Springer Series on

Wave Phenomena

6

Editor: Leopold B. Felsen

Springer Series on
Wave Phenomena

Editors: L. M. Brekhovskikh   L. B. Felsen   H. A. Haus
Managing Editor: H. K. V. Lotsch

Yu. A. Kravtsov   Yu. I. Orlov

Geometrical Optics of Inhomogeneous Media

With 108 Figures

Springer-Verlag
Berlin Heidelberg New York London
Paris Tokyo Hong Kong Barcelona

Professor Dr. Yury A. Kravtsov
General Physics Institute, USSR Academy of Sciences, Vavilova 38,
SU-117942 Moscow, USSR

Present address: ZIL Institute, Antozavodskaja 16, SU-109 068 Moscow, USSR

Dr. Yuri Ilich Orlov †

Series Editors:

Professor Leonid M. Brekhovskikh, Academician

P. P. Shirsov Institute of Oceanology, Academy of Sciences of the USSR, Krasikowa Street 23,
SU-117218 Moscow, USSR

Professor Leopold B. Felsen, Ph. D.

Department of Electrical Engineering, Polytechnic University, Route 110,
Farmingdale, NY 11735, USA

Professor Hermann A. Haus

Department of Electrical Engineering & Computer Science, MIT,
Cambridge, MA 02139, USA

Managing Editor: Helmut K. V. Lotsch

Springer-Verlag, Tiergartenstrasse 17,
D-6900 Heidelberg, Fed. Rep. of Germany

ISBN-13: 978-3-642-84033-3   e-ISBN-13: 978-3-642-84031-9
DOI: 10.1007/ 978-3-642-84031-9

Library of Congress Cataloging-in-Publication Data. Kravt͡sov, I͡Uriĭ Aleksandrovich. [Geometricheskai͡a optika neodnorodnykh sred. English] Geometrical optics of inhomogeneous media / Yu. A. Kravtsov, Yu. I. Orlov. p. cm. – (Springer series on wave phenomena ; 6) Translation of: Geometricheskai͡a optika neodnorodnykh sred. Includes bibliographical references. ISBN 0-387-51944-0 (U.S.) 1. Optics, Geometrical. I. Orlov, I͡U. I. (I͡Uriĭ Il ʼich) II. Title. III. Series. QC381.K713 1990 535–dc20 89-26299

The text was prepared using the PS™ Technical Word Processor.

2154/3150-543210 – Printed on acid-free paper

Preface

This monograph is concerned with the fundamentals of up-to-date geometrical optics treated as an approximate method of wave theory.

Geometrical optics has changed dramatically over the last two decades. Primarily, it has acquired a number of novel disciplines: space-time geometrical optics, the quasi-isotropic approximation, the modern theory of caustics related to catastrophe theory, and perturbation techniques for rays, to name only a few. Another acquisition is the reliable boundaries of applicability for geometrical optics, based upon the concept of the Fresnel volume for a ray. These recent additions to the field are the focus of discussion in the book.

We did not attempt to separate study-oriented and illustrative material from that intended for professionals, but rather we spread it throughout the text to facilitate for the reader the mastering of this attractive, intuitively appealing and efficient ray method.

In preparing the manuscript we used a set of lecture notes devised for All-Union Schools on Diffraction and Wave Propagation, published in Russian. Sections 2.1-4,6 and 10 result from joint efforts of both authors. The other material of the book we wrote separately. I contributed Sects. 2.5,9 and 3.17 and Chap.4; Yu.I. Orlov prepared the rest. Unfortunately, he could not take part in the preparation of the English edition, as he died in 1982 at the age of 41, on the verge of what would have been great achievements considering his strong and original talent.

The majority of the authors' results presented were obtained with the encouraging support of S.M. Rytov and E.N. Vasilyev, to whom I owe my sincere gratitude. I also wish to thank M.G. Edelev for his translation, and N.S. Orlova for her help in preparing the English text. I would also like to express my deep gratitude to Prof. L.B. Felsen of the Polytechnic University, and Dr. H. Lotsch of Springer-Verlag for making the publication of this book possible.

Moscow, USSR
January 1990Yury A. Kravtsov

Contents

1. Introduction

Geometrical optics is understood in two different ways. As a way to conduct ray analysis it is confined to the techniques of image formation by ray tracing. In a wider, wavelike sense it is a method for approximately defining wave fields. In this monograph we shall stick to the wave-field rather than the ray-tracing analysis: here rays form only a geometrical backbone supporting the wave field spanned by them.

Historically, these interpretations of geometrical optics have seen two periods of development. The initial "ray period" had its basic ideas formulated already in the works of *Hamilton* [1.1], which influenced considerably the development of classical mechanics (for a discussion of *Hamilton's* method, see [1.2-5]). Ray-tracing analysis is essentially utilized in optical-instrument design. The advancements in this field have been summarized in well-known monographs [1.6-13], also containing historical accounts of the ray period.

The modern "wave period" of geometrical optics was initiated by *Debye* [1.14] who influenced the ray formulation in wave-optics theory (among them is the quasi-classic representation in quantum mechanics). The results of *Debye* were extended by *Sobolev* [1.15], *Rytov* [1.16, 17], *Luneburg* [1.18], *Courant* [1.19], and his coworkers *Keller* et al. [1.20], and *Babich* and *Buldyrev* [1.21], to name only the most profound contributions. The history of this period is reiterated in books [1.19, 21, 22] and reviews [1.20, 23].

The method of geometrical optics is of immense importance in wave-field analysis. It is simple and intuitively appealing for a wide variety of wave phenomena (electromagnetic, elastic, and sound waves; waves in plasma and liquids; quantum-mechanical phenomena; etc.). It works well as long as the wavelength is small compared with the characteristic dimension of the problem at hand.

Geometrical optics is a probing tool for a multitude of applications. Even a concise survey of them would exceed the scope of this volume. The method is especially popular in radio science and optics, plasma physics, wave propagation through the earth's atmosphere, solar corona, and interstellar space, as well as with radar, radio navigation, and remote-sensing techniques. The list of applications might spread even further if we recall the counterparts in acoustics (geometrical acoustics), seismology (geometrical seismology), quantum mechanics (the WKB method), and hydrodynamics, to name only a few. Without any risk of overemphasizing, geometrical optics and its modifications are applied to perform most high-frequency field computations, especially at an initial stage of estimating.

Of tremendous importance is the use of ray optics in analyzing wave fields in inhomogeneous media. Attention is drawn to applications concerned with the emitting and propagating of waves in laboratory, ionospheric and cosmic plasmas; in planetary atmospheres, in a solar corona, in the turbulent ocean, and in the earth's crust; and in various media with random inhomogeneities.

Although the ray method has long been a powerful tool both in theoretical and applied physics, thus far it has not enjoyed any systematic examination of its recent achievements in the literature, only some of the key topics having been outlined in monographs. (By these we mean books either dealing with physical problems [1.24-30] or discussing generalizations of the ray method [1.21,31]; the relevant mathematical fundamentals were presented in [1.2,18,19,32]). We attempt to fill the gap by elucidating - systematically and completely insofar as possible - the state of the art in geometrical optics and by demonstrating applications of the method to wave problems of modern physics. For each topic discussed, our evaluation will involve four aspects:

(i) Whereever necessary, the "wave" origin of all the geometrical-optics notions will be emphasized. For example, the wave treatment for the Fermat principle is given, and the concept of a ray is defined from the wave standpoint.

(ii) We shall pay much attention to demonstrating how to solve equations of geometrical optics and apply the solution to constructing an approximate wave field.

(iii) The illustrative examples - collected in accordance with the main objective of the book from studies in optics, acoustics, and radio-wave propagation - are used to demonstrate applications of the method to various physical and technical problems, thereby serving as a guide to a novice in the field.

(iv) Intending to give a complete picture, we shall briefly review papers proposing applications and generalizations of geometrical optics that could not be discussed here even in minor detail. Occasionally, however, we will have to confine ourselves merely to referring to the original papers and reviews when touching upon such important topics as the geometrical optics of open resonators and optical waveguides, nonlinear and random media, and some others.

The presentation of references provided us some headache. On one hand, did we aim at a balanced coverage of original and recent works but, on the other hand, we took advantage of the opportunity to acquaint the reader from abroad with the most signifacant publications on geometrical optics and related topics, which until recently have been available only in Russian. As a consequence, the Russian part of the literature listed represents a kind of overview of Soviet publications on the subject matter.

2. The Scalar Wave Field

This chapter is concerned with general aspects of the geometrical optics of a scalar wave field. We first illustrate how the equations of geometrical optics (including space-time equations) are derived and then consider the main properties of rays, caustics, and ray fields. At the end of the chapter we formulate the uniformly valid conditions of applicability. These conditions pave the way for the solution of a number of complicated problems, which can never be solved by ordinary geometrical optics, among them are estimates of the caustic-zone width and of fields at caustics and in the focal spot.

2.1 Equations of Geometrical Optics

2.1.1 Mathematical Background

Consider monochromatic waves described by the Helmholtz (reduced wave) equation

$$\Delta u + k_0^2 n^2(\mathbf{r}) u(\mathbf{r}) = 0 , \qquad (2.1.1)$$

where $k_0 = \omega/c$ is the wave number, c is the characteristic velocity of the wave, ω is the angular frequency [the time dependence is given by $e^{-i\omega t}$], and the refractive index $n(\mathbf{r})$ defines the medium in which the wave propagates.

In electrodynamics, (2.1.1) describes the behavior of electromagnetic waves, excluding polarization, in any isotropic medium. The square of the refractive index is taken to represent the dielectric permittivity of the medium, $n^2(\mathbf{r}) = \epsilon(\mathbf{r})$, while here and henceforth the magnetic permeability of the medium is assumed to be unity.

Equation (2.1.1) governs acoustic waves, if we set $k_0 = \omega/\tilde{c}$ and $n(\mathbf{r}) = \tilde{c}/c(\mathbf{r})$, where $c(\mathbf{r})$ is the local velocity of the sound wave, and $\tilde{c}$ is the characteristic velocity of sound in the space under consideration. For the stationary Schrödinger equation, defining the wave function of a particle of mass m and energy E in the field of the potential $U(\mathbf{r})$, K_0 should be replaced by $(2mE)^{1/2}/\hbar$, and $n(\mathbf{r})$ by $[1-U(\mathbf{r})/E]^{1/2}$ so that $k^2 n^2(\mathbf{r}) = 2m(E-U)\hbar^2$, $\hbar = h/2\pi$ being Planck's constant.

For the simplest case of a plane wave traveling in the direction of the unit vector 1 in a homogeneous medium of constant n the solution of (2.1.1) has the form

$$u(\mathbf{r}) = \mathcal{A}e^{i\Psi(\mathbf{r})} \ , \quad \Psi(\mathbf{r}) = k_0 n(\mathbf{r}\cdot\mathbf{l}) \ ,$$

where $\mathcal{A}$ is the constant amplitude of the wave, and $\Psi(\mathbf{r})$ is its phase such that the wave vector $\mathbf{k} = \nabla\Psi$ is constant and directed in $\mathbf{l}$, i.e., $\mathbf{k} = n\mathbf{l}k_0$. If the properties of the wave or the medium change only slowly over the wavelength, then we may assume that the wave field will also change slowly in space, and hence may locally be represented by an "almost plane" wavefront. This assumption presents the essence of the ray method for monochromatic waves.

Having assumed that (2.1.1) yields, as a solution, the almost plane wave

$$u(\mathbf{r}) = \mathcal{A}(\mathbf{r})e^{i\Psi(\mathbf{r})} \ , \tag{2.1.2}$$

we imply that the amplitude $\mathcal{A}(\mathbf{r})$ and the local wave vector $\mathbf{k}(\mathbf{r}) = \nabla\Psi(\mathbf{r})$ *vary insignificantly over the wavelength* $\lambda(\mathbf{r}) = 1/|\mathbf{k}(\mathbf{r})|$ in that medium, or mathematically

$$\lambda|\nabla\mathcal{A}| \ll |\mathcal{A}| \ , \quad \lambda|\nabla k_i| \ll |k_j| \ . \tag{2.1.3a}$$

Here k_j are the components of the wave vector $\mathbf{k}$, and $\lambda = \lambda/2\pi$. A similar assumption should be ascribed to the refractive index $n(\mathbf{r})$

$$\lambda|\nabla n| \ll n \ . \tag{2.1.3b}$$

Since in a medium of refractive index n the wavelength λ is n times shorter than that in free space, i.e., $\lambda = \lambda_0/n$, the inequalities (2.1.3a, b) are equivalent to

$$\frac{\lambda_0}{n} \ll \frac{|\mathcal{A}|}{|\nabla\mathcal{A}|} \equiv L_1 \ ,$$

$$\frac{\lambda_0}{n} \ll \frac{|k_j|}{|\nabla k_j|} \equiv L_2 \ , \tag{2.1.3c}$$

$$\frac{\lambda_0}{n} \ll \frac{n}{|\nabla n|} \equiv L_3 \ ,$$

where $\lambda_0 = \lambda_0/2\pi = c/\omega$.

Alternatively, these conditions, stating that $\mathcal{A}(\mathbf{r})$, $\mathbf{k}(\mathbf{r})$, and $n(\mathbf{r})$ vary insignificantly within a region of the order of λ, may be united in the single inequality

$$\mu = 1/kL = 1/k_0 nL = \lambda/L \ll 1 \ , \tag{2.1.4}$$

where μ is the parameter of smallness in the method of geometrical optics, and L is the smallest of the characteristic lengths of $\mathcal{A}$, $\mathbf{k}$, and n, i.e., $L =$

$\min(L_1, L_2, L_3)$. The characteristic length of the problem, L, is here taken to be the distance over which the increment of a certain variable is comparable with it in value, say, $|n(r+L)-n(r)| \simeq n(r)$. Substituting the term linear in L from the Taylor series for the increment $n(r+L)$ yields $L|\nabla n| \simeq n$, then an estimate for L is $L \simeq n/|\nabla n|$, which is the same as L_3 in (2.1.3c). For plane-wave propagation in a homogeneous medium, the quantities $\mathcal{A}$, $\mathbf{k}$, and n are strictly constant, corresponding to $L = \infty$ and $\mu = 0$ so that (2.1.4) becomes trivial.

2.1.2 Field Expansion in a Dimensionless Parameter

The basic equations can be derived in several ways. A procedure which seems to be most systematic is that developed by *Rytov* [2.1, 2], who applied a series expansion in the small dimensionless parameter $\mu \simeq 1/k_0 L$. With the dimensionless variables $x_1 = k_0 x$, $y_1 = k_0 y$, and $z_1 = k_0 z$, (2.1.1) takes the form

$$\Delta_1 u + n^2 u = 0 , \quad \Delta_1 \equiv \frac{\partial^2}{\partial x_1{}^2} + \frac{\partial^2}{\partial y_1{}^2} + \frac{\partial^2}{\partial z_1{}^2} , \tag{2.1.5}$$

and the characteristic length L, over which $\mathcal{A}$, k, and n vary significantly, converts into the dimensionless parameter $k_0 L = 1/\mu$. We can introduce it into (2.1.5) by letting $r_2 = \mu r_1 = r/L$ and

$$n = n(r/L) = n(r_1/k_0 L) = n(\mu r_1) = n(r_2) , \quad |\nabla_2 n| \simeq 1 , \tag{2.1.6}$$

where $\nabla_2 = \partial/\partial r_2$. Introducing r_2 transforms (2.1.5) into

$$\Delta_2 u(r_2) + \frac{n^2(r_2)}{\mu^2} u(r_2) = 0 , \quad \Delta_2 \equiv \frac{\partial^2}{\partial x_2{}^2} + \frac{\partial^2}{\partial y_2{}^2} + \frac{\partial^2}{\partial z_2{}^2} . \tag{2.1.7}$$

The refractive-index dependence on r_2 in (2.1.6, 7) reflects a slow (within the wavelength λ) variation of n with x, y, and z, since the increment $|\delta r| \simeq \lambda$ for $L \gg \lambda$ gives rise only to a small variation $\delta n \simeq |\delta r \nabla n| \simeq \lambda|\nabla_2 n|/L \simeq \lambda/L = \mu \ll 1$.

The amplitude is assumed to vary slowly, i.e. $\mathcal{A} = \mathcal{A}(\mu r_1) = \mathcal{A}(r_2)$. It is convenient to write the phase in the form $\Psi(r) = \Psi_1(\mu r_1)/\mu = \Psi_1(r_2)/\mu$ so that its gradient, i.e., the wave vector $\nabla\Psi = \nabla_1 \Psi_1(\mu r_1)/\mu = \nabla_2 \Psi_1(r_2)$ would also be a slow function of coordinates, here $\nabla_{1,2} \equiv \partial/\partial r_{1,2}$. Thus, we have arrived at representing the field by the product of a slowly varying amplitude and a rapidly oscillating function

$$u = \mathcal{A}(\mu r_1)e^{i\Psi_1(\mu r_1)/\mu} = \mathcal{A}(r_2)e^{i\Psi_1(r_2)/\mu} . \tag{2.1.8a}$$

Now the wave problem has been reduced to seeking two functions, $\mathcal{A}$ and Ψ_1. Substituting (2.1.8a) into (2.1.7) yields

$$\Delta_2 u + \frac{n^2}{\mu^2}\, u \equiv \tag{2.1.8b}$$

$$\left\{ \frac{1}{\mu^2}[n^2 - (\nabla_2 \Psi_1)^2]\mathcal{A} + \frac{i}{\mu}(2\nabla_2 \mathcal{A}\nabla_2 \Psi_1 + \mathcal{A}\Delta_2 \Psi_1) + \Delta_2 \mathcal{A} \right\}\exp(i\Psi_1/\mu) = 0 \; .$$

A widely used approach consists of expanding $\mathcal{A}$ in powers of μ to represent the field (2.1.8a) as a ray series[1]

$$u(\mathbf{r}_2) = e^{i\Psi_1(\mathbf{r}_2)/\mu} \sum_{m=0}^{\infty} (\mu/i)^m\, \mathcal{A}_m(\mathbf{r}_2) \; , \tag{2.1.9}$$

where the fraction μ/i appears to simplify the derivation. Substituting (2.1.9) into (2.1.8b) and equating coefficients of equal power in μ yields

$$
\begin{aligned}
(\mu^{-2}) \qquad & (\nabla_2 \Psi_1)^2 = n^2 \; , \\
(\mu^{-1}) \qquad & 2(\nabla_2 \mathcal{A}_0 \nabla_2 \Psi_1) + \mathcal{A}_0 \Delta_2 \Psi_1 = 0 \; , \\
(\mu^{0}) \qquad & 2(\nabla_2 \mathcal{A}_1 \nabla_2 \Psi_1) + \mathcal{A}_1 \Delta_2 \Psi_1 = -\,\Delta_2 \mathcal{A}_0 \; , \\
\text{-----------------------} \qquad & \\
(\mu^{m-1}) \qquad & 2(\nabla_2 \mathcal{A}_m \nabla_2 \Psi_1) + \mathcal{A}_m \Delta_2 \Psi_1 = -\,\Delta_2 \mathcal{A}_{m-1}
\end{aligned}
\tag{2.1.10}
$$

(the respective powers of μ are shown in the leftmost column).

Rewriting these equations in the dimensional variables x,y,z and denoting the phase $\Psi = \Psi_1/\mu$ by $k_0\psi$ yields

$$(\nabla\psi)^2 = n^2 \; , \tag{2.1.11}$$

$$
\begin{aligned}
& 2(\nabla\mathcal{A}_0 \nabla\psi) + \mathcal{A}_0 \Delta\psi = 0 \; , \\
& 2(\nabla\mathcal{A}_1 \nabla\psi) + \mathcal{A}_1 \Delta\psi = -\,\mathrm{L}\Delta\mathcal{A}_0 \; , \\
& \text{-------------------} \\
& 2(\nabla\mathcal{A}_m \nabla\psi) + \mathcal{A}_m \Delta\psi = -\,\mathrm{L}\Delta\mathcal{A}_{m-1} \; .
\end{aligned}
\tag{2.1.12}
$$

The function ψ has come to be known as the *eikonal* and the respective equation (2.1.11) as the eikonal equation.[2] The eikonal ψ has the dimension of length, and an alternative name - the optical path of the

---

[1] Alternative derivations of the geometrical-optics equations can be based on the path integral [2.3-5], by the averaged Lagrangian [2.6-9], and other techniques. We have adopted here the simplest and most intuitively appealing approach of expanding the amplitude in powers of μ.

[2] Hamilton termed ψ the point characteristic [2.11]. The name **eikonal** (which is the Greek for image) was suggested by *Bruns* [2.12,13]. The mechanical analog of the eikonal function is the action [2.14,15]. The quantity $\Psi = k_0\psi$ defines the phase of a field in the zero-order approximation of geometrical optics. The higher terms of (2.1.9) add to the phase only small corrections of order μ^m, $m = 1,2$, etc.

6

wave. The equations for the amplitudes $\mathcal{A}_0$, $\mathcal{A}_1$, etc. are called the approximations to the transport equation of the zeroth, first, etc. order.

Geometrical optics evaluates the field by solving (2.1.11,12). These equations are simpler than the initial Helmholtz equation (2.1.1), firstly, because they are *first*-order partial differential equations and, secondly, because they can be reduced to *ordinary* differential equations (Sects. 2.2, 3).

2.1.3 Field Expansion in Inverse Wave Numbers

The eikonal and transport equations could also be derived by formally expanding the field in inverse powers of the wave number k_0, which is a dimensional parameter, i.e.,

$$u(\mathbf{r}) = \sum_{m=0}^{\infty} \frac{A_m(\mathbf{r})}{(ik_0)^m} e^{ik_0 \psi(\mathbf{r})} . \tag{2.1.13}$$

This type of ray series is known as the *Debye expansion* in deference to its initiator [2.10].

Substituting (2.1.13) into (2.1.1) and equating the coefficients of the same powers in k_0 leads us to the eikonal equation (2.1.11) and to the transport equations in A_m that differ from (2.1.12) only in that L is absent from the right-hand sides:

$$2(\nabla A_0 \nabla \psi) + A_0 \Delta \psi = 0 , \tag{2.1.14a}$$
$$2(\nabla A_1 \nabla \psi) + A_1 \Delta \psi = - \Delta A_0 , \tag{2.1.14b}$$
$$\text{-----------------}$$
$$2(\nabla A_m \nabla \psi) + A_m \Delta \psi = - \Delta A_{m-1} . \tag{2.1.14c}$$

This implies that the coefficients $\mathcal{A}_m$ in (2.1.9) and A_m in (2.1.13) are interrelated by

$$A_m = \mathcal{A}_m / L^m . \tag{2.1.15}$$

All $\mathcal{A}_m$ are of the same dimension, coinciding with that of the field u [because (2.1.9) is expanded in the dimensionless parameter μ]; the dimensions of the amplitudes A_m are seen to differ, with $[A_m] = [u]/L^m$.

Although (2.1.9, 13) are equivalent, the use of the dimensionless parameter $\mu = 1/kL$ is of certain methodological advantage. The point is that the smallness of μ required for the ray method to be valid can be achieved not only by increasing k (shorter wavelength), but also by raising L, which comes in effect when a ray crosses into a more smoothly varying medium. A formal expansion in powers of $1/k_0$ would overlook this possibility.

On the other hand, the Debye expansion provides conciser equations. In what follows we shall preferably use (2.1.13) keeping in mind, however, that we may formally obtain (2.1.9) from (2.1.13) by inserting $k_0 L = \mu^{-1}$ in place of k_0, which is a useful tool in qualitative analysis.

2.1.4 Initial Conditions for the Eikonal and Amplitude Equations

Consider an initial surface Q, parametrically defined as

$$r = r^0(\xi,\eta) \,, \qquad (2.1.16)$$

where ξ and η are curvilinear coordinates on Q. In the specific case of the plane $z = 0$ taken as Q, the role of ξ and η may be identified with the coordinates x^0 and y^0 of this plane.

Let an initial field $u^0(\xi,\eta)$ be given on Q by a series similar to (2.1.13)

$$u^0(\xi,\eta) = \sum \frac{A_m^0(\xi,\eta)}{(ik_0)^m} \exp[ik_0\psi^0(\xi,\eta)] \,. \qquad (2.1.17)$$

The initial values, implied by the superscript ^{0}, immediately follow from the representation of the initial field, namely

$$\psi|_Q = \psi^0(\xi,\eta) \,, \quad A_m|_Q = A_m^0(\xi,\eta) \,. \qquad (2.1.18)$$

Frequently it is only the zeroth-order amplitude that is prescribed to be nonzero, i.e.,

$$A_0^0 = A_0^0(\xi,\eta), \quad \text{with} \quad A_m^0 = 0 \ \text{ for } m \geq 1 \,, \qquad (2.1.19)$$

as, say, for a plane wave incident on an inhomogeneous medium. On the other hand, for the problem of a wave striking an interface, generally all the amplitudes A_m^0 corresponding to the refracted and reflected waves are nonzero, even if for the incident wave $A_m^0 = 0$ ($m \geq 1$). Under these circumstances the initial values of A_m are not specified und must rather be evaluated from boundary conditions (Sect.2.5).

2.1.5 Asymptotic Nature of the Ray Series

Both (2.1.9, 13) are usually asymptotic series. The difference between the exact solution and its truncated sum

$$u_M = \exp(i\Psi_1/\mu) \sum_{m=0}^{M} (\mu/i)^m \mathscr{A}_m = \exp(ik_0\psi) \sum_{m=0}^{M} \frac{A_m}{(ik_0)^m} \qquad (2.1.20)$$

vanishes as $\mu \to 0$ as μ^{M+1}

$$|u - u_M| \underset{\mu\to 0}{=} O(\mu^{M+1}) \,. \qquad (2.1.21)$$

8

Recall that the condition $\mu \ll 1$ is due to the fact that the wavelength λ is small compared with the characteristic dimension of the problem, L, allowing one to say that the ray expansion provides a *short-wavelength asymptotics* to the field. Another name is *high-frequency asymptotics* because $k = \omega n/c \gg L^{-1}$ at high frequencies, which corresponds to a small λ.

For μ other than zero the ray series diverge as a rule, i.e., $|u-u_M| \to \infty$ as $M \to \infty$.

The asymptotic behavior of both (2.1.9 and 13) has been proved for many problems. For other situations, (2.1.9, 13) may supply only a formal solution to (2.1.1) because this behavior has not yet been proved in general.

For the majority of physical applications, a satisfactory analysis results from discussing the zero-order geometrical-optics approximation by letting

$$u \simeq u_0 = A_0 \exp(ik_0 \psi) = A_0 \exp(i\Psi_1/\mu) \tag{2.1.22}$$

with $\mathscr{A}_0 = A_0$ by virtue of (2.1.15). The lack of attention paid in the literature to higher-order terms of the ray expansion is primarily due to computational difficulties encountered in deriving them (Sect.2.1.3). Therefore a higher-order approximation is worth of attempting when it promises to describe the entire phenomenon as in problems of reflection either from a weak interface (Sect.2.1.5) or close to the Brewster angle (Sect.3.1.2).

2.2 Rays and the Eikonal

2.2.1 The Method of Characteristics

The eikonal equation (2.1.11) is a nonlinear, partial differential equation of the first order belonging to the Hamilton-Jacobi variety.

In the general case the Hamilton-Jacobi equation has the form

$$\mathscr{H}\left[\frac{\partial\psi}{\partial q_1}, \frac{\partial\psi}{\partial q_2}, ..., \frac{\partial\psi}{(\partial q_n)}; q_1, q_2, ..., q_n \right] = 0 , \tag{2.2.1a}$$

or

$$\mathscr{H}(p_j, q_j) = 0 , \quad p_j = \frac{\partial\psi}{\partial q_j} . \tag{2.2.1b}$$

Here $\psi = \psi(q_1, q_2 ..., q_n)$ is the function to be determined, q_j are arbitrary coordinates ($j = 1, 2, ..., n$), and p_j are the associated "momenta".

This first-order partial differential equation can be reduced to a characteristic set of ordinary differential equations solvable by integration [2.16, 17], i.e.

$$\frac{dq_j}{\partial\mathscr{H}/\partial p_j} = -\frac{dp_j}{\partial\mathscr{H}/\partial q_j} = \frac{d\psi}{\Sigma_{j=1}^{n} p_j(\partial\mathscr{H}/\partial p_j)} . \tag{2.2.2}$$

On equating the ratios in (2.2.2) with the differentials of the parametric variable $d\tau$, the characteristic system can be also written[3] as

$$\frac{dq_j}{d\tau} = \frac{\partial \mathcal{H}}{\partial p_j} \, , \tag{2.2.3}$$

$$\frac{dp_j}{d\tau} = \frac{\partial \mathcal{H}}{\partial q_j} \, , \tag{2.2.4}$$

$$\frac{d\psi}{d\tau} = \sum_{j=1}^{n} p_j \, \frac{\partial \mathcal{H}}{\partial p_j} \, . \tag{2.2.5}$$

The solution $q_j = q_j(\tau)$, $p_j = p_j(\tau)$, and $\psi = \psi(\tau)$ of the characteristic system (2.2.3-5) are referred to as the characteristic strip in the theory of differential equations [2.16, 17]. In physics, the family of the 2n functions $q_j = q_j(\tau)$ is said to constitute the *characteristic* of (2.2.1).[4]

The characteristics (2.2.3,4) are given here in the canonic Hamiltonian form. They were initially suggested by W.R. Hamilton to handle problems of light-ray propagation. It was only later that they were extended to mechanics where p_j and q_j in (2.2.1,3-5) are called the generalized momenta and coordinates, and the function $\mathcal{H}(p_j, q_j)$ is termed the Hamiltonian of the system [2.11, 12]. We shall adhere to this terminology, keeping in mind that in wave problems no meaning of energy is attached to $\mathcal{H}(p_j, g_j)$.

Let q_j ($j = 1,2,3$) be Cartesian coordinates and $\mathcal{H} = \mathcal{H}(\mathbf{p}, \mathbf{r})$, then the characteristic system (2.2.3-5) turns into vector form

$$\frac{d\mathbf{r}}{d\tau} = \frac{\partial \mathcal{H}}{\partial \mathbf{p}} \, , \quad \frac{d\mathbf{p}}{d\tau} = -\frac{\partial \mathcal{H}}{\partial \mathbf{r}} \, , \tag{2.2.7}$$

$$\frac{d\psi}{d\tau} = \mathbf{p} \, \frac{\partial \mathcal{H}}{\partial \mathbf{p}} \, , \quad \mathbf{p} = \nabla \psi \, . \tag{2.2.8}$$

Here and in what follows $\partial f / \partial \mathbf{a}$ should be understood as the vector $\nabla_{\mathbf{a}} f$ of the components $\{\partial f / \partial a_i\}$.

---

[3] We recall the procedure leading to the characteristic equations (2.2.2) or (2.2.3,4). For a vanishing hypersurface in the phase space $\{p_j, q_j\}$

$$d\mathcal{H} = \sum_{j=1}^{n} \left[\frac{\partial \mathcal{H}}{\partial p_j} \, dp_j + \frac{\partial \mathcal{H}}{\partial q_j} \, dq_j \right] = 0 \, , \tag{2.2.6}$$

which can be satisfied by (2.2.3,4) or, what is the same, by the first two equalities in (2.2.2); Eq.(2.2.5) can be obtained by differentiating along the characteristics (2.2.3,4).

[4] Recognizing that the eikonal ψ is itself a characteristic of the Helmholtz equation (2.1.1), these characteristics are often called bicharacteristics [2.16].

2.2.2 Ray Equations and the Eikonal

Geometrical optics treats a ray as the coordinate space projection $q_j = q_j(\tau)$ of the characteristic of the eikonal equation (2.1.11). For brevity, the characteristics $q_j = q_j(\tau)$, $p_j = p_j(\tau)$ in the phase space $\{p_j, q_j\}$ are at times also called a ray, and the respective equations are termed the *ray equations.*

As follows from the method of characteristics, the use of rays reduces the eikonal problem to a quadrature one. To illustrate this, assume we know the solution $\mathbf{r} = \mathbf{r}(\tau)$, $\mathbf{p} = \mathbf{p}(\tau)$ to the ray equations written, for instance, in the form (2.2.7). Then integrating (2.2.8) yields

$$\psi = \psi^0 + \int_{\tau^0}^{\tau} \mathbf{p}\, \frac{\partial \mathscr{H}}{\partial \mathbf{p}}\, d\tau \ . \tag{2.2.9}$$

Here $\psi^0 = \psi^0(\tau^0)$ is the initial value of the eikonal at $\tau = \tau^0$, and the integral is taken along the ray $\mathbf{r} = \mathbf{r}(\tau)$, $\mathbf{p} = \mathbf{p}(\tau)$.

The parameter τ varying along the ray can readily be related to the arc length σ. On the one hand, from (2.2.7) we have $d\mathbf{r}^2 = (\partial\mathscr{H}/\partial\mathbf{p})^2 d\tau^2$, on the other, $d\mathbf{r}^2 = d\sigma^2$. Consequently

$$d\tau = \frac{d\sigma}{|\partial\mathscr{H}/\partial\mathbf{p}|} \ . \tag{2.2.10}$$

Consider some specific differential ray equations for a definite Hamiltonian $\mathscr{H}(\mathbf{p}, \mathbf{r})$. The ray equations transform into a convenient form with the eikonal equation (2.1.11) written as

$$\mathscr{H} = \tfrac{1}{2}\,[\mathbf{p}^2 - n^2(\mathbf{r})] = 0\ , \tag{2.2.11}$$

where $\mathbf{p} = \nabla\psi$. Then from (2.2.7) we get the ray equations in the Hamiltonian form

$$\frac{d\mathbf{r}}{d\tau} = \mathbf{p}\ , \tag{2.2.12a}$$

$$\frac{d\mathbf{p}}{d\tau} = \tfrac{1}{2}\,\nabla n^2(\mathbf{r})\ . \tag{2.2.12b}$$

From (2.2.12a) it is readily seen that the vector $d\mathbf{r}/d\tau$, being tangent to the ray, is parallel to the momentum $\mathbf{p}$, i.e., to the eikonal gradient $\nabla\psi$. In other words, in an isotropic medium the rays are normal to the surfaces of equal eikonal, $\psi = $ constant, that is, to the wavefronts.

For the eikonal ψ we have from (2.2.9), assuming $\mathbf{p}^2 = n^2$,

$$\psi = \psi^0 + \int_{\tau^0}^{\tau} \mathbf{p}^2\, d\tau = \psi^0 + \int_{\tau^0}^{\tau} n^2[\mathbf{r}(\tau)]d\tau \ . \tag{2.2.13}$$

For the Hamiltonian (2.2.11), the parameter τ is connected with the arc length by the relation

$$d\tau = \frac{d\sigma}{p} = \frac{d\sigma}{n} \, , \tag{2.2.14}$$

immediately following from (2.2.8). Therefore (2.2.13) is equivalent to the more familiar expression

$$\psi = \psi^0 + \int_{\sigma^0}^{\sigma} n \, d\sigma \, , \tag{2.2.15}$$

where $d\sigma$ is the arc length on the ray.

For another form of the ray equations, the Hamiltonian is

$$\mathcal{H}(\mathbf{p},\mathbf{r}) = p - n(\mathbf{r}) = 0 \, , \quad p = \sqrt{\mathbf{p}^2} \, . \tag{2.2.16}$$

In this case $\partial\mathcal{H}/\partial\mathbf{p} = \mathbf{p}/p$ so that $|\partial\mathcal{H}/\partial\mathbf{p}| = 1$, and according to (2.2.10) the parameter τ is the arc length σ, i.e., $\tau = \sigma$. The ray equations corresponding to (2.2.16) are derived from (2.2.7)

$$\frac{d\mathbf{r}}{d\sigma} = \frac{\mathbf{p}}{n} \, , \quad \frac{d\mathbf{p}}{d\sigma} = \nabla n \, , \tag{2.2.17}$$

with the eikonal found from (2.2.15).

Inserting the unit vector $\mathbf{l} = \mathbf{p}/p = \mathbf{p}/n$, being tangent to the ray, into (2.2.17) yields

$$\frac{d\mathbf{r}}{d\sigma} = \mathbf{l} \, , \quad \frac{d\mathbf{l}}{d\sigma} = \frac{\nabla n}{n} - \mathbf{l}\left(\frac{\nabla n}{n}\cdot\mathbf{l}\right) \equiv \nabla_{\perp}\ln(n) \, , \tag{2.2.18}$$

where $dn/d\sigma = \mathbf{l}\nabla n$, and the notation $\nabla_{\perp} = \nabla - \mathbf{l}(\mathbf{l}\cdot\nabla)$ has been introduced for the transverse (with respect to the ray) gradient. The eikonal (2.2.15) remains valid for this form of the ray equations.

The systems (2.2.12, 17) or (2.2.18) can be combined into one single second-order differential equation if we eliminate the momentum $\mathbf{p}$ or the unit vector $\mathbf{l}$. The resultant equation becomes

$$\frac{d^2\mathbf{r}}{d\tau^2} = \frac{1}{2}\nabla n^2 \quad \text{or} \quad \frac{d}{d\sigma}\left(n\frac{d\mathbf{r}}{d\sigma}\right) = \nabla n \, , \tag{2.2.19}$$

frequently encountered in the literature.[5] The first of them is similar to the Newton equation of motion of a unit mass in the force field $U(\mathbf{r})$ [2.14]

---

[5] These ray equations are equivalent to the Lagrangian equations of motion in mechanics.

$$\frac{d^2 \mathbf{r}}{dt^2} = - \nabla U(\mathbf{r}) \ .$$

This analogy permits one to borrow results of path computations from mechanics, and vice versa.

2.2.3 Curvature and Torsion of Rays

Let us first define the curvature K and the torsion κ of a ray in a homogeneous medium. For this purpose we invoke the Frenet reference frame [2.18] associated with the ray (defined at every point by the tangent, normal, and binormal to the curve)

$$\frac{d\mathbf{l}}{d\sigma} = K\boldsymbol{\nu} \ , \tag{2.2.20a}$$

$$\frac{d\boldsymbol{\nu}}{d\sigma} = - K\mathbf{l} - \kappa\mathbf{b} \ , \qquad \frac{d\mathbf{b}}{d\sigma} = \kappa\boldsymbol{\nu} \ . \tag{2.2.20b}$$

Here $\boldsymbol{\nu}$ is the unit vector of the principal normal to the ray, and $\mathbf{b} = \mathbf{l} \times \boldsymbol{\nu}$ is the unit vector of the binormal.

From a comparison of (2.2.20a and 18) it follows that the principal normal $\boldsymbol{\nu}$ is parallel to the vector $\nabla_\perp \ln(n)$

$$K\boldsymbol{\nu} = \nabla_\perp \ln(n) = \nabla_\perp n/n$$

$$= \mathbf{l} \times \left(\frac{\nabla n}{n} \times \mathbf{l} \right) = \frac{\nabla n}{n} - \mathbf{l} \cdot \left(\mathbf{l} \cdot \frac{\nabla n}{n} \right) \ . \tag{2.2.21}$$

As can be readily seen from (2.2.21), in an inhomogeneous medium a ray bends toward increasing refractive index.[6] The curvature of the ray is

$$K = \left| \mathbf{l} \times \left(\frac{\nabla n}{n} \times \mathbf{l} \right) \right| = \left| \frac{\nabla n}{n} \times \mathbf{l} \right| = \left| \frac{\nabla n}{n} \right| \sin\theta \ , \tag{2.2.22}$$

where θ is the angle between the ray ($\mathbf{l}$) and the vector ∇n. The radius of curvature, ρ, is the inverse of (2.2.22), i.e., $\rho = 1/K$. In a homogeneous medium, where n is constant and $\nabla n = 0$, the curvature is zero everywhere; that is, the rays are straight lines.

The normal $\boldsymbol{\nu}$ and binormal $\mathbf{b}$ can be expressed in terms of $\mathbf{l}$, n, and ∇n with the help of (2.2.21)

$$\boldsymbol{\nu} = \frac{1}{K} \nabla_\perp \ln(n) \ ,$$

---

[6] We recall that the curvature of a spatial curve is always positive, so $\boldsymbol{\nu}$ points into the concave segment of the curve.

$$\mathbf{b} = \mathbf{l} \times \boldsymbol{\nu} = \frac{1}{K} \ \mathbf{l} \times \frac{\nabla n}{n} = (\text{curl}\,\mathbf{l})/K \ . \tag{2.2.23}$$

The last equation accounts for the identity

$$\text{curl}\,\mathbf{l} = \mathbf{l} \times \nabla n / n \ , \tag{2.2.24}$$

which follows from

$$\text{curl}(\mathbf{l}n) = \text{curl}\,\mathbf{p} = \text{curl}\,\nabla\psi = 0 \ .$$

From (2.2.20b) the torsion of a ray, κ, is

$$\kappa = -\,\mathbf{b}\cdot\frac{d\boldsymbol{\nu}}{d\sigma} = \boldsymbol{\nu}\cdot\left(\frac{d\mathbf{b}}{d\sigma}\right) = \boldsymbol{\nu}\cdot(\mathbf{l}\cdot\nabla)\cdot\mathbf{b} \ , \tag{2.2.25}$$

where $d/d\sigma = \mathbf{l}\cdot\nabla$ is the derivative taken along the ray.

Substituting the expressions for $\mathbf{b}$ and $\boldsymbol{\nu}$ from (2.2.13) into (2.2.25) yields

$$\kappa = \frac{1}{K}\,(\mathbf{l}\cdot\nabla)\mathbf{b}\cdot\nabla_\perp \ln(n) = \frac{1}{K}\,\mathbf{b}\cdot(\mathbf{l}\cdot\nabla)\,\nabla_\perp \ln(n) \ . \tag{2.2.26}$$

Recalling the formula for $(\mathbf{l}\cdot\nabla)\cdot\mathbf{b}$ from vector algebra we can recast (2.2.25) into the form

$$2\kappa = \boldsymbol{\nu}\cdot\text{curl}\,\boldsymbol{\nu} + \mathbf{b}\cdot\text{curl}\,\mathbf{b} \ , \tag{2.2.27}$$

which will be needed in Chap.4.

The torsion κ vanishes for the rays that are plane curves as is the case in homogeneous, plane-stratified and spherically stratified media, and in two-dimensional problems. In all the other situations, including media that possess axial symmetry, the torsion is different from zero with the possible exclusion of some singled-out directions.

2.2.4 Initial Conditions for Rays. Ray Coordinates

A solution of the ray equation calls for some initial conditions. It is natural to impose such conditions at the origin of the ray[7]

$$\mathbf{r}(\tau^0) = \mathbf{r}^0(\zeta, \eta) \ , \tag{2.2.28}$$

i.e., the point on the surface where the ray emerges, and where the initial conditions (2.1.18) for the wave amplitude and the eikonal are given.

---

[7] Later we will mostly use the ray equations in the form (2.2.12). The variable τ does not appear explicitly on the right-hand sides of (2.2.12); therefore the solution $\mathbf{r}(\tau)$, in fact, depends only upon the difference τ-τ^0. This leads us to set $\tau^0 = 0$, but we shall instead assume τ^0 to be an arbitrary smooth function of ξ and η.

14

Besides this source point (2.2.28) we need to prescribe the direction of the outgoing ray or, which is the same, the initial value of the momentum $\mathbf{p}$ on Q.

Two equations for the components of the initial momentum $\mathbf{p}(r^0) = \mathbf{p}^0(\xi,\eta)$ can be obtained by differentiating the eikonal distribution $\psi(r^0) = \psi^0(\xi,\eta)$ with respect to both ξ and η

$$\mathbf{p}^0 \frac{\partial \mathbf{r}^0}{\partial \xi} = \frac{\partial \psi^0}{\partial \xi} , \quad \mathbf{p}^0 \frac{\partial \mathbf{r}^0}{\partial \eta} = \frac{\partial \psi^0}{\partial \eta} . \tag{2.2.29}$$

As a third equation we may take the eikonal equation (2.1.11) written for the initial surface Q

$$(\mathbf{p}^0)^2 = n^2(\mathbf{r}^0) . \tag{2.2.30}$$

Eqs. (2.2.29-30) define the three components of the initial momentum $\mathbf{p}^0$. Generally, the expressions for p_i^0 are fairly involved. We demonstrate them only for the simplest case of the eikonal specified on the plane $z = 0$. Letting $x^0 = \xi$, $y^0 = \eta$, and $z^0 = 0$, we have

$$p_x^0 = \frac{\partial \psi^0}{\partial \xi} , \quad p_y^0 = \frac{\partial \psi^0}{\partial \eta} , \quad p_z^0 = \pm\sqrt{\epsilon^0 - (\partial\psi^0/\partial\xi)^2 - (\partial\psi^0/\partial\eta)^2} . \tag{2.2.31}$$

Here $\epsilon^0 = n^2(\mathbf{r}^0)$, and the sign is chosen according to how the wave propagates – a plus sign if the wave travels in the positive z direction and a minus sign otherwise.

The solution of the differential equations (2.2.12), derived subject to the initial conditions (2.2.28-29), can be represented in the form

$$\mathbf{r} = \mathbf{R}(\xi,\eta,\tau) , \quad \mathbf{p} = \mathbf{P}(\xi,\eta,\tau) . \tag{2.2.32}$$

Here, the parameters ξ and η label the rays leaving the surface Q, while τ (or arc length σ) indicates a point on a fixed ray. The parameter set ξ,η,τ (or, equivalently, ξ,η,σ) is referred to as *ray coordinates* (Fig.2.1). In the general case, they are nonorthogonal.[8]

2.2.5 Ray Families and Phase Fronts

The equation $\mathbf{r} = \mathbf{R}(\xi,\eta,\tau)$ of (2.2.32) defines a family (or *congruence*) of ray produced by a given field distribution on the initial surface $\mathbf{r} = \mathbf{r}^0(\xi,\eta)$ (Fig.2.1). The equation of a ray family relates ray coordinates with Cartesian coordinates. If the Jacobian $\mathcal{D} = \partial(x,y,z)/\partial(\xi,\eta,\tau)$ is different from zero in the domain under consideration, the equation $\mathbf{r} = \mathbf{R}(\xi,\eta,\tau)$ can

---

[8] The set ξ,η,ψ may also be taken as ray coordinates [2.19]. Then the coordinate surface ψ = constant is a wave front, and the lines ξ = constant and η = constant are rays. Thus the coordinate ψ is orthogonal to ξ ad η, whereas ξ and η generally are nonorthogonal.

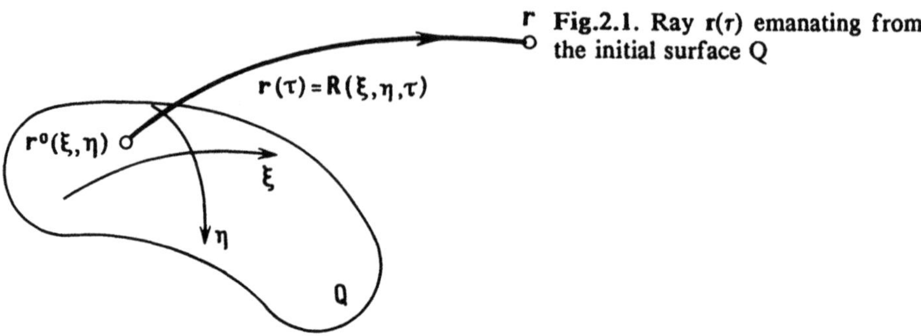

Fig.2.1. Ray $r(\tau)$ emanating from the initial surface Q

uniquely be solved for the ray coordinates ξ, η, τ corresponding to a given point of observation: $\xi = \xi(r)$, $\eta = \eta(r)$, $\tau = \tau(r)$.

The initial conditions for the eikonal $\psi = \psi^0(\xi, \eta)$ on the initial surface Q define a family of spatial phase fronts orthogonal to the family of rays $r = R(\xi, \eta, \tau)$. The set of equations governing the phase fronts has the form

$$\psi(\xi, \eta, \tau) = \psi^0(\xi, \eta) + \int_{\tau^0}^{\tau} n^2 [R(\xi, \eta, \tau)] d\tau = \text{const.} = \psi_0 , \qquad (2.2.33)$$

$$r = R(\xi, \eta, \tau) .$$

If we find $\tau = \tau_{ph}(\xi, \eta, \psi_0)$ from the first equation in (2.2.33) for a fixed ψ_0 and substitute it into the second equation in (2.2.33), then we obtain the following vector equation of the phase front

$$r = R[\xi, \eta, \tau_{ph}(\xi, \eta, \psi_0)] \equiv r_{ph}(\xi, \eta, \psi_0) . \qquad (2.2.34)$$

The phase fronts (2.2.34) can be diverse and fairly complicated in shape (Sect.2.4).

It should be emphasized that the initial conditions (2.2.28,29) are specified simultaneously for a family of rays emerging from the surface Q, rather than for separate paths as it is done in classical mechaniscs for trajectories of individual poit masses. This feature reflects the wave character of the problem at hand – to a wave field there corresponds a family or a bundle of rays rather than a single path (to contrast this difference from mechanics, the term "a coherent bundle of rays" had even be coined [2.12]). This stresses the importance of studying the whole family of rays – the ray pattern – in order to evaluate the distribution of a wave field.

In physical space, a family of rays $r = R(\xi, \eta, \tau)$ may, in general, be placed in a fairly complicated manner (Chap.3). However, in phase space $\{p, r\}$, or $\{p_j, q_j\}$, each point may be associated with a single phase trajectory passing through it; i.e., the phase trajectories do not intersect.[9] In essence this property follows from the theorem of a unique solution to the system of ordinary differential equations with initial conditions [2.17].

2.2.6 The Fermat Principle

The theory of the Hamilton-Jacobi equations (2.2.1) is intimately related to the classical calculus of variations [2.16], as the solution of (2.2.1) is equivalent to finding the functions that extremize the functional

$$\int \mathbf{p}(\partial \mathcal{H}/\partial \mathbf{p})\mathrm{d}\tau \, ,$$

i.e., reduces its first variations to zero

$$\delta \int_{\tau_1}^{\tau_2} \mathbf{p}\left(\frac{\partial \mathcal{H}}{\partial \mathbf{p}}\right) \mathrm{d}\tau \equiv \delta \int_{\tau_1}^{\tau_2} \sum_{j=1}^{n} p_j \frac{\partial \mathcal{H}}{\partial p_j} \, \mathrm{d}\tau = 0 \, . \tag{2.2.35}$$

In mechanics this corresponds to the least-action principle of Hamilton, and in geometrical optics to the Fermat principle.

When the Hamiltonian $\mathcal{H}(\mathbf{p}, \mathbf{r})$ is given by (2.2.11 or 16), the Fermat principle states that the functional

$$\psi(a, b) = \int_{\mathbf{r}_a}^{\mathbf{r}_b} n^2 \mathrm{d}\tau = \int_{\mathbf{r}_a}^{\mathbf{r}_b} n \mathrm{d}\sigma \, , \tag{2.2.36}$$

defined on the rays connecting the end points $\mathbf{r}_a$ and $\mathbf{r}_b$, is stationary (i.e., extremal) on the paths satisfying (2.2.12 or 17). The stationarity condition can be written by equating to zero the first variation of the eikonal

$$\delta\psi(a, b) = 0 \, . \tag{2.2.37}$$

The ray equations (2.2.19) then play the role of Euler's equations for the extremal of (2.2.36), which are geodesics in the non-Euclidean space of the metric [2.85]

$$\mathrm{d}s^2 = n^2(x, y, z)\cdot(\mathrm{d}x^2 + \mathrm{d}y^2 + \mathrm{d}z^2) \, .$$

In the approach under consideration, the Fermat principle is a corrolary to the eikonal equation (2.1.11).

An alternative approach to geometrical optics may be based on the Fermat principle (2.2.37) rather than the eikonal equation (2.1.11), which then follows from (2.2.37). As a proof, if the integrals in (2.2.36) are taken along the extremals (rays), then ψ can be shown [2.16, 19] to satisfy the eikonal equation (2.1.11).

---

[9] The 3D hypersurface produced by the phase trajectories (2.2.32) in the 6D phase space $\{\mathbf{p}, \mathbf{r}\}$ defines the Lagrangian manifold [2.20]. With smooth initial conditions, this manifold is also smooth. All singularities of a ray pattern (multi-path propagation, caustics, focuses, etc.) appear in projecting the Lagrangian manifold $\{\mathbf{P}(\xi,\eta,\tau), \mathbf{R}(\xi,\eta,\tau)\}$ into the coordinate space.

Elementary textbooks treat the Fermat principle as the requirement of the minimum time it takes for a wave to travel from $\mathbf{r_a}$ to $\mathbf{r_b}$. This treatment is valid only for nondispersive media since in the presence of dispersion the propagation time $\Delta t(a,b)$ is generally no longer proportional to $\psi(a,b)$ (Sect.2.7).

An up-to-date treatment of Fermat's principle (2.2.37) is based on the Huygens-Fresnel-Kirchhoff principle of *interference* of the wave disturbance, originated from a primary wave at the point a and travelling to the point b through any virtual paths. Such paths need not necessarily be solutions of characteristic equation. With this interference of the secondary wavelets, only those paths will contribute significantly to the resultant field at b for which the phase shift $\psi(a,b)$ deviates from the extremal value $\psi(a,b)$ for not more than $\lambda_0/2$. The contribution of all the other paths will be negligible as the interfering waves with drastically different phases practically cancel each other. Thus only those virtual paths "survive" the interference that are consistent with the eikonal $\psi(a,b)$, i.e., obey the Fermat principle.

The representation by means of interfering waves corresponding to different paths was used by *Feynman* in formulating of quantum mechanics via path integrals [2.3,4]. We do not intend to describe the path-integration technique as it will not be needed later, but we will recur to this technique of secondary wave interference (Sect.2.10) in formulating the conditions sufficient for the applicability of ray method. We will also discuss the physical concept of a "ray".

2.2.7 Ray Equations in Curvilinear Coordinates

The ray equations can frequently be written in a more convenient form in curvilinear coordinates. A curvilinear coordinate transformation could be applied to (2.2.12 or 17), but would be fairly cumbersome. A straightforward procedure results from the use of characteristic equations in the canonic Hamiltonian form that are left unaltered upon transformation to any curvilinear coordinates.

Orthogonal curvilinear coordinates q_i are usually of greatest interest; in these coordinates the eikonal equation (2.1.11) has the form

$$\mathscr{H}(p_i,q_i) = \frac{1}{2}\left[\sum_{j=1}^{3}\frac{p_j^2}{h_j^2(q_i)} - n^2(q_i)\right] = 0 , \qquad (2.2.38)$$

where $p_j = \partial\psi/\partial q_j$ are the generalized components of the momentum, and h_j are the Lame coefficients of the coordinates q_j ($j = 1,2,3$). Substituting the Hamiltonian (2.2.38) into the equations of characteristics (2.2.3,4) yields the ray equations

$$\frac{dq_i}{d\tau} = h_i^{-2} p_i \ , \qquad \frac{dp_i}{d\tau} = n \frac{\partial n}{\partial q_i} + \sum_{j=1}^{3} h_j^{-3} \frac{\partial h_j}{\partial q_i} p_j^2 \ , \qquad (2.2.39)$$

with the eikonal defined by (2.2.13).

Using in place of p_j the momentum components in curvilinear coordinates

$$\tilde{p}_i = \frac{1}{h_i} p_i = \frac{1}{h_i} \frac{\partial \psi}{\partial q_i} \qquad (2.2.40)$$

changes the ray equations (2.2.39) to still another form

$$\frac{dq_i}{d\tau} = \frac{1}{h_i} \tilde{p} \ ,$$

$$\frac{d\tilde{p}_i}{d\tau} = \frac{1}{h_i} n \frac{\partial n}{\partial q_i} + \frac{1}{h_i} \sum_{j \neq i}^{3} \frac{\tilde{p}_i}{h_j} \left[\tilde{p} \frac{\partial h_j}{\partial q_i} - \tilde{p}_j \frac{\partial h_i}{\partial q_j} \right] . \qquad (2.2.41)$$

At times, (2.2.41) becomes more convenient if converted to direction cosines of the ray in curvilinear coordinates by

$$\tilde{p}_i = n\cos\alpha_i \ , \qquad \sum_{i=1}^{3} \cos^2\alpha_i = 1 \ . \qquad (2.2.42)$$

Then

$$\frac{d\tilde{p}_i}{d\tau} = - n\sin\alpha_i \frac{d\alpha_i}{d\tau} + \cos\alpha_i \sum_{j=1}^{3} \frac{\partial n}{\partial q_j} \frac{dq_j}{d\tau}$$

$$= - n\sin\alpha_i \frac{d\alpha_i}{d\tau} + n\cos\alpha_i \sum_{j=1}^{3} \frac{1}{h_j} \frac{\partial n}{\partial q_j} \cos\alpha_j \ . \qquad (2.2.43)$$

Writing (2.2.41) in spherical coordinates r,θ,ϕ, letting $q_1 = r$, $q_2 = r$, $q_2 = \theta$, $q_3 = \phi$, and noting that $h_1 = 1$, $h_2 = r$, and $h_3 = r\sin\theta$ transforms (2.2.41) into

$$\frac{dr}{d\tau} = \tilde{p}_r \ , \qquad \frac{d\theta}{d\tau} = \frac{1}{r} \tilde{p}_\theta \ , \qquad \frac{d\phi}{d\tau} = \frac{\tilde{p}_\theta}{r\sin\theta} ,$$

$$\frac{d\tilde{p}_r}{d\tau} = n \frac{\partial n}{\partial R} + \frac{1}{r} \tilde{p}_\theta^2 + \frac{1}{r} \tilde{p}_\phi^2 \ ,$$

$$\frac{d\tilde{p}_\theta}{d\tau} = \frac{1}{r}\left[n\frac{\partial n}{\partial \theta} - \tilde{p}_r\tilde{p}_\theta + \cot\theta\tilde{p}_\phi{}^2\right],$$

$$\frac{d\tilde{p}_\phi}{d\tau} = \frac{1}{r\sin\theta}\left[n\frac{\partial n}{\partial \phi} - \tilde{p}_r\tilde{p}_\phi\sin\theta - \tilde{p}_\theta\tilde{p}_\phi\cos\theta\right], \tag{2.2.44}$$

where $\tilde{p}_r = \partial\psi/\partial r$, $\tilde{p}_\theta = r^{-1}(\partial\psi/\partial\theta)$, and $\tilde{p}_\phi = (1/r\cdot\sin\theta)(\partial\psi/\partial\phi)$ are related by the eikonal equation

$$\tilde{p}_r{}^2 + \tilde{p}_\theta{}^2 + \tilde{p}_\phi{}^2 = n^2 . \tag{2.2.45}$$

We can eliminate $\tilde{p}_\phi$ between (2.2.45 and 36) and so reduce the number of the unknowns. This set of equations defines the ray path $r=r(\tau)$, $\theta=\theta(\tau)$, $\phi=\phi(\tau)$ and its direction $\tilde{p}_r=\tilde{p}_r(\tau)$, $\tilde{p}_\theta=\tilde{p}_\theta(\tau)$, $\tilde{p}_\phi=\tilde{p}_\phi(\tau)$ parametrically. A similar system in the two-dimensional case ($\tilde{p}_\phi\equiv0$) has the form

$$\frac{dr}{d\tau} = \tilde{p}_r , \qquad \frac{d\tilde{p}_r}{d\tau} = n\frac{\partial n}{\partial r} + \frac{1}{r}\tilde{p}_\theta{}^2 ,$$

$$\frac{d\theta}{d\tau} = \frac{1}{r}\tilde{p}_\theta , \qquad \frac{d\tilde{p}_\theta}{d\tau} = \frac{1}{r}\left[n\frac{\partial n}{\partial \theta} - \tilde{p}_r\tilde{p}_\theta\right]. \tag{2.2.46}$$

Introducing the "angle of refraction" α by

$$p_r = n\cos\alpha , \quad p_\theta = n\sin\alpha$$

reduces (2.2.46) to the three equations

$$\frac{dr}{d\tau} = n\cos\alpha , \qquad \frac{d\theta}{d\tau} = \frac{n}{r}\sin\alpha ,$$

$$\frac{d\alpha}{d\tau} = \frac{1}{r}\left[\frac{\partial n}{\partial \theta}\cos\alpha - \frac{\partial(nr)}{\partial r}\sin\alpha\right]. \tag{2.2.47}$$

The equation for $d\theta/d\tau$ can also be eliminated by virtue of (2.2.45). If $\alpha \neq \pi/2$ then dividing (2.2.47) by the first relation of (2.2.47) yields

$$\frac{d\theta}{dr} = \frac{1}{r}\tan\alpha , \qquad \frac{d\alpha}{dr} = \frac{1}{nr}\left[\frac{\partial n}{\partial \theta} - \frac{\partial(rn)}{\partial r}\frac{d\theta}{dr}\right]. \tag{2.2.48}$$

This system results in a ray path expressed as $\alpha = \alpha(r)$, $\theta = \theta(r)$.

2.2.8 Other Types of Ray

For a long time geometrical-optical rays had been the only type of ray known. The situation changed when *Keller* proposed his *geometrical theory*

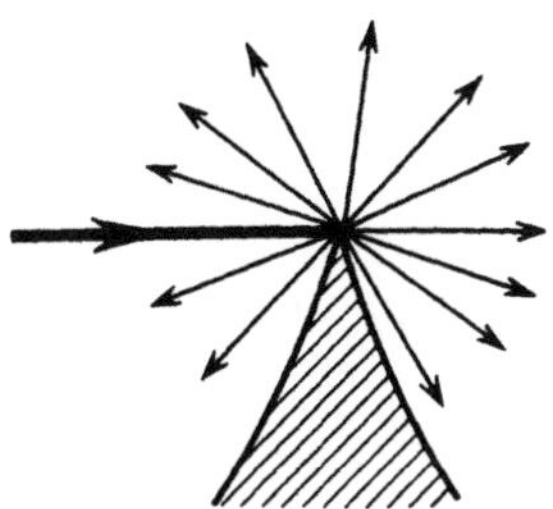

Fig.2.2. Edge rays

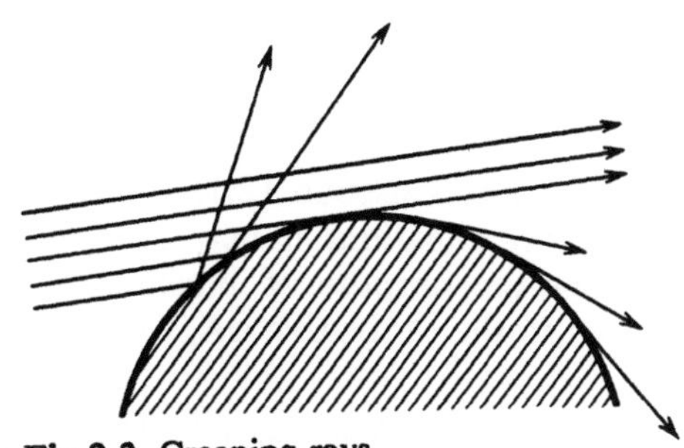

Fig.2.3. Creeping rays

of diffraction [2.19, 21-24], which uses diffracted rays as well as usual geo-metrical-optical rays to describe diffraction effects in the geometrical shadow behind the obstacle. The two major categories of diffracted rays include *edge (vertex) rays* which emanate from the edges and vertices on illuminated obstacles (Fig.2.2), and *creeping rays* which creep around the surface of smooth obstacles (Fig.2.3).

Another specific category of diffracted rays appears on a refracting interface, this gives rise to the diffracted rays corresponding to *lateral waves* (Fig.2.4) [2.23-25]. We should also mention *complex rays*. They may be associated with the complex eikonal and describe the wave field in a caustic shadow, the field of a Gaussian wave bundle, or the surface and leaking waves, to name only a few [2.23, 26, 27].

Having been introduced, these rays can be handled in much the same manner as ordinary geometrical-optics rays. More specifically, their paths are described by (2.2.12) and the field on these rays obeys the laws of geo-metrical optics (Sect.2.3). The applicability of various ray methods can be assessed by virtue of the criteria discussed in Sect.2.10.

Space-time rays appearing in problems of nonstationary wave propa-gation or in those of more general nature existing in anisotropic nonsta-tionary media with temporal and spatial dispersion of waves are considered in detail in Sects.2.7 and 4.4. Various types of space-time rays (complex, vertex, and others) related to the geometrical theory of diffraction were discussed in [2.28].

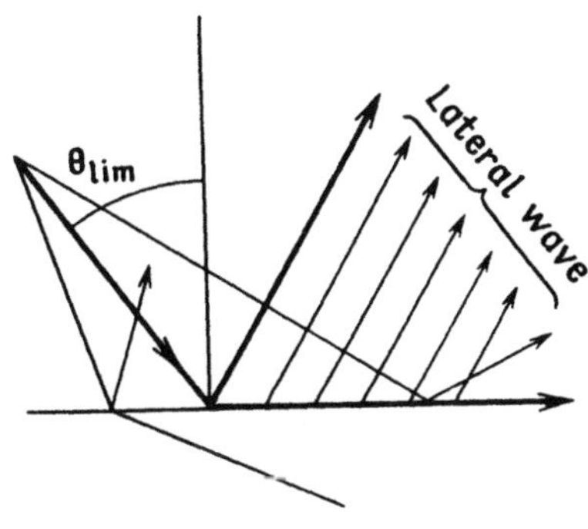

Fig.2.4. Lateral wave rays

2.3 Wave Amplitude

2.3.1 Formal Solution of the Transport Equation

The transport equation (2.1.14) contains partial derivatives, but with the help of ray equations it may be reduced to an ordinary differential equation. For the purpose of illustration, the quantity $(\nabla A_0 \nabla \psi)$ in (2.1.14) is reduced to the derivative $dA_0/d\tau$

$$(\nabla A_0 \nabla \psi) = \frac{dA_0}{d\sigma}|\nabla \psi| = \frac{dA_0}{d\tau}\frac{d\tau}{d\sigma}n = \frac{dA_0}{d\tau} , \qquad (2.3.1)$$

because $\nabla \psi = n\mathbf{l}$, $\mathbf{l}\cdot\nabla = d/d\sigma$, and $d\tau = d\sigma/n$. Then the transport equation becomes

$$2\frac{dA_0}{d\tau} + A_0\Delta\psi = 0 \qquad \text{or} \qquad (2.3.2)$$

$$\frac{dA_0^2}{d\tau} + A_0^2\Delta\psi = \frac{dA_0^2}{d\tau} + (A_0^2)\mathrm{div}\mathbf{p} = 0 . \qquad (2.3.3)$$

Integration gives

$$A_0^2(\tau) = A_0^2(\tau^0)\exp\left(-\int_{\tau^0}^{\tau}\mathrm{div}\mathbf{p}\,d\tau\right), \qquad (2.3.4)$$

where $A_0(\tau^0) = A_0^0$ is the initial value of the wave amplitude defined by initial conditions on the surface Q.

To derive $\mathrm{div}\mathbf{p} = \Delta\psi$ in (2.3.4) we invoke the Liouville formula which yields when the ray equation $d\mathbf{r}/d\tau = \mathbf{p}$ has the solution $\mathbf{r} = \mathbf{r}(\xi,\eta,\tau)$,

$$\mathrm{div}\mathbf{p} = \frac{d}{d\tau}\ln\mathscr{D}(\tau) , \qquad \text{where} \qquad (2.3.5)$$

$$\mathscr{D}(\tau) = \frac{\partial(x,y,z)}{\partial(\xi,\eta,\tau)} = \begin{vmatrix} \dfrac{\partial x}{\partial\xi} & \dfrac{\partial x}{\partial\eta} & \dfrac{\partial x}{\partial\tau} \\[2mm] \dfrac{\partial y}{\partial\xi} & \dfrac{\partial y}{\partial\eta} & \dfrac{\partial y}{\partial\tau} \\[2mm] \dfrac{\partial z}{\partial\xi} & \dfrac{\partial z}{\partial\eta} & \dfrac{\partial z}{\partial\tau} \end{vmatrix} \qquad (2.3.6)$$

is the Jacobian of the transformation from the Cartesian coordinates x,y,z to the ray coordinates ξ,η,τ. It can be expressed in terms of

$$\mathbf{r}_\xi \equiv \frac{\partial\mathbf{r}}{\partial\xi} , \quad \mathbf{r}_\eta \equiv \frac{\partial\mathbf{r}}{\partial\eta} , \quad \mathbf{r}_\tau \equiv \frac{\partial\mathbf{r}}{\partial\tau} = \mathbf{p} , \qquad (2.3.7)$$

as the scalar triple product

$$\mathscr{D}(\tau) = (\mathbf{r}_\xi \times \mathbf{r}_\eta){\cdot}\mathbf{r}_\tau = (\mathbf{r}_\xi, \mathbf{r}_\eta, \mathbf{r}_\tau) = (\mathbf{r}_\xi \times \mathbf{r}_\eta){\cdot}\mathbf{p} \quad . \tag{2.3.8}$$

Substituting (2.3.5) into (2.3.4) ultimately results in

$$A_0 = A_0^0 \sqrt{\frac{\mathscr{D}(\tau^0)}{\mathscr{D}(\tau)}} \equiv A_0^0/\sqrt{\mathscr{J}} \quad . \tag{2.3.9}$$

The quantity

$$\mathscr{J} = \frac{\mathscr{D}(\tau)}{\mathscr{D}(\tau^0)} = \frac{(\mathbf{r}_\xi \times \mathbf{r}_\eta){\cdot}\mathbf{p}}{(\mathbf{r}_\xi, \mathbf{r}_\eta, \mathbf{p})_{\tau=\tau^0}} = \frac{(\mathbf{r}_\xi \times \mathbf{r}_\eta){\cdot}\mathbf{p}}{(\mathbf{r}_\xi, \mathbf{r}_\eta, \mathbf{p})^0} \tag{2.3.10}$$

is called the ray divergence (the geometrical meaning of this definition will be discussed later in Sect.2.3.3).

If a family of rays is described by the equation $\mathbf{r} = \mathbf{r}(\xi, \eta, \sigma)$ where σ is the arc length, then

$$\mathscr{D}(\tau) = (\mathbf{r}_\xi \times \mathbf{r}_\eta){\cdot}\mathbf{r}_\tau = (\mathbf{r}_\xi \times \mathbf{r}_\eta){\cdot}\mathbf{r}_\sigma \frac{\mathrm{d}\sigma}{\mathrm{d}\tau} = n\mathscr{D}(\sigma) \quad , \qquad \text{and} \tag{2.3.11}$$

$$\mathscr{J} = \frac{n\mathscr{D}(\sigma)}{n^0\mathscr{D}(\sigma^0)} = \frac{n(\mathbf{r}_\xi \times \mathbf{r}_\eta){\cdot}\mathbf{l}}{n^0(\mathbf{r}_\xi, \mathbf{r}_\eta, \mathbf{l})^0} \quad , \tag{2.3.12}$$

where n^0 is the initial value of the refractive index (at $\sigma = \sigma^0$), and $\mathbf{l} = \mathbf{p}/n = \mathbf{r}_\sigma$ is the unit vector tangent to the ray.

2.3.2 Rays and the Direction of Energy Flow

From the Helmholtz equation (2.1.1) we can readily derive that with a real refractive index $n(\mathbf{r})$ the quantity

$$\mathbf{I} = \frac{1}{2ik_0} (u^* \nabla u - u\nabla u^*) \tag{2.3.13}$$

is conserved, i.e.,

$$\mathrm{div}\,\mathbf{I} = 0 \quad . \tag{2.3.14}$$

Therefore $\mathbf{I}$ may be adopted as a measure of energy flux.[10]

---

[10] If we neglect polarization effects and think of u being the strength of electromagnetic field, then we obtain for the Poynting vector $\mathbf{S} = c\mathbf{I}/8\pi$. For sound waves, with u being the velocity potential, the energy-flux vector (the Umov vector) is $\mathbf{S} = \rho\omega^2\mathbf{I}/2c$, where ρ is the density of the medium, and c the velocity of sound. In quantum mechanics, $\mathbf{I}$ (appropriately normalized) is the probability density flow.

Substituting (2.1.13) into (2.3.13) and assuming a real amplitude A_m [if A_0 is real then by virtue of the transport equation (2.1.14) the amplitudes of all the subsequent approximations are real, too] yields

$$I = pA_0^2 + k_0^{-2}\left[p(A_1^2 - 2A_0A_1) + A_1\nabla A_0 - A_0\nabla A_1 + ...\right] . \qquad (2.3.15)$$

This leads to the important conclusion that the leading term in a ray expansion describes the flux of energy directed along the rays

$$I \simeq pA_0^2 \equiv I_0 .$$

The transverse to the ray component of an energy flux appears only upon account of the first and higher terms, and, in view of (2.3.15), is of the second order in $1/k_0$.[11]

2.3.3 Conservation of Energy Flux in a Ray Tube

The ray-optics approximation yields the energy conservation (2.3.14) in the form

$$\mathrm{div}I_0 = \mathrm{div}(pA_0^2) = 0 , \qquad (2.3.16)$$

which is equivalent to the transport equation (2.1.14a) since $p = \nabla\psi$ and $\mathrm{div}p = \Delta\psi$. It can be used to connect the amplitude variation of the leading term A_0 with the cross section da of an indefinitely narrow ray tube formed by a "pencil" of closely spaced rays (Fig.2.5). On integrating (2.3.16) over the volume bounded by the ray tube we have by the divergence theorem

$$\int_V \mathrm{div}I_0 \, dV = \oint_\Sigma I_0 da ,$$

where Σ is the surface of the confining ray tube. The integral over the side walls turns out to be zero as here da is at right angles with p so that $I_0 \cdot da = 0$, whereas the integral over the section ends gives

$$I_0 \cdot da = \mathrm{const.} = I_0^0 \cdot da^0 . \qquad (2.3.17)$$

Observing here that $I_0 = A_0^2 p = A_0^2 n$ yields the zeroth-approximation amplitude

$$A_0 = A_0^0(\xi,\eta)\sqrt{\frac{n^0 da^0}{nda}} . \qquad (2.3.18)$$

---

[11] It would be more accurate to speak of smallness with respect to the dimensionless parameter $\mu = 1/kL$.

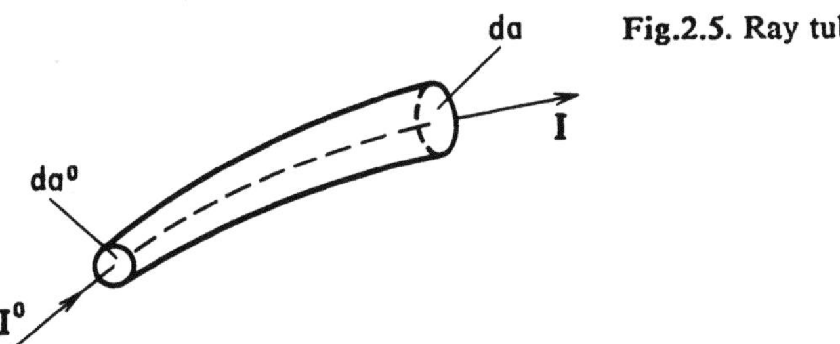

Fig.2.5. Ray tube

In one-dimensional problems with da = da^0 = constant,

$$A_0 = A_0^0 \sqrt{n^0/n} \ , \tag{2.3.19}$$

i.e., the amplitude varies inversely as $\sqrt{n}$. In a homogeneous medium, where $n = n^0$ = constant, the amplitude varies with the cross section of the ray tube only; i.e., when the rays diverge or converge,

$$A_0 = A_0^0 \sqrt{da^0/da} \ . \tag{2.3.20}$$

The general expression (2.3.18) looks essentially like $A_0 = A_0^0/\sqrt{\mathcal{J}}$ in (2.3.9) with $\mathcal{J}$ being equal to the product of the true divergence of rays da/da^0 and the ratio n/n^0, i.e.,

$$\mathcal{J} = nda/n^0 da^0 \ . \tag{2.3.21}$$

So $\mathcal{J}$ may appropriately be named the generalized divergence, but for brevity we shall refer to it as the simply divergence.

Formula (2.3.9) as well as (2.3.17, 18) suggests that in ray optics the field amplitude decreases along a ray if the divergence increases, and vice versa when the rays converge.

One can readily establish the equivalence of (2.3.21) and (2.3.10) or (2.3.12) for the divergence $\mathcal{J}$. The elementary volume of a ray tube $dV = dxdydz = d\sigma da$ being expressed in terms of $\mathscr{D}(\tau)$ or $\mathscr{D}(\sigma)$ becomes

$$dV = \mathscr{D}(\tau)d\xi d\eta d\tau = \mathscr{D}(\tau)d\xi d\eta \frac{d\sigma}{n} = \mathscr{D}(\sigma)d\xi d\eta d\sigma \ , \tag{2.3.22}$$

whence

$$da = \frac{dV}{d\sigma} = n^{-1} \mathscr{D}(\tau)d\xi d\eta = \mathscr{D}(\sigma)d\xi d\eta \ . \tag{2.3.23a}$$

Similarly

$$da^0 = \frac{1}{n^0} \mathscr{D}(\tau^0)d\xi d\eta = \mathscr{D}(\sigma^0)d\xi d\eta \ . \tag{2.3.23b}$$

Substituting (2.3.23) into (2.3.21) leads to a formula equivalent to (2.3.10 or 12). Note in passing the relationship between the Jacobians $D(\tau)$ and $D(\sigma)$ and the relative normal section of the ray tube $da/d\xi d\eta$

$$D(\sigma) = D(\tau)/n = da/d\xi d\eta \; . \tag{2.3.24}$$

2.3.4 The Field due to a Point Source in an Inhomogeneous Medium

For a point source the initial conditions are given on a sphere of small radius $R_0 \ll L$ within which the medium can still be deemed homogeneous. Let a source be emitting a directional spherical wave having a normalized radiation function $G(\theta, \phi)$, where θ and ϕ are the colatitude and longitude which we will use as ray coordinates. Then

$$A^0(\theta,\phi) = B^0 G(\theta,\phi)/R_0 \; , \tag{2.3.25}$$

where B^0 is a coefficient related to the strength of the source. Inserting (2.3.25) into (2.3.18) and noting that at small distances from the source $da^0 = R_0{}^2 d\Omega$, with $d\Omega = \sin\theta d\theta d\phi$ being the differential solid angle, we obain for the amplitude A an expression independent of R_0, i.e.,

$$A = B^0 G(\theta, \phi) \sqrt{\frac{n^0}{n} \frac{d\Omega}{da}} \; . \tag{2.3.26a}$$

This formula takes on another form if we substitute $\mathscr{D}(\tau)d\theta d\phi/n$ for da, by virtue of (2.3.24),

$$A = B^0 G(\theta, \phi) \sqrt{\frac{n^0 \sin\theta}{\mathscr{D}(\tau)}} \; . \tag{2.3.26b}$$

2.3.5 The Resultant Field in the Ray-Optics Approximation

Combining the expressions for the eikonal (2.2.9) and the zeroth-approximation amplitude (2.3.9) we write the resultant field along a given ray as

$$u(\tau) = \frac{u^0(\xi,\eta)}{\sqrt{\mathscr{J}}} \exp\left(ik_0 \int_{\tau^0}^{\tau} n^2 d\tau\right), \qquad \text{where} \tag{2.3.27}$$

$$u^0(\xi,\eta) = A_0^0(\xi,\eta) \exp\left[ik_0\psi^0(\xi,\eta)\right] \tag{2.3.28}$$

is the initial field. Equation (2.3.27) suggests that the field on a ray depends only on the initial field at the point $r^0(\xi,\eta)$ where the ray leaves surface Q. This indicates the *local nature* of the wave disturbance transfer described by the leading term of the geometrical-optics expansion. Various

nonlocal phenomena, that is, deviations from the ray-optics law, are reerred to as *diffractional*. The validity of the locality principle of geometrical
optics will be discussed in Sect.2.10.

For a majority of applications it is desirable to know the field at a
fixed point r rather than along the ray. For this purpose we should first
find the ray travelling through this point, that is, solve the vector ray
equation

$$\mathbf{R}(\xi,\eta,\tau) = \mathbf{r} , \tag{2.3.29}$$

which is equivalent to the system of three equations in the ray parameters
ξ, η, and τ. The respective values of $\xi(\mathbf{r})$ and $\eta(\mathbf{r})$ define the point where
the ray leaves the initial surface Q

$$\mathbf{r}^0 = \mathbf{r}^0[\xi(\mathbf{r}),\eta(\mathbf{r})] , \tag{2.3.30}$$

whereas the quantity $\tau(\mathbf{r})$ determines the "distance" from the initial point $\mathbf{r}^0$
to the point in observation $\mathbf{r}$. After this we may compute the ray field by
(2.3.27).

It may so happen that (2.3.30) allows several sets of parameter $\xi_\nu(\mathbf{r})$,
$\eta_\nu(\mathbf{r})$, $\tau_\nu(\mathbf{r})$ ($\nu = 1,2,...$). This implies that several rays emanating from different points on the initial surface Q come to this point. Each of them
arrives with its own amplitude and phase defined by (2.2.9) and (2.3.9).
With $\xi = \xi_\nu(\mathbf{r})$, $\eta = \eta_\nu(\mathbf{r})$, and $\tau = \tau_\nu(\mathbf{r})$, then the resultant field will be
given as the sum

$$u(\mathbf{r}) = \sum_\nu \frac{u^0(\xi_\nu,\eta_\nu)}{\sqrt{|\mathscr{J}_\nu|}} \exp(\mathrm{i}k_0 \int_{\tau_\nu^0}^{\tau_\nu} n^2 \mathrm{d}\tau + \delta\Psi_\nu) , \tag{2.3.31}$$

where all ν-indexed values correspond to the νth ray. The term $\delta\Psi_\nu$ known
as the phase shift at the caustic is added to account for the change of sign
by the ray divergence $\mathscr{J}$ at the caustic (Sect.2.4.7).

To sum up, let us retrace the computational sequence that we have
followed in deriving the geometrical-optics field:
i) Solve the ray equations (2.2.12) with the initial conditions specified on
    the initial surface Q.
ii) Find the rays incoming to the observation point $\mathbf{r}$, i.e., derive from
    (2.3.29) parameters ξ, η, and τ.
iii) Determine the eikonals by (2.2.9) and the amplitudes by (2.3.9), sub-
    ject to the initial conditions corresponding to the initial points of the
    rays.
iv) Determine the caustic phase shift $\delta\Psi$ as shown in Sect.2.4.7.
v) Compute the total field by (2.3.31).

2.3.6 The Field Amplitudes of Higher-Order Approximations

The transport equation (2.1.14c) in the amplitude A_m can be written as

$$2 \frac{dA_m}{d\tau} + A_m \frac{d\ln \mathscr{D}(\tau)}{d\tau} = - \Delta A_{m-1} \ . \tag{2.3.32}$$

This equation can readily be solved by quadrature

$$A_m = \frac{A_m^0}{\sqrt{\mathscr{D}}} - \frac{1}{2\sqrt{\mathscr{D}(\tau)}} \int_{\tau^0}^{\tau} \Delta A_{m-1}(\tau') \sqrt{\mathscr{D}(\tau')} \ d\tau' \ . \tag{2.3.33}$$

For a real A_0, all A_m will also be real-valued. An analysis of higher-order approximations for a homogeneous medium can be found in [2.22-24].

As follows from (2.3.33) the *local nature* of wave-disturbance transfer *no longer holds* in higher-order approximations of geometrical optics. To prove this, the Laplacian ΔA_{m-1} in (2.3.33) implies that A_m depends not only upon the values of A_{m-1} on the ray norm, but also upon the values of A_{m-1} in the vicinity of the ray. The amplitudes of higher approximations, therefore, describe *diffraction effects*.

2.3.7 Accounting for Weak Absorption

In an absorbing medium the square of the refractive index is a complex-valued quantity $n^2 = \epsilon = \epsilon' + i\epsilon''$ (in electrodynamics ϵ is the complex permittivity). If the absorption is weak, $\epsilon'' \ll \epsilon'$, we may account for it by a minor modification in the ordinary computation procedure. To this end we will assume ϵ'' to be a small quantity of the order of $\mu \ll 1$. Then, accurate to the terms of order μ^2, we have for $n = \sqrt{\epsilon} = n' + in''$

$$n' \simeq \sqrt{\epsilon'} \ , \quad n'' = \epsilon''/2\sqrt{\epsilon'} \ . \tag{2.3.34}$$

At $\epsilon'' \simeq \mu \ll 1$, among the first-order terms in (2.1.10) there is a term $\epsilon'' A_0$. In translating from dimensionless to dimensional variables, we obtain the new term $k_0 \epsilon'' A_0$ in the transport equation for the zeroth-approximation amplitude

$$2(\nabla A_0 \nabla \psi) + A_0 \nabla \psi + k_0 \epsilon'' A_0 = 0 \ . \tag{2.3.35}$$

The eikonal equation becomes

$$(\nabla \psi)^2 = \epsilon' \ . \tag{2.3.36}$$

An alternative derivation of (2.3.35) can be obtained by invoking the Debye approximation technique, i.e., manipulating ϵ'' into a a quantity of the order of $1/k_0$. It would suffice to let $\epsilon'' = \kappa/k_0$ assuming κ to be the quantity of zeroth order in $1/k_0$. Upon equating to zero the coefficients of like powers of k_0, (2.1.14) acquires the term $\kappa A_0 = k_0 \epsilon'' A_0$, i.e., we arrive at (2.3.34).

The term $k_0 \epsilon'' A_0$ in (2.3.34) causes an additional factor to appear in the solution of (2.3.9) to describe the attenuation of the wave

$$A_0 = \frac{A_0{}^0(\xi,\eta)}{\sqrt{\mathscr{J}}}\,\mathscr{P}\,, \qquad \text{where} \tag{2.3.37}$$

$$\mathscr{P} = \exp\!\left(-\frac{k_0}{2}\int_{\tau^0}^{\tau}\epsilon''d\tau\right) = \exp\!\left(-k_0\int_{\sigma^0}^{\sigma}n''d\sigma\right). \tag{2.3.38}$$

The resultant zeroth-approximation field becomes

$$u = \frac{u^0(\xi,\eta)}{\sqrt{\mathscr{J}(\tau)}}\,\exp\!\left(ik_0\int_{\tau^0}^{\tau}\epsilon'd\tau - \frac{k_0}{2}\int_{\tau^0}^{\tau}\epsilon''d\tau\right). \tag{2.3.39}$$

The strong-absorption case would call for a geometrical-optical approximation in the complex domain [2.26] accounting for absorption along the ray path.

2.4 Caustics

2.4.1 Fundamental Properties

Caustic surfaces or simply caustics are the envelopes of ray families. The Jacobian $\mathscr{D}(\tau)$ vanishes at caustics, i.e.,

$$\mathscr{D}(\tau) = 0 \quad \text{at the caustic .} \tag{2.4.1}$$

This corresponds to the main feature of the envelope: at $\mathscr{D}(\tau) = 0$ the equation of the ray family $\mathbf{r} = \mathbf{R}(\xi,\eta,\tau)$ cannot be resolved uniquely for ξ, η, and τ, corresponding to an infinite narrowing of the rays (Fig.2.6).

The position of the caustics is defined by $\mathbf{r} = \mathbf{R}(\xi,\eta,\tau)$ subject to (2.4.1), i.e., by the system of equations

$$\mathbf{r} = \mathbf{R}(\xi,\eta,\tau)\,, \quad \mathscr{D}(\xi,\eta,\tau) = 0\,, \tag{2.4.2}$$

where it is taken into account that the Jacobian (2.3.6) is a function of coordinates ξ and η. Eliminating from (2.4.2) the parameter τ we may transfer the equation of the caustic into its vector parametric form

$$\mathbf{r} = \mathbf{R}[\xi,\eta,\tau(\xi,\eta)] = \mathbf{r}_c(\xi,\eta)\,, \tag{2.4.3}$$

where $\tau = \tau(\xi,\eta)$ is to be found from $\mathscr{D}(\xi,\eta,\tau) = 0$.

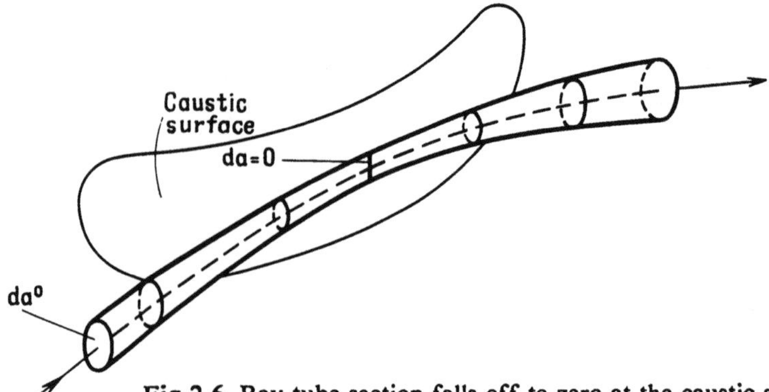

Fig.2.6. Ray tube section falls off to zero at the caustic surface

According to (2.4.3) the position of a point on the caustic can be defined by the same coordinates as on the initial surface. Occasionally it appears more convenient to introduce on a caustic its "own" two curvilinear caustic coordinates which along with a third parameter giving the deviation from the caustic, define the position of a point in space.

In one-dimensional problems where the field is described by the ordinary differential equation

$$d^2u/dz^2 + k_0^2 \epsilon(z)u = 0 , \qquad (2.4.4)$$

caustics have their analogs in the turning point $z = z_{tp}$ where the coefficient $\epsilon(z)$ vanishes. The name "turning point" was borrowed from mechanics where at $\epsilon(z) = 0$ the kinetic and potential energies balance each other, so the particle reverses its motion. Trajectories of rays in space also have their turning points (Chap.3).

Smooth caustics containing no singularities are termed *simple* or *nonsingular*. In the vicinity of a simple caustic two rays intersect tangent to the caustic (Fig.2.7a). If caustics involve singularities such as cusps, swallowtails, etc. these singularities are accompanied by congestion of a large number of rays. So in the neighborhood of a cusp, specifically inside a caustic cusp (Fig.2.7b) three rays intersect.

Each transition through a nonsingular region of a caustic causes a pair of rays to appear or disappear, as the real-valued roots of (2.3.29) appear in pairs as **r** varies. Figure 2.7 illustrates an application of this simple rule by showing a nonsingular caustic and a caustic with a single cusp point.

The importance of caustic surfaces for wave problems is that they define a family of rays as a whole, and lead to a qualitative assessment of the field behavior throughout the space. If the ray direction tangent to the caustic is available, then the family of these rays can be reconstructed from the shape of the caustic. This procedure can be especially simple for rays in a homogeneous medium where the rays are straight lines tangent to caustics.

Regions into which rays cannot penetrate form geometrical or caustic shadows. In the geometrical-optics approximation, to a caustic shadow

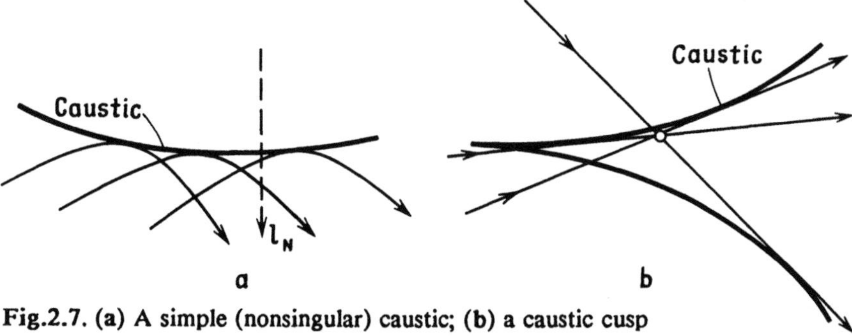

Fig.2.7. (a) A simple (nonsingular) caustic; (b) a caustic cusp

there corresponds a zero field. Actually the field there is other than zero and is of diffractional origin (in the simplest case, the field in a shadow vanishes exponentially away from caustics). Under certain conditions it can be derived by the complex generalization of geometrical optics [2.26].

2.4.2 Wave-Field Focusing on Caustics

Visible light focused into the vicinity of caustics can often be observed with a naked eye (light spot at the focal point of a lens, shallow-water bottom dappled with sunlight, etc. - a number of interesting examples have been listed in [2.29].) At other frequencies, the field focussed on caustics is to be detected instrumentally by acoustic probes, radio receivers, etc.

Close to a caustic, the cross section of a ray tube and the associated divergence of rays $\mathcal{J} = \mathcal{D}/\mathcal{D}^0$ decrease (Fig.2.6). By virtue of (2.3.9) this increases the amplitude A_0. At a caustic $D(\tau) = 0$ so A_0 tends to infinity. In geometrical optics the amplitudes A_m of the higher approximations ($m \geq 1$) are infinite at caustics, and there they have singularities of order higher than A_0 due to the presence of ΔA_{m-1} in (2.3.33).

The Helmholtz equation and other fundamental equations of wave theory do not permit infinite values of the field other than at sources. Therefore the singularity of the amplitudes A_0, A_1, etc. implies that the geometrical-optics solution *does not apply* both at caustics and in their vicinity.

Exact and approximate solutions of wave theory that work on caustics indicate that close to the caustics the field does concentrate, or focusses,[12] though the values of a field are, of course, finite. By way of example, Fig.2.8 illustrates the behavior of the square of the magnitude of the resultant field $|u|^2$ in the vicinity of a nonsingular caustic as a function of the normal distance from the caustic, l_N. In the bright region ($l_N > 0$) where two rays intersect, they give rise to interference of the two waves. Closer to the caustic the geometrical-optics field tends to infinity (dashed line) whereas the exact wave function remains finite. In the shadow region ($l_N < 0$) the field decays exponentially.

---

[12] The intensity can indeed be very great, sufficient to burn a piece of paper, for example. It is for this reason that such surfaces are called caustics.

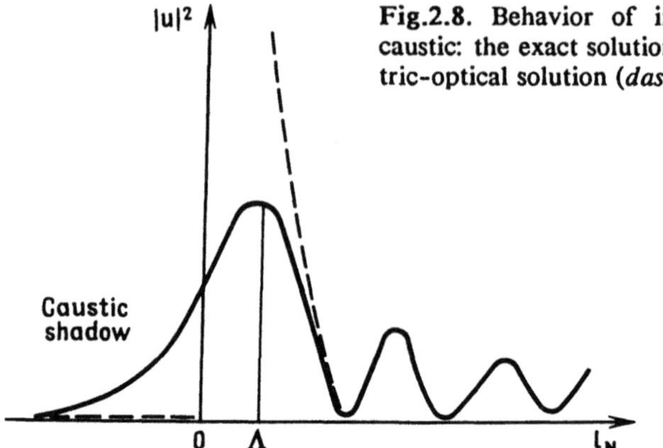

Fig.2.8. Behavior of intensity $|u|^2$ near a simple caustic: the exact solution (*solid line*) and the geometric-optical solution (*dashed*)

The phenomenon of focussing is especially strong within a certain finite region which we shall call the *caustic zone*. We can roughly estimate this region by the distance Λ from the caustic to the first interference maximum (Fig.2.8). Section 2.10 discusses a simple way to determine the width of this region, and estimates the field intensity at a caustic. Furthermore, it treats other topics associated with focusing, such as index and factor of focusing, observability of caustics, etc.

2.4.3 Types of Caustics

The aforementioned nonsingular caustic and a caustic with one cusp are the simplest types of caustics known. Real caustics can have a much more complicated shape. They are bound to appear, for example, in optical systems of mirrors and lenses with aberrations [2.30-32] and in propagating electromagnetic, sound, elastic, and other waves in graded index media. It appears that even in media of relatively simple structure (say plane stratified), caustics can possibly form, having several cusps, pockets, etc. [2.25, 33-38]. Complex-shaped caustics appear also in nonuniform (range-dependent) waveguides [2.25, 39] and in inhomogeneous random media [2.40, 41].

In spite of the fact that caustics have been investigated for a comparatively long time, it is only lately that they have been classified. This classification relies on *catastrophe theory* (R. Thom), which studies singularities of differentiable mappings (diffeomorphisms). The fundamental results of this theory were obtained by *Arnold* and co-workers [2.42-45] and others [2.46-47]. Discussions bearing relavance to wave problems may be found, for example, in [2.29, 42, 48-55]. Various specific caustics will be discussed in Chap.3 in relation to particular problems. Until later we confine ourselves to listing the most recurrent caustics.

The method of geometrical optics deals with the mapping of points of the initial surface Q: $r = r^0(\xi, \eta)$ onto three-dimensional space $\{x, y, z\}$. The mapping is performed by the functions

$$x = X(\xi, \eta, \tau) , \quad y = Y(\xi, \eta, \tau) , \quad z = Z(\xi, \eta, \tau) , \qquad (2.4.5)$$

describing a family of trajectories $\mathbf{r} = \mathbf{R}(\xi, \eta, \tau)$. These functions are analytic in ξ, η, and τ if the parameters of the medium $n = n(x, y, z)$ and the initial conditions, i.e., the functions $\mathbf{r} = \mathbf{r}^0(\xi, \eta)$ and $\psi^0 = \psi^0(\xi, \eta)$, are analytic.

In the extended space $\{x, y, z, \xi, \eta, \tau\}$ the ray family (2.4.5) forms a three-dimensional surface. Denoting for brevity $\boldsymbol{\xi} = (\xi, \eta, \tau)$, the equation of this surface can be written

$$F(\mathbf{r}, \boldsymbol{\xi}) = 0 , \qquad\qquad (2.4.6)$$

with the function $F(\mathbf{r}, \boldsymbol{\xi})$ being analytic in all its arguments. Its mapping onto the coordinate space $\mathbf{r} = \{x, y, z\}$ is not necessarily unique, however. The boundaries of the multivalued regions to be mapped are the singularities of this mapping. These singularities should be identified as caustics, for the Jacobian $\mathscr{D}(\tau)$ vanishes on them.[13] We may consider further the ray family (2.4.5) that is embedded in a space of still higher dimension by adding to the three parameters ξ, η, τ a set of others w', w'', ... They characterize, e.g., the initial eikonal $\psi^0 = \psi_{(\xi, \eta)}{}^0$, the shape of the intial surface $\mathbf{r} = \mathbf{r}^0(\xi, \eta)$, and the profile of the refrative index $n = n(\mathbf{r})$. As such extending parameters one may use, for example, the Taylor expansion coefficients for the functions $\psi^0(\xi, \eta)$, $\mathbf{r}^0(\xi, \eta)$, $n(\mathbf{r})$. If $\mathbf{w} = (\xi, \eta, \tau, w', w'', ...)$ is the totality of all N parameters of the problem, then for the surface $F(\mathbf{r}, \mathbf{w})$ the mapping in question will be done from the (3+N)-dimensional space into the coordinate space $\mathbf{r} = \{x, y, z\}$.

Catastrophe theory analyzes singularities by describing the local behavior of the functions $F(\mathbf{r}, \mathbf{w})$ in the neighborhood of a certain point $(\mathbf{r}_0, \mathbf{w}_0.)$. Let $\tilde{\mathbf{r}} = (\tilde{x}, \tilde{y}, \tilde{z}) = \mathbf{r} - \mathbf{r}_0$, $\tilde{\mathbf{w}} = \mathbf{w} - \mathbf{w}_0$ be local coordinates in the vicinity of the point $(\mathbf{r}_0, \mathbf{w}_0)$. An analytic function $F(\mathbf{r}, \mathbf{w})$ in the vicinity of this point may be represented by its Taylor expansion which we shall write symbolically as

$$F(\mathbf{r}, \mathbf{w}) = F(\mathbf{r}_0, \mathbf{w}_0) + P_1(\tilde{\mathbf{r}}, \tilde{\mathbf{w}}) + P_2(\tilde{\mathbf{r}}, \tilde{\mathbf{w}}) + ... = 0 , \qquad (2.4.7)$$

where $P_n(\tilde{\mathbf{r}}, \tilde{\mathbf{w}})$ is a homogeneous polynomial in $\tilde{\mathbf{r}}$, $\tilde{\mathbf{w}}$ of degree n.

The classification of singularities is based on reducing the analytic function $F(\mathbf{r}, \mathbf{w})$ by means of a smooth transformation of variables (translation, rotation of axes, and change of scale) to one of the *normal forms*, i.e., to one of the simplest standard polynomials in one, two, etc. parameters. These normal forms correspond to *structurally stable singularities*,[14] and

---

[13] As such an extended space we may also view phase space $\{p, r\}$ in which the phase trajectories $\mathbf{r} = \mathbf{R}(\xi, \eta, \tau)$, $\mathbf{p} = \mathbf{P}(\xi, \eta, \tau)$ form a smooth three-dimensional surface (a Lagrangian manifold) devoid of singularities (Sect.2.2). Caustics occur when this surface is projected onto the configuration space, so they are the singularities of the projections of the Lagrangian manifolds [2.20, 43, 50, 51].

[14] Stable singularities retain their structure when disturbed in the sense that the pertubed singularity is diffeomorphic, i.e, connected to its undisturbed counterpart by a one-to-one continuous differentiable transformation.

Table 2.1. Seven elementary caustics

Co-dimension	Type of catastrophe	Symbol	Focusing index
1	Fold (nonsigular caustic)	A_2	1/6
2	Cusp	A_3	1/4
	Swallowtail	A_4	3/10
3	Hyperbolic umbilic	D_4^-	1/3
	Elliptic umbilic	D_4^+	1/3
	Butterfly	A_5	2/3
4	Parabolic umbilic	D_5	3/8

the number m of free parameters is reerred to as the *codimension*.[15] This classification is mainly due to the studies of *Arnold* [2.42, 43] and others [2.29]. For $m \leq 4$, polynomials of the type (2.4.7) to reduce to R. Thom's seven standard forms (seven elementary catastrophes). These are summarized in Table 2.1 along with their notation, A_k, D_k (k = m+1) in *Arnold's* classification, and with the focusing degree which we shall discuss in Sect.2.10.

Table 2.1 deserves a brief comment. With the codimension $m \leq 4$ the normal forms of the polynomials (2.4.7) contain one or two *state* (or *internal*) variables ζ_1 and ζ_2 to be eliminated on projecting, and a number of *control* (or *external*) variables v_1, v_2, v_3, and v_4 equal to the codimension of the problem. In actual problems all these variables are linear combinations of the initial coordinates r and parameters w. To make the representation more vivid we shall treat v_1 and v_2 as local coordinates in the x, y plane.

For the simplest singularity $(m=1)$[16] the normal form is

$$F_1(\zeta, v_1) = 3\zeta^2 + v_1 = 0 , \qquad (2.4.8)$$

which is known as a *fold*. The image of the surface $F_1(\zeta, v_1) = 0$ in the (ζ, $v_1=x$, $v_2=y$) space is shown in Fig.2.9. Projecting the surface (2.4.8) onto the plane $v_1=x$, $v_2=y$ we obtain a simple (nonsingular) caustic, stressed by a heavy straight line at the bottom of the figure.

The complex singularity next in order is known as a *cusp* $(m=2)$. The corresponding surface[17] is

---

[15] It should be recognized that the dimension of the physical (phase) space does not impose any constraints on the codimension m. So already in a two-dimensional phase space, for example, there could occur a caustic with an arbitrary number of cusps, pockets, etc. [2.33-37,39-41].

[16] In the vicinity of a nonsingular point, F can be reduced to the form $F = \zeta + v_1$. In this case the surface $F=0$ can be uniquely projected on an arbitrary axis v_1, say $v_1 = x$.

[17] There also exists the dual cup differing from (2.4.10) by the sign of ζ^3. The dual singularities of this type were considered, for instance, in [2.49].

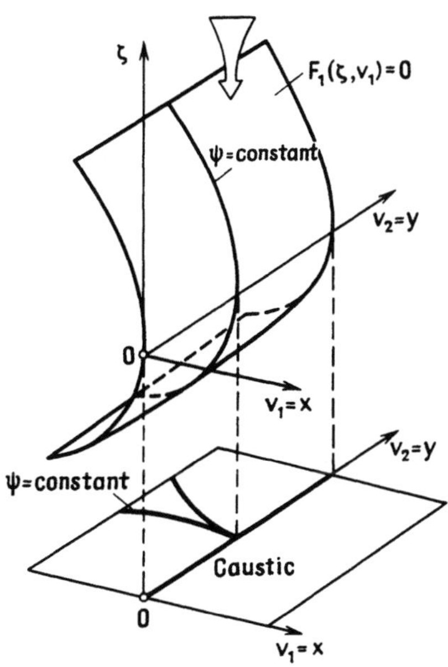

Fig.2.9. A_2 type of caustic, fold, occurs as a result of the ray surface F being projected on the coordinate plane $v_1 = x$, $v_2 = y$

$$F_2(\zeta, v_1, v_2) = 4\zeta^3 - 2v_2\zeta + v_1 = 0 \; . \tag{2.4.9}$$

Its projection onto the plane $v_1 = x$, $v_2 = y$ has the form of a cusp, as is seen from Fig.2.10. It is described by a semicubic parabola $27v_1^2 = 8v_2^3$.

For $m = 3$, we have already three normal forms. One of them, the *swallowtail*, is defined by the function

$$F_3(\zeta, v_1, v_2, v_3) = 5\zeta^4 + 3v_3\zeta^2 + 2v_1\zeta + v_2 = 0 \; . \tag{2.4.10}$$

This three-dimensional hypersurface in the space $\{\zeta, v_1, v_2, v_3\}$ can no longer be visualized graphically, but its singularities are mapped into a surface of characteristic form $\Phi(v_1, v_2, v_3) = 0$ in the $\{v_1, v_2, v_3\}$ space (Fig.2.11a). Figure 2.11b shows sequential sections of this surface by the planes $v_3 = $ constant. For $v_3 < 0$ the caustic is seen to form a pocket produced by two cusps connected by a common branch. At the critical point $v_3 = 0$ the caustic bifurcates (i.e., suffers a qualitative change): the pocket contracts into a point and then, for $v_3 > 0$, disappears entirely.

In this problem the variable v_3 may be a third spatial coordinate, but it may also define the position of the source, the gradient of a refractive index, the curvature of an initial phase front, etc. ([2.33-37,41] and Chap.3). Generally speaking, caustics observed in physical space are the projections of the sections resulting when a hypersurface of singularities in the space $\{v_1, v_2,...\}$ is cut by various planes, say, by a plane $x = c_1 v_1 + c_2 v_2 + ... = $ constant.

The other two types of caustic for $m=3$ are described by the hypersurface $F(v_1, v_2, v_3; \zeta_1, \zeta_2) = 0$ which contains the internal variables $\zeta_{1,2}$ and separates into the two equations $F_1 = 0$, $F_2 = 0$. So, for the relevant catastrophe, called the *hyperbolic umbilic*, D_4^- we have

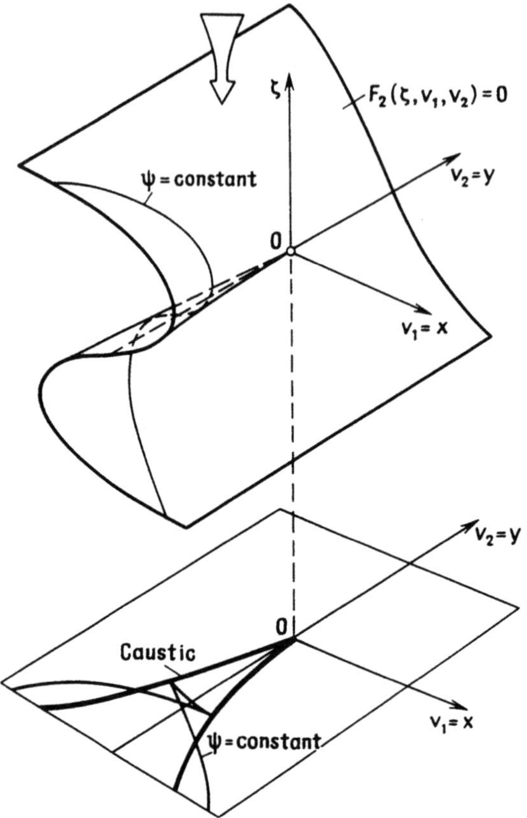

Fig.2.10. The caustic cusp results by projecting the fold on the plane $v_1 = x$, $v_2 = y$

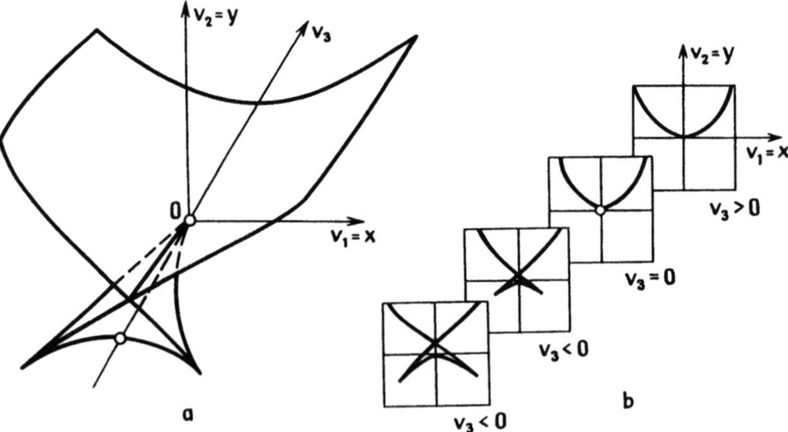

Fig.2.11. (a) The swallowtail (A_4 singularity) in the space v_1, v_2, v_3; (b) cross sections of the bifurcation set of the swallowtail by planes $v_3 =$ constant

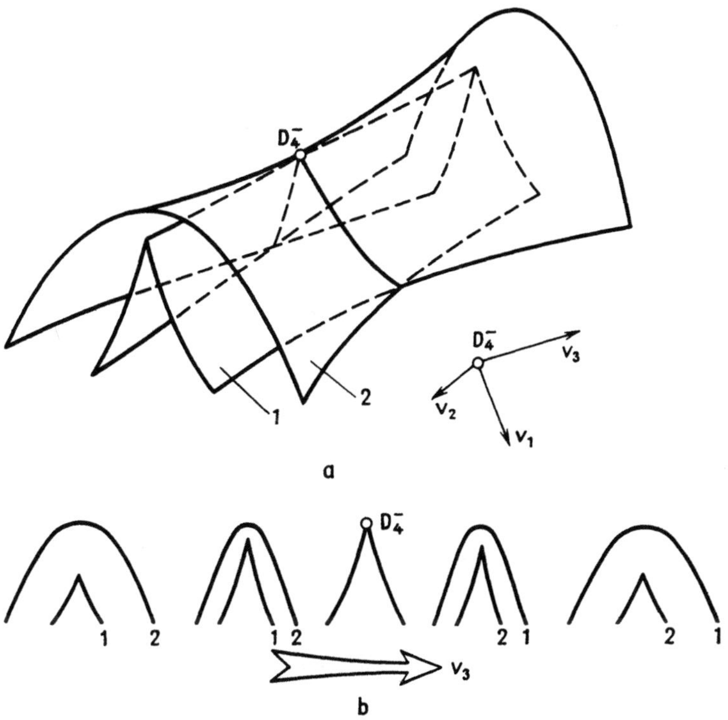

Fig.2.12. (a) The hyperbolic umbilic (D_4^- singularity) in the space v_1, v_2, v_3; (b) cross sections of the bifurcation set of the hyperbolic umbilic

$$F_1(v_i;\zeta_1,\zeta_2) = 3\zeta_1{}^2 + v_3\zeta_2 - v_1 = 0 \ ,$$

$$F_2(v_i;\zeta_1,\zeta_2) = 3\zeta_2{}^2 + v_3\zeta_1 - v_2 = 0 \ .$$

Eliminating from the above equations ζ_1 and ζ_2 by means of the condition $(\partial F_1/\partial\zeta_1)(\partial F_2/\partial\zeta_2) - (\partial F_1/\partial\zeta_2)^2 = 0$, we may obtain the equation of the caustic $\Phi(v_1,v_2,v_3) = 0$. Its geometry is depicted in Fig.2.12a, and its cross sections by planes $v_3 = $ constant in Fig.2.12b.

Similarly, for the *elliptic umbilic* D_4^+ we have

$$F_1(v_i;\zeta_1,\zeta_2) = 3(\zeta_1{}^2 - \zeta_2{}^2) - v_3\zeta_1 - v_1 = 0 \ ,$$

$$F_2(v_i;\zeta_1,\zeta_2) = 6\zeta_1\zeta_2 + 2v_3\zeta_2 + v_2 = 0 \ .$$

The caustic of the elliptic umbilic catastrophe and its sections are shown in Fig.2.13.

More detailed sketches picturing elementary catastrophes and, consequently, other types of caustic, listed in Table 2.1, can be found in [2.29, 42, 46]; more complicated singularities of codimension $m \geq 5$ were considered in [2.50, 51].

We will deal with a variety of singularities of caustics in Chap.3. Specifically, caustic pockets which are the sections of a swallowtail will

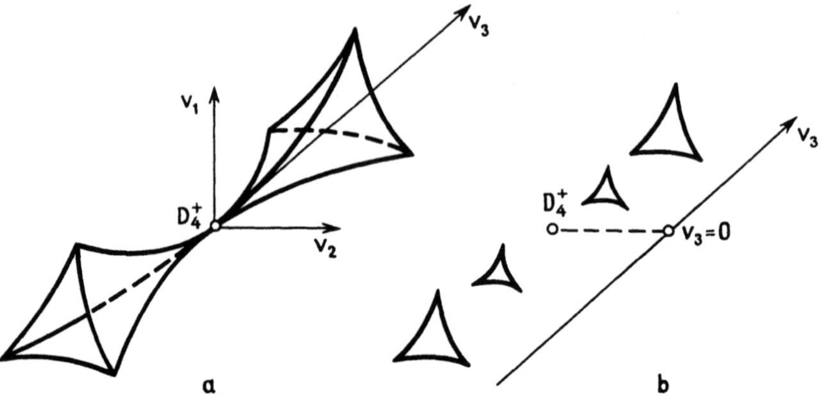

Fig.2.13. (a) The elliptic umbilic ($D_4{}^+$ singularity) in the space v_1, v_2, $v_{3;}$ (b) cross sections of the elliptic umbilic by planes $v_3 = $ constant

come up in Sect.3.3 (Fig.3.19) and various cross sections of a butterfly-type caustic in Sect.3.1 (Fig.3.2) and Sect.3.3 (Fig.3.26), to name only a few.

2.4.4 Structurally Stable and Unstable Caustics in Physical Problems

The mathematical notion of "structural stability" should not be identified with the physical stability of caustics. The structural stability is associated with the variation of control (external) variables v_1, v_2, ..., v_m only, i.e., those parameters which enter the normal form of the function F_m. It is this principle that enables caustics to retain their structure - for example, a pocket form of the caustic in Fig.2.11b - under a small variation of the control variables.

Physical perturbations (that is, additions δF to the function F) may, however, involve arbitrarily high orders of the coordinates x_i and the parameters w_i and so almost always upset structural stability (the function $F+\delta F$, in general, reduces to a normal form other than F). Thus, a weak sinusoidal perturbation of a phase front may cause many pockets to appear on the smooth caustic.

It would now be natural to dwell on the structurally stable singularities in physical theories. The importance of these singularities is determined by how well they are able to model valuable features of real objects, with the "goodness of fit" criterion derived from the physical or technical statement of the problem.

Dealing with caustics, a model could be considered as feasible if the disturbance of any initial condition or parameter, characteristic of this problem, brings about only a small disturbance of the wave field. These disturbances may be expected to be small if *the perturbations of caustic surface are small compared with the width of the caustic zone* (estimates will be given in Sect.2.10). If one or another disturbing factor causes a caustic to protrude beyond the limits of an initial caustic zone, then the model should be made more sophisticated by including this factor among the internal (state) variables of the problem.

From the viewpoint of the introduced criterion, applied physicists may be interested not only in structurally stable caustics (simple caustics, cusps, pockets, etc.), but also in *structurally unstable objects* - similar to an ideal focus or a plane wave,[18] if they supply suitable models to the wave phenomena under consideration. This pragmatic approach is, certainly, characteristic of physics in general rather than of wave optics alone.

2.4.5 Other Types of Caustics

The caustics we have discussed already, related to diffeomorphisms, do not exhaust all caustics of any physical value.

First of all, we note the so-called *broken* caustics which occur when the problem fails to be analytic. They appear, for example, in screening a part of a ray bundle by a stop or in a sudden change of the gradient of the refractive index $n(r)$ [2.56-60].

The geometrical theory of diffraction deals with diffracted rays shed by edges, vertices of smooth surfaces which are themselves caustics for these rays [2.21, 22]. Besides, diffracted rays can form complicated caustics on smooth surfaces [2.61]. Caustics of diffracted rays and, incidentally, broken caustics can be found close to the penumbra region, giving them the name penumbra caustics [2.58, 62].

Finally, we should mention complex caustics which are the envelopes of complex rays [2.26]. A specific case of such caustics embraces penetration caustics, i.e., the envelopes associated with wave fields penetrating through obstacles [2.63, 64].

2.4.6 Singularities of Phase Fronts

The eikonal of a wave, ψ, is an analytic function of the parameters ξ, η, and τ, so that $\psi = \psi(\xi, \eta, \tau)$, and therefore can be thought of as an analytic function on the surface $F(r, \xi)$ in the space $\{r, \xi\}$. On this surface the lines ψ = constant will have singularities exactly on caustics, upon projecting them into the $\{x, y, z\}$ space. So projecting a fold from $\{\xi, v_1, v_2\}$ space onto the plane $v_1 = x$, $v_2 = y$ yields for the lines ψ = constant a cusp orienting at right angles to the caustic (Fig.2.9); the orthogonality follows from the fact that the rays are tangent to the caustic and normal to the wave fronts).

---

[18] A collimated beam supporting a plane wave is defined by the equations $x = \xi$, $y = \eta$, $z = \tau$. Obviously these will not change if polynomials of arbitrary degree m (m = 1,2,_) with zero coefficients of v_1, v_2, _, v_m are added. This agglomerate will be structurally unstable since arbitrarily small nonzero coefficients of v_1, v_2, _, v_m give rise to arbitrarily complex singularities. A similar argument applies to a spherical wave which corresponds to an ideal focus of rays. Berry [2.29], ascribed an infinite codimension, m = ∞ to such mathematical objects. An infinite codimension might be associated with any real object, yet in modelling, i.e., in idealized description of a physical object, we may safely retain a finite number of the most valuable parameters.

Upon projecting onto the coordinate plane, the lines ψ = constant form pockets which are the cross sections of a swallowtail (Fig.2.10).

The aforementioned examples illustrate that the singularity of the phase fronts has a codimension one unit higher than that of the respective caustic. This principle is of a general nature and stems from the theory of sigularities in differentiable mapping. In the particular case three functions (2.4.5) should be augmented by the function $\psi = \psi(\xi,\eta,\tau) = \psi(\xi)$ to consider the hypersurface $\Phi(\psi,\mathbf{r},\xi) = 0$ in the extended space $\{\psi,\mathbf{r},\xi\}$. Projecting this surface into $\{\psi,\mathbf{r}\}$ the space displays singularities of diffeomorphisms. The cross sections of these surfaces by the plane ψ = constant will yield the possible forms of phase front in coordinate space [2.48,50].

2.4.7 Phase Shifts at Caustics

A ray grazing a caustic suffers a phase shift, already noted in Sect.2.3. We shall explain the nature of this shift with reference to a simple caustic.

Let the Jacobian be positive, $\mathscr{D}(\tau) > 0$ at an initial point on the ray. It becomes negative upon touching the caustic (being zero on the caustic) and therefore (2.3.9) acquires the factor $\sqrt{-1}$ which can be interpreted either as $i = \exp(i\pi/2)$ or as $-i = \exp(-i\pi/2)$. This factor (the sign will be defined later) corresponds to the phase shift that the ray suffers at the caustic.

That the phase of a wave changes by $-\pi/2$ upon touching a nonsingular caustic and by $-\pi$ upon passing through a three-dimensional focus has been known for some time from specific examples to be solved exactly. However, a uniformly valid rule on phase retardation at a caustic is of comparatively recent origin. *Lewis* [2.65] derived this rule from an asymptotic consideration of rigorous integral representation of a field in a homogeneous medium, and *Maslov* and *Fedoryuk* [2.20] obtained it by the canonic-operator technique. Without giving much consideration to the derivation[19] we shall only illustrate how a caustic phase shift can be evaluated.

Consider the matrix

$$
\hat{\mathscr{D}} = \begin{bmatrix}
\dfrac{\partial x}{\partial \xi} & \dfrac{\partial x}{\partial \eta} & \dfrac{\partial x}{\partial \tau} \\[2ex]
\dfrac{\partial y}{\partial \xi} & \dfrac{\partial y}{\partial \eta} & \dfrac{\partial y}{\partial \tau} \\[2ex]
\dfrac{\partial z}{\partial \xi} & \dfrac{\partial z}{\partial \eta} & \dfrac{\partial z}{\partial \tau}
\end{bmatrix} .
\tag{2.4.11}
$$

At the point where a ray is tangent to a caustic, $\det\hat{\mathscr{D}}$ vanishes and simultaneously one, two, or three eigenvalues of the matrix $\hat{\mathscr{D}}$ may also become zero. Let q be the number of eigenvalues vanishing on the caustic. Then

---

[19] The caustic phase shift can also be derived by means of a uniformly asymptotic representation of the field which is valid in the neighborhood of caustics [2.23,25,66-68], and by means of complex rays which in the same manner as in the method of phase integrals enable the caustic to be encircled in complex domain (the last method was used in [2.69] to produce the phase shift $-\pi/2$ at a nonsingular caustic).

upon touching the caustic the field amplitude on the ray undergoes a phase change of $\delta\Psi = -\pi q/2$, namely

$$A_0(\tau) = \frac{A_0{}^0}{\sqrt{|\mathscr{J}|}}\, e^{(-i\pi q)/2} = \frac{A_0{}^0}{\sqrt{|\mathscr{J}|}}\, e^{i\delta\Psi} \; . \tag{2.4.12}$$

The quantity $\delta\Psi$ is reerred to as the caustic phase shift.

In a one-dimensional problem, q may assume only the value of unity so that $\delta\Psi = -\pi/2$ (we ignore in the discussion the trivial case of $q = 0$ corresponding to the absence of a caustics). Two-dimensional problems suggest two possibilities: $q = 1$ and $q = 2$ for which $\delta\psi = -\pi/2$ and $-\pi$, respectively. A simple calculation may indicate that in a homogeneous medium only one of two eigenvalues may vanish, so that here for a two-dimensional caustic of any shape $q = 1$ and $\delta\Psi = -\pi/2$. In three-dimensional problems q may be 1, 2, or 3. A value of $q = 1$ with the associated phase shift $-\pi/2$ is realized on nonsingular segments of caustics. The value $q = 2$ with the phase shift $\delta\psi = -\pi$ corresponds to the twofold degenerate matrix $\hat{\mathscr{D}}$ and is observed in a three-dimensional focus or in a 3-D (axially symmetric) cusp.

In a situation where a ray is tangent to several caustics, the phase shift equals the sum of the phase shifts the ray acquired at all the caustics, i.e.,

$$\delta\Psi = \sum_j \delta\Psi_j = -\frac{\pi}{2}\sum_j q_j \; . \tag{2.4.13}$$

The quantity Σq_j is called the *index of a trajectory* or Maslov's index [2.20].

The caustic phase shift affects the interference pattern under multipath propagation. Specifically, an account of $\delta\psi$ is needed to determine the position of interference maxima and minima, to obtain the frequencies of modes in an open resonator [2.19, 70-72], to calculate phase velocities of normal waves in refraction (graded-index) waveguides [2.25], etc.

2.5 Reflection and Refraction of Waves at Interfaces

2.5.1 The Locality Principle in Wave Reflection

The method of geometrical optics enables solution of the most complicated problem of diffraction - that of wave reflection from a curvilinear interface - in a comparatively simple manner with reference to the principle of locality. Indeed, the zeroth approximation assumes that at each point of a boundary the reflection proceeds in such a manner as if the incident wave were plane and the curvilinear boundary were replaced by a tangent plane (Fig.2.14). This principle, therefore, enables us to use as a zero-order approximation, the formulas for plane wave reflection on plane boundaries,

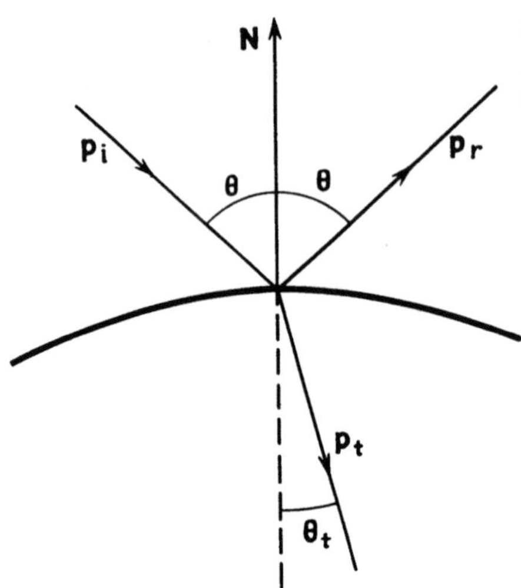

Fig.2.14. Incident, reflected, and refracted (transmitted) rays at the interface between two media

in particular, the Fresnel formulas in acoustics and electrodynamics. We shall discuss this topic in more detail below following [2.24, 73].

2.5.2 Relations for Rays and Eikonals

Let $u_i = A_i \exp(ik_0 \psi_i)$ be a wave incident on the boundary Q between media of the refractive indices n_1 and n_2, giving rise to the reflected wave $u_r = A_r \exp(ik_0 \psi_r)$ and refracted (transmitted) wave $u_t = A_t \exp(ik_0 \psi_t)$. At the interface Q, the rapidly oscillating exponetial functions $\exp(ik_0 \psi_i)$, $\exp(ik_0 \psi_r)$, and $\exp(ik_0 \psi_t)$ must be equal (otherwise, to meet boundary conditions we would have to abandon the main assumption of geometrical optics, namely that of slowly varying amplitudes). Consequently, at the interface, i.e., for $r \in Q$, there should be

$$\psi_r = \psi_t = \psi_i \ . \tag{2.5.1}$$

These relationships serve as initial conditions for the eikonals of the reflected and refracted waves leaving the interface.

According to (2.5.1) the tangential to the Q components of the vectors $p_i = \nabla \psi_i$, $p_r = \nabla \psi_r$, and $p_t = \nabla \psi_t$ should be equal, namely,

$$(p_i)_t = (p_r)_t = (p_t)_t \ , \quad p_t = p - N(p \cdot N) \ , \tag{2.5.2}$$

where N is the unit normal to the interface. We define the plane of incidence for an incident ray as that containing p_i and N (Fig.2.14). Then from (2.5.2) it follows that the vectors p_r and p_t also belong to this plane; that is, all three of them – p_i, p_r, and p_t – are coplanar. For the normal components of these vectors, which are simultaneously the normal derivatives of the eikonal ψ_i, ψ_r, and ψ_t we have

$$(p_i)_N = - \sqrt{n_1^2 - (p_i)_t^2} \ ,$$

$$(p_r)_N = - \sqrt{n_1^2 - (p_i)_t^2} \ , \qquad r \in Q \tag{2.5.3}$$

$$(p_t)_N = - \sqrt{n_2^2 - (p_i)_t^2} \ .$$

Denoting the angles of incidence, reflection and refraction by θ, θ_r, and θ_t, respectively, transforms (2.5.2) into

$$n_1 \sin\theta = n_1 \sin\theta_r = n_2 \sin\theta_t \ , \tag{2.5.4}$$

implying that the specular-reflection law

$$\theta_r = \theta \tag{2.5.5}$$

and Snell's law of refraction

$$\sin\theta / \sin\theta_t = n_2 / n_1 \tag{2.5.6}$$

are obeyed.

With account of these equalities the normal deriviatives of the eikonals assume the form

$$\partial\psi_r / \partial N = - \ \partial\psi_i / \partial N = n_1 \cos\theta \ ,$$

$$\partial\psi_t / \partial N = - n_2 \cos\theta_t = - \sqrt{n_2^2 - n_1^2 (\sin\theta)^2} \ , \quad \text{for} \quad r \in Q \ . \tag{2.5.7}$$

The relations (2.5.2-7) give all the necessary information to derive the rays and eikonals of reflected and refracted waves far from the interface. Note that in deriving these relations we have nowhere used an explicit form of boundary conditions. To find the amplitudes A_r and A_t, on the other hand, requires a definite form of the boudnary conditions. We shall proceed along the lines of [2.24].

2.5.3 Reflection Formulas for Amplitude

A comparatively large number of problems of practical value obey the boundary conditions

$$\kappa_1 (u_i + u_r) = \kappa_2 u_t \ ,$$

$$\nu_1 \left[\frac{\partial u_i}{\partial N} + \frac{\partial u_r}{\partial N} \right] = \nu_2 \frac{u_t}{\partial N} \ , \tag{2.5.8}$$

where $u_i + u_r$ describes the field in the primary (subscript "1") medium, while u_t that in the secondary (subscript "2"). Conditions of the type (2.5.8) can be satisfied, for example, by acoustic fields. If u is the potential of a sound field, $v = \nabla u$ being the velocity of the particles in the sound wave and $P = \rho \partial u / \partial t$ the pressure, then (2.5.8) follows from the requirement that the pressure and the normal component of the velocity $v_N = -\partial u / \partial N$ should be continuous at the boundary with $\kappa_1,2 = \rho_1,2$ and $\nu_1,2 = 1$. These relations also contain, as a specific case, the Dirichlet boundary conditions $(u_i + u_r)_Q = 0$ realized as $\kappa_2 \to 0$, and those of Neumann $\partial/(\partial N)(u_i + u_r) = 0$, as $\nu_2 \to 0$.

When we represent each of the fields u_i, u_r, and u_t as a series in powers of $(1/ik_0)$ such as

$$u_r = \sum_{m=0}^{\infty} \frac{A_m^r}{(ik_0)^m} \exp(ik_0 \psi_r) \ . \tag{2.5.9}$$

Substituting (2.5.9) into (2.5.8) and equating the coefficients of equal powers in k_0 yields

$$\kappa_1(A_m^i + A_m^r) = \kappa_2 A_m^t \ ,$$

$$\nu_1 \left[A_m^i \frac{\partial \psi_i}{\partial N} + \frac{\partial A_{m-1}^i}{\partial N} + A_m^r \frac{\partial \psi_r}{\partial N} \right] = \nu_2 \left[A_m^t \frac{\partial \psi_t}{\partial N} + \frac{\partial A_{m-1}^t}{\partial N} \right] , \tag{2.5.10}$$

all A_m with $m < 0$ are assumed zero, i.e., $A_{-1} = A_{-2} = ... = 0$, as is usually done in ray expansions. Accounting for (2.5.7) and solving (2.5.10) for A_m^r and A_m^t, we arrive at the set of recursive equations

$$A_m^r = \Gamma A_m^i + \frac{1}{1+Y} \delta_{m-1} \ ,$$

$$A_m^t = D A_m^i + \frac{\kappa_1}{\kappa_2(1+Y)} \delta_{m-1} \ , \qquad \text{where} \tag{2.5.11}$$

$$\Gamma = \frac{1-Y}{1+Y} \ , \quad D = \frac{2\kappa_1}{\kappa_2(1+Y)} \ , \tag{2.5.12}$$

$$Y = \frac{\kappa_1 \nu_2 n_2 \cos\theta_t}{\kappa_2 \nu_1 n_1 \cos\theta} \ , \tag{2.5.13}$$

$$\delta_{m-1} = \frac{1}{n_1 \cos\theta} \left[\frac{\nu_2}{\nu_1} \frac{\partial A_{m-1}^t}{\partial N} - \frac{\partial A_{m-1}^r}{\partial N} \right] . \tag{2.5.14}$$

With (2.5.11) we can compute in sequence, having started with the zeroth approximation (m=0), the amplitudes of the reflected and refracted waves at the boundary Q. With these amplitudes as initial values we then shall be able to deduce by (2.3.9) the amplitudes of the refracted and reflected fields far from the interface.

As a zero approximation, (2.5.11) yields the familiar formula for the reflected field

$$A_0^r = \frac{1 - Y}{1 + Y} A_0^i = \Gamma A_0^i ,$$
(2.5.15)

and the formula for the refracted field

$$A_0^t = \frac{2\kappa_1}{\kappa_2(1 + Y)} A_0^i = D A_0^i .$$
(2.5.16)

Here Γ is the local (amplitude) reflection coefficient, and D the local transmission coefficient. In (2.5.15, 16) they assume Fresnel values and substantiate the locality principle discussed earlier in this section.

When a plane wave is incident on a plane boundary between homogeneous media, the amplitudes A_0^i, A_0^r, and A_0^t are constant at the interface, and all δ_{m-1} in (2.5.11) are zeros. In this case the geometrical-optics approximation gives an exact Fresnel solution to the problem of wave reflection and refraction. In the general case, this approximation yields results the more accurate the less the curvature of the wave fronts and the interface, and the closer both media are to being homogeneous. The applicability of the geometrical-optics approximation (2.5.15, 16) at an interface will be discussed in Sect.3.2 [2.25, 74]

2.5.4 Reflection from Weak Interfaces

By virtue of (2.4.4) we may find the field reflected from the so-called *weak* boundaries which render discontinuous not the parameters n, κ, and ν proper, but their derivatives of mth order, e.g.,

$$\partial^m n_1/\partial N^m = \partial^m n_2/\partial N^m \quad \text{for} \quad m < M , \quad \mathbf{r} \in Q ,$$
(2.5.17)

but

$$\partial^M n_1/\partial N^M \neq \partial^M n_2/\partial N^M \quad \text{for} \quad \mathbf{r} \in Q$$
(2.5.18)

(an ordinary interface corresponds, from this viewpoint, to the discontinuous derivatives of zeroth order). For weak interfaces we have $Y = 1$, $\kappa_1 = \kappa_2$, and $\nu_1 = \nu_2$, the reflection coefficient vanishes ($\Gamma = 0$) and the transmission coefficient is unity (D = 1), so relations (2.5.11) assume the form

$$A_m^r = \delta_{m-1}/2 ,$$

$$A_m^t = A_m^i + \delta_{m-1}/2 , \qquad \text{where} \qquad (2.5.19)$$

$$\delta_{m-1} = \frac{1}{n_1 \cos\theta} \frac{\partial}{\partial N} (A_{m-1}^t - A_{m-1}^i - A_{m-1}^r) .$$

The quantities δ_{m-1} equal zero for $m < M$, so that

$$A_m^r = 0, \quad A_m^t = A_m^i , \quad m = 0,1,...,M-1 , \qquad (2.5.20)$$

but δ_{M-1} is no longer zero, which results in the nonzero amplitude A_m^r of the reflected field

$$A_M^r = \frac{1}{2n_1 \cos\theta} \frac{\partial}{\partial N}(A_{M-1}^t - A_{M-1}^i) . \qquad (2.5.21a)$$

It implies that a reflected wave has appeared, and is described by the formula

$$u_r = \frac{1}{(ik_0)^M} e^{ik_0 \psi_r} \{A_M^r + O(1/k_0)\} , \qquad (2.5.21b)$$

where the leading term is of the order k_0^{-M} (or μ^M if expanded in the dimensionless parameter μ). Thus, at weak interfaces reflected waves occur only in the Mth term of the ray expansion.

One could conclude that in media whose parameters possess all their derivatives ($M \to \infty$) no reflected field would be present, and as $M \to \infty$ the ray expansion (2.1.20) would tend to an exact solution. Actually, the reflected or, more correctly, scattered field is nonzero even in a medium of infinitely differentiable parameters. However, in contrast to (2.20), the scattered field is *exponentially* small [2.70], i.e.,

$$u_{sc} \propto \exp(-1/\mu) = \exp(-k_0 nL) . \qquad (2.5.22)$$

This is the reason why this field cannot be computed in the framework of geometrical optics; indeed, as $kL \to \infty$ the field u_{sc} decays faster than any finite power $\mu^M = (1/kL)^M$, and consequently cannot be represented by the series (2.1.20).

The inability of ray expansions to account for exponentially small scattered fields is another proof that the truncated sums of (2.1.20) converge asymptotically to an exact solution, as we have already mentioned in Sect.2.1.

2.5.5 The Geometrical Optics of Surface Waves

We recalled that an interface may give rise to surface waves, either electromagnetic or acoustic. Their field decays exponentially away from the sur-

face. If the properties of such a boundary vary slowly, then the wave field
may be described by geometrical optics. The respective modification - surface geometrical optics - deals with surface rays and caustics, and with the
surface eikonal and transport equations [2.19, 24, 75].

Similar problems occur for the creeping waves associated with rays
diffracted at a smooth body [2.19, 57, 76, 77].

2.6 Reciprocity of Rays and Caustics

2.6.1 The Reciprocity Theorem

At certain conditions, the fields described by the scalar equation (2.1.1)
obey only the *theorem of reciprocity*. According to this theorem, a field
$u_1(r)$ being excited by source $q_1(r)$ and satisfying the inhomogeneous
equation

$$\Delta u_1 + k_0{}^2 n^2 u_1 = q_1(r) \tag{2.6.1}$$

and a field $u_2(r)$ being excited by source $q_2(r)$ obey the relationship

$$\int_V u_1(r) q_2(r) dV = \int_V u_2(r) q_1(r) dV . \tag{2.6.2}$$

Write the reciprocity theorem (2.6.2) for a particular case with the fields u_1
and u_2 being excited by point sources of unit strength:

$$q_1(r) = \delta(r - r') , \quad q_2(r) = \delta(r - r'') . \tag{2.6.3}$$

In this case

$$u_1(r'') = u_2(r') ; \tag{2.6.4}$$

that is, the field $u_1(r'')$ created by the point source q_1 at point $r = r''$ turns
out to be equal to the field $u_2(r')$ created by soruce q_2 at point $r = r'$.

2.6.2 Reciprocity Relations for Rays and Caustics

If the fields u_1 and u_2 are represented by their ray expansions, relations of
the type (2.6.4) must hold for any power of $1/k_0$. Specifically, as the zero-order approximation of geometrical optics we obtain the *ray reciprocity
theorem* [2.78]

$$A_1(r'') \exp[ik_0 \psi(r', r'')] = A_2(r') \exp[ik_0 \psi(r'', r')] , \tag{2.6.5}$$

where by virtue of (2.2.9 and 2.3.26)

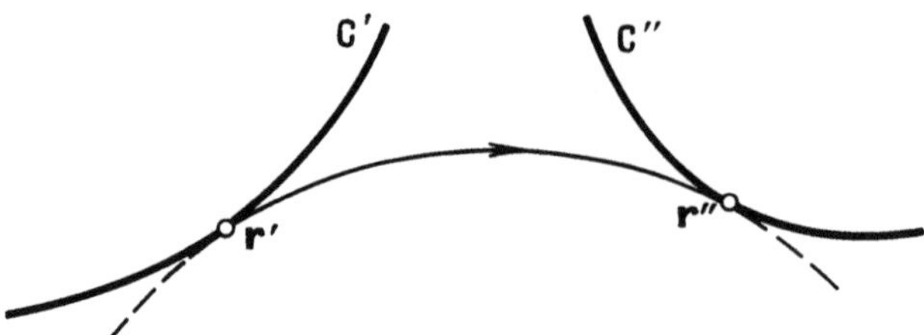

Fig.2.15. Illustrating the reciprocity theorem for caustics

$$\psi(\mathbf{r'},\mathbf{r''}) = \int_0^{\tau(\mathbf{r'},\mathbf{r''})} n^2 d\tau \ , \quad \psi(\mathbf{r''},\mathbf{r'}) = \int_0^{\tau(\mathbf{r''},\mathbf{r'})} n^2 d\tau \ , \tag{2.6.6}$$

$$A_1(\mathbf{r''}) = - \frac{1}{4\pi} \left. \sqrt{\frac{n' d\Omega_1}{n'' da_1}} \right|_{\mathbf{r}=\mathbf{r''}}, \quad A_2(\mathbf{r'}) = - \frac{1}{4\pi} \left. \sqrt{\frac{n'' d\Omega_2}{n' da_2}} \right|_{\mathbf{r}=\mathbf{r'}}. \tag{2.6.7}$$

From (2.6.5) it follows that $\psi(\mathbf{r'},\mathbf{r''}) = \psi(\mathbf{r''},\mathbf{r'})$; that is, the phase shifts for traveling in both the directions (from $\mathbf{r'}$ to $\mathbf{r''}$) and back (from $\mathbf{r''}$ to $\mathbf{r'}$) are the same. Obviously, the functionals $\psi(\mathbf{r'},\mathbf{r''})$ and $\psi(\mathbf{r''},\mathbf{r'})$ reach their extrema on one and the same ray connecting the end points $\mathbf{r'}$ and $\mathbf{r''}$ (elementary geometrical optics treats this as *reversibility of rays paths*).

Further, the equality of the amplitudes $A_1(\mathbf{r''})$ and $A_2(\mathbf{r'})$ (at equal elementary solid angles $d\Omega_1$ and $d\Omega_2$) leads us to the reciprocity relationship for ray-tube cross sections:

$$n^2(\mathbf{r''}) \, da_1(\mathbf{r''}) = n^2(\mathbf{r'}) da_2(\mathbf{r'}) \ . \tag{2.6.8}$$

If several rays join points $\mathbf{r'}$ and $\mathbf{r''}$, the reciprocity relations hold for all of them.

We may draw a useful conclusion for the configuration of caustics with reference to Fig.2.15. If $da_1(\mathbf{r''}) = 0$, that is, if point $\mathbf{r''}$ lies on caustic C' formed by the rays leaving point $\mathbf{r'}$, then caustic C'' for the rays leaving $\mathbf{r''}$ traces through $\mathbf{r'}$ since $da_2(\mathbf{r'}) = 0$ by virtue of (2.6.8). Both caustics C' and C'' are tangent to one and the same "reference" ray connecting $\mathbf{r'}$ and $\mathbf{r''}$ [2.79].

Similar reciprocity arguments can be applied to singular sections of caustics, which must retain the nature of the singularity in corresponding points [2.36]. For example, if $\mathbf{r''}$ is at the apex of a cusp of caustic C'', then $\mathbf{r'}$ will be a cusp point of caustic C'. This reciprocity theorem for caustics helps us in many occasions to imagine the shape of a caustic without conducting a special study [2.35, 36, 38].

2.7 Space-Time Geometrical Optics

2.7.1 The Wave Equation for Media
with Temporal (Frequency) Dispersion

This section is concerned with the application of the method of geometrical
optics to nonstationary wave fields, arbitrarily dependent on time. This
type of problem occurs in propagating modulated or pulsed waves, traveling sources, reflected from moving interfaces, etc.

A wide number of wave problems can be described (exactly or
approximated to some extent) by the equation

$$\Delta u(t,r) - \frac{1}{c^2}\frac{\partial}{\partial t^2}\hat{M}[u(t,r)] = 0, \qquad (2.7.1)$$

in which the linear integral operator

$$\hat{M}[u] = \int_{-\infty}^{t} \tilde{\epsilon}(t-t',r)u(t',r)dt' \qquad (2.7.2)$$

characterizes the temporal (frequency) dispersion[20] of the medium. The
operator is seen to depend not only upon the field at a given moment of
time t, but also upon the field $u(t',r)$ at the earlier time $t' < t$. The relation
$v(t,r) \equiv M[u(t,r)]$ is essentially the constitutional equation of the problem,
the nucleus $\tilde{\epsilon}(t-t',r)$ of M being derived by considering microprocesses in
the substance under investigation ($\tilde{\epsilon}$ corresponds to the permittivity in electrodynamics). For $t-t' < 0$, the nucleus $\tilde{\epsilon}(t-t',r)$ vanishes implying that the
casual principle holds.

Provided the field $u(t,r)$ satisfies the wave equation (2.7.1), its spectral
amplitude

$$u(\omega,r) = \int_{-\infty}^{\infty} u(t,r)e^{i\omega t}dt \qquad (2.7.3)$$

obeys, as can be readily verified, the Helmholtz equation

$$\Delta u(\omega,r) + \frac{\omega^2}{c^2}n^2(\omega,r)u(\omega,r) = 0, \qquad (2.7.4)$$

where $n^2(\omega,r)$ is defined as the Fourier transform[21] of $\tilde{\epsilon}(t-t',r)$:

---

[20] A geometrical optics of media with spatial dispersion is constructed (for the scalar
problem statement) in [2.80].

[21] Here and in (2.7.12) below we have taken account of $\epsilon(\Delta t,r) = 0$ for $\Delta t < 0$. Therefore
the lower limit of integration in (2.7.5,12) is set to zero.

$$n^2(\omega,r) \equiv \epsilon(\omega,r) = \int_0^\infty \tilde{\epsilon}(t-t',r)e^{i\omega(t-t')}d(t-t') . \qquad (2.7.5)$$

Consequently, if an exact solution $u(\omega,r)$ to the Helmholtz equation (2.7.4) is found, then the solution $u(t,r)$ of the wave equation (2.7.1) can be derived by taking the Fourier transform:

$$u(t,r) = \frac{1}{2\pi} \int_{-\infty}^\infty u(\omega,r)e^{-i\omega t}d\omega . \qquad (2.7.6)$$

Thus a consideration of nonstationary wave processes (pulse or modulated waves) can, in principle, be reduced to a solution of a sinusoidal wave problem. Taking the transform in (2.7.6) appears, as a rule, too cumbersome a process, so it proves often convenient and intuitively appealing to build up the ray approximation directly for the field $u(t,r)$ rather than for its spectral amplitude $u(\omega,r)$. This is a problem central to space-time geometrical optics, enjoying intensive development in recent years [2.8, 9, 19, 65, 73, 81-87].

The method has two intimately related modifications, one of which deals with the propagation of quasi-monochromatic waves (quasi-monochromatic *wave packets* are treated in Sects. 2.7.2-8) and the other with the propagation of field discontinuities (singularities) in nondispersive media (this theory will be the subject of Sect. 2.7.9).

2.7.2 Necessary Conditions for the Geometrical-Optics Applied to Quasi-Monochromatic Wave Packets

Geometrical optics is based on the representation of nonmonochromatic waves as the field

$$u(t,r) = A(t,r)\exp[i\phi(t,r)] \qquad (2.7.7)$$

with a slowly varying amplitude $a(t,r)$ and rapidly changing phase $\phi(t,r)$. It behaves locally as that of plane monochromatic waves in a homogeneous and stationary medium. Denote by

$$\omega = \omega(t,r) \equiv -\frac{\partial\phi(t,r)}{\partial t} , \quad k = k(t,r) \equiv \nabla\phi(t,r) , \qquad (2.7.8)$$

respectively, the instantaneous frequency $\partial\phi/\partial t$, and local wave vector $\nabla\phi$, to which we shall subsequently refer as "frequency" and "wave vector", respectively. The quantities ω, k, and A are strictly constant for a plane monochromatic wave $u = A\exp(ik\cdot r - i\omega t)$ traveling in a homogeneous and stationary medium, but in the general case these quantities significantly vary in time and space over their characteristic time interval T and dimension L.

In order that a wave may be deemed plane and monochromatic we ought to require that. along with (2.1.4) for spatially smooth parameters of both the wave and the medium, the wave variation should be slow:

$$\bar{\tau}/T \simeq 1/\bar{\omega}T \ll 1 \, , \tag{2.7.9}$$

where $\bar{\tau} \simeq 1/\bar{\omega}$ is the characteristic period of oscillations, and T is the characteristic time interval of nonstationarity. We will also require that

$$\tau_0/T \ll 1 \, , \tag{2.7.10}$$

where τ_0 is the characteristic interval of temporal dispersion, i.e., the interval over which the nucleus $\tilde{\epsilon}(t\text{-}t',\mathbf{r})$ significantly varies in the argument t-t'. Condition (2.7.12) enables us to assume that the wave is monochromatic within the interval $|\Delta| = |t\text{-}t'| \leq \tau_0$, essential for integration in (2.7.2). Taken together, inequalities (2.1.4) and (2.7.9, 10) form the *necessary conditions for the applicability* of geometrical optics in dispersive media:

$$\mu = \max\{\lambda/L, \bar{\tau}/T, \tau_0/T\} \ll 1 \, . \tag{2.7.11}$$

It should be recognized that these inequalities allow both weak ($\tau_0 \leq \bar{\tau}$) and strong ($\tau_0 > \bar{\tau}$) temporal dispersion to be considered.

2.7.3 Differential Form of the Constitutive Equation (2.7.2)

In order to perform a systematic derivation of the equation of space-time geometrical optics we should expand the field (2.7.7) in the small parameter μ.[22] To make our derivation more concise we shall not pursue explicit expansions in μ, but instead we shall check the orders of those quantities dropped and retained.

Since we have assumed that the amplitude A and frequency ω change over intervals $|\Delta t| = |t\text{-}t'| \leq \tau_0$ only insignificantly, we may develop A(t') and $\phi(t')$ as Taylor series in $\Delta t = t\text{-}t'$.

$$A(t') \sim A(t) - \left[\frac{\partial A(t)}{\partial t}\right]\Delta t + ... \, ,$$

$$\phi(t') \sim \phi(t) - \frac{\partial \phi}{\partial t}\Delta t + \frac{1}{2}\frac{\partial^2 \phi}{\partial t^2}(\Delta t)^2 + ... \, .$$

In view of (2.7.11) the quantity $|(\partial A/\partial t)\Delta t| \leq A\tau_0/T$ is small compared with A, and the quantity $|\partial^2 \phi/\partial t^2 (\Delta t)^2| \leq |(\partial \omega/\partial t)\tau_0^2| \simeq \bar{\omega}\,\tau_0^2/T \simeq (\tau_0/\bar{\tau})(\tau_0/T)$ is small compared with unity, so that

---

[22] An example of such a derivation for the "plasma" law of dispersion can be found in [281].

$$u(t',r) = A(t',r)\exp(i\phi(t',r))$$

$$\simeq \exp[i\phi(t,r) + i\omega(t,r)\Delta t]$$

$$\cdot \left[A(t,r) - \frac{\partial A(t,r)}{\partial t}\Delta t + \frac{i}{2}A(t,r)\frac{\partial^2 \phi(t,r)}{\partial t^2}(\Delta t)^2 + ... \right] . \quad (2.7.12)$$

Substituting (2.7.12) into 2.7.2) we obtain the constitutive equation (2.7.2) in *differential form*

$$\hat{M}[u(t,r)] = e^{i\phi(t,r)}\left[\epsilon(\omega)A + i\frac{\partial\epsilon(\omega)}{\partial\omega}\frac{\partial A}{\partial t} - \frac{iA}{2}\frac{\partial^2\epsilon(\omega)}{\partial\omega^2}\frac{\partial^2\phi}{\partial t^2} + ... \right],$$

$$(2.7.13)$$

which accounts for the definition of $\epsilon(\omega) = \epsilon(\omega,r)$ given by (2.7.5) and the equation that follows from it:[23]

$$\int_0^\infty (\Delta t)^m \tilde{\epsilon}(\Delta t,r)e^{i\omega\Delta t}d(\Delta t) = (-i)^m \frac{\partial^m \epsilon(\omega,r)}{\partial\omega^m} . \quad (2.7.14)$$

Upon differentiating (2.7.13) twice with respect to t we have

$$\frac{\partial^2}{\partial t^2}\hat{M}[u(t,r)] = -\omega^2 \epsilon A e^{i\phi}$$

$$+ ie^{i\phi}\left[-2\omega\frac{\partial}{\partial t}(A\epsilon) + A\epsilon\frac{\partial^2\phi}{\partial t^2} - \omega^2\frac{\partial A}{\partial t^2}\frac{\partial\epsilon}{\partial\omega} + \frac{A\omega^2}{2}\frac{\partial^2\phi}{\partial t^2}\frac{\partial^2\epsilon}{\partial\omega^2} \right] + ...$$

$$\equiv -\omega^2\epsilon A e^{i\phi} + ie^{i\phi}\hat{\Gamma}A + ... \quad (2.7.15)$$

All the quantities lumped into $\hat{\Gamma}A$ are by virtue of (2.7.11) small compared with the zero-order term $\omega^2\epsilon A$; for example $|\partial\phi^2/\partial t^2| = |\partial\omega/\partial t| \simeq \bar{\omega}/T \ll \omega^2$ since $1/\omega T \ll 1$.

2.7.4 Eikonal and Transport Equations

Letting

$$A = A_0 + iA_1 + i^2 A_2 + ... , \quad (2.7.16)$$

where $|A_1| \simeq \mu|A_0|$, $|A_2| \sim \mu A_1$, etc., we substitute (2.7.15,16) into

---

[23] For the integrals of type (2.7.13) to converge the nucleus $\epsilon(\Delta t,r)$ must diminish suffi- ciently fast (faster than any power of $1/\Delta t$, for instance, exponentially) as Δt increases.

(2.7.1) and equate to zero the sums of similar order in μ. This gives in the zeroth order the eikonal equation ($k=\nabla\phi$, $\omega=-\partial\phi/\partial t$)

$$\mathscr{H}(\omega, k, r) \equiv k^2 - \frac{\omega^2}{c^2}\epsilon(\omega, r) = 0 \ , \tag{2.7.17}$$

and in the first order the transport equations for the zero approximation amplitude A_0, denoted below simply by A,

$$2\nabla A \nabla\phi + A \nabla\phi - \hat{\Gamma}A/c^2 = 0 \ . \tag{2.7.18}$$

Multiplying this equation by A we restate it in the form

$$\nabla A^2 \nabla\phi + A^2 \Delta\phi + \frac{1}{2c^2}\frac{\partial}{\partial t}\left[A^2 \frac{\partial(\omega^2\epsilon)}{\partial\omega}\right] = 0 \ . \tag{2.7.19}$$

The eikonal equation (2.7.17) is formally the same as the equation of dispersion for a plane time-harmonic wave in a homogeneous medium of permittivity ϵ. This reflects the most important feature of the geometrical-optics field (2.7.7), which behaves locally similar to the field of a plane monochromatic wave. With this in insight the eikonal equation (2.7.17) is often referred to as the *local dispersion equation*.

By the content, though, the eikonal equation (2.7.17) markedly differs from the dispersion equation as the former is not simply an algebraic relation between frequency ω and the components of wave vector k, rather a *differential equation* in the unknown function $\phi(t, r)$.

2.7.5 Space-Time Rays

The eikonal equation (2.7.17) belongs to the Hamilton-Jacobi variety and can be solved by the method of characteristics (Sect. 2.2). We write the equation of characterstics in the canonic Hamiltonian form[24]

$$\frac{dr}{d\varsigma} = \frac{\partial\mathscr{H}}{\partial k} \ , \quad \frac{dk}{d\varsigma} = -\frac{\partial\mathscr{H}}{\partial r} \ , \quad \frac{dt}{d\varsigma} = -\frac{\partial\mathscr{H}}{\partial\omega} \ , \quad \frac{d\omega}{d\varsigma} = \frac{\partial\mathscr{H}}{\partial t} \ , \tag{2.7.20}$$

where the functional $\mathscr{H} = \mathscr{H}(\omega, k, r)$ defined by (2.7.17) plays the role of a Hamiltonian, and ς is a parameter varying along the characteristic.

The solution $\{t(\varsigma), r(\varsigma)\}$ of (2.7.20) describes *space-time rays* in the four-dimensional space $\{t, r\}$ which are, as will be shown below, the world lines of wave packets. Spatial rays (or simply rays) $r = r(\varsigma)$ are the projections of space-time rays into the three-dimensional (coordinates) space r. In turn, space-time rays $\{t(\varsigma), r(\varsigma)\}$ can be regarded as the projection of the

---

[24] One of the four equations (2.7.20) is the corollary of the other three since the variables t, r_j, k_j, and ω are combined by the equation $\mathscr{H} = 0$.

8-D phase trajectory $\{t(\varsigma),\ r(\varsigma),\ \omega(\varsigma),\ k(\varsigma)\}$, satisfying (2.7.20) into the 4-D space $\{t, r\}$.

Equation (2.7.20) can be rewritten as

$$\frac{dr}{dt} = -\frac{\partial\mathcal{H}}{\partial k}\bigg/\frac{\partial\mathcal{H}}{\partial\omega} \equiv g\ ,$$

$$\frac{dk}{dt} = \frac{\partial\mathcal{H}}{\partial r}\bigg/\frac{\partial\mathcal{H}}{\partial\omega}\ , \qquad \frac{d\omega}{dt} = -\frac{\partial\mathcal{H}}{\partial t}\bigg/\frac{\partial\mathcal{H}}{\partial\omega}\ . \tag{2.7.21}$$

Using here the Hamiltonian (2.7.17) yields

$$\frac{dr}{dt} = g\ , \qquad \frac{dk}{dt} = \omega\left(\frac{\partial\omega n}{\partial\omega}\right)^{-1}\nabla n\ , \qquad \frac{d\omega}{dt} = 0\ , \tag{2.7.22}$$

where the quantity

$$g = -\frac{\partial\mathcal{H}}{\partial k}\bigg/\frac{\partial\mathcal{H}}{\partial\omega} = cl\left(\frac{\partial\omega n}{\partial\omega}\right)^{-1}\ , \qquad l = \frac{k}{k}\ , \tag{2.7.23}$$

is the *group velocity* of the wave. In an isotropic medium the group velocity g is directed along the vector $k = \nabla\phi$; i.e., in the normal to the phase front $\phi(t, r) = $ constant at any fixed t.[25] When dispersion is absent, the permittivity ϵ does not depend on frequency, i.e., $\epsilon = \epsilon(r) = n^2(r)$, and the group velocity, as is seen from (2.7.23), coincides with the phase velocity

$$g = \frac{c}{n}\,l = v_{ph}l = v_{ph}\ . \tag{2.7.24}$$

In accordance with (2.7.22) the frequency ω remains unchanged along a space-time trajectory

$$\omega = \text{const} = \omega^0\ . \tag{2.7.25}$$

and is defined by the initial condition which will be discussed in Sect.2.7.6.

---

[25] Introducing the coordinate $x_4 = ct$ and the fourth component of the wave vector $k_4 = -\omega/c$, we could consider the four-component quantity

$$G = \gamma(g, c) = \gamma\left(\frac{dr}{dt}, c\right) = \gamma c\left(\frac{\partial\mathcal{H}}{\partial k}, \frac{\partial\mathcal{H}}{\partial k_4}\right)\bigg/\frac{\partial\mathcal{H}}{\partial k_4}\ ,$$

where $\gamma = (1 - g^2/c^2)^{-1/2}$, as a 4-tuple of group velocity, tangent to the world line $\{t, r(t)\}$ and normal to the dispersion surface $\mathcal{H} = $ constant (see [2.81,83,84] which consider specific types of dispersion and present a graphic method of ray tracing, and which systematically analyzes the role of the 4-tuple group velocity in electrodynamics).

54

If we recognize that the frequency is constant along a ray, and introduce arc length σ along the ray as a parameter by the expression $d\sigma = g\, dt$, then we can easily transform the first two equations in (2.7.22) to the form

$$dr/d\sigma = 1 \, ,$$

$$dp/d\sigma = \nabla n \, , \quad (p = \cdot kc/\omega) \tag{2.7.26}$$

characteristic of the ray equations (2.2.17) for monochromatic waves. The respective group time of wave propagation is then

$$t = t^0 + \int_{\sigma^0}^{\sigma} d\sigma/g \, . \tag{2.7.27}$$

According to (2.7.26) a wave of an instantaneous frequency ω proceeds in the same trajectory as a monochromatic wave of this frequency, but the following train of a slightly different frequency $\omega_1 = \omega + \delta\omega$ will proceed through another path corresponding to the frequency ω_1. This creates, in a dispersive medium, a situation when the wave trains of larger group velocity can overrun other trains (this will be discussed in more detail in Sect. 3.9). In the absence of dispersion, space-time rays become frequency-independent so that the world lines of wave trains at different frequencies coincide.

2.7.6 Initial Conditions

In contrast to the geometrical optics of monochromatic waves, space-time geometrical optics allows *nonstationary* initial conditions to be specified, such as those for modulated (specifically, pulse) oscillations. Moreover, initial conditions may now be given on *moving surfaces* (moving sources, reflection from moving interfaces, etc.). In the general case initial conditions for a field may be specified on a three-dimensional hypersurface Σ in a four-dimensional space $\{t, r\}$. We write the equation for Σ parametrically as

$$r = r^0(\xi) \, , \quad t = t^0(\xi) \, , \tag{2.7.28}$$

where $\xi = \{\xi_1, \xi_2, \xi_3\}$ are curvilinear coordinates on Σ, which are the *ray coordinates* in our consideration.

In particular, if initial conditions are specified on a stationary surface Q, then two ray coordinates, as in Sects. 2.1-3, are provided by the curvilinear coordinates on Q, and the third can be taken as the initial moment of time t^0, then

$$r_\Sigma = r^0(\xi, \eta) \, , \quad t_\Sigma = t^0 \, , \quad \xi_1 = \xi \, , \quad \xi_2 = \eta \, , \quad \xi_3 = t^0 \, . \tag{2.7.29}$$

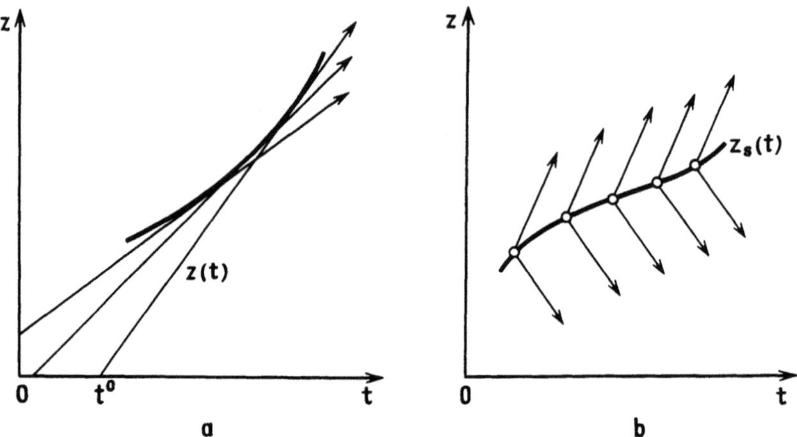

Fig.2.16. Behavior of space time rays in the vicinity of (a) the space time caustic and (b) a moving source

A point source moving along a path $\mathbf{r} = \boldsymbol{\rho}(t)$ may have, as the ray coordinates, the initial time t^0 for the ray and the angular coordinates on an infinitesimally small sphere containing the source:

$$\mathbf{r}_\Sigma = \boldsymbol{\rho}(t^0) , \quad t_\Sigma = t^0 , \quad \xi_1 = \theta , \quad \xi_2 = \nu , \quad \xi_3 = t^0 \tag{2.7.30}$$

(θ and v are the angular coordinates on the sphere).

Let the initial field be given on Σ in the form corresponding to the approximation of space–time geometrical optics:

$$u_\Sigma = A(t^0, \mathbf{r}^0)\exp[i\phi(t^0, \mathbf{r}^0)] = A^0(\xi)\exp[i\phi^0(\xi)] . \tag{2.7.31}$$

On differentiating the initial eikonal $\phi(t^0, \mathbf{r}^0) = \phi^0(\xi)$ with respect to ξ_j ($j = 1, 2, 3$) we obtain

$$\frac{\partial\phi^0}{\partial\xi_j} = \left(\nabla\phi\frac{\partial\mathbf{r}}{\partial\xi_j} + \frac{\partial\phi}{\partial t}\frac{\partial t}{\partial\xi_j}\right)\Big|_\Sigma = \mathbf{k}^0\frac{\partial\mathbf{r}^0}{\partial\xi_j} - \omega^0\frac{\partial t^0}{\partial\xi_j} . \tag{2.7.32}$$

The three equations of (2.7.32) along with the eikonal equation (2.7.17) present the initial conditions for the frequency $\omega_\Sigma = \omega^0(\xi)$ and the wave vector $\mathbf{k}_\Sigma = \mathbf{k}^0(\xi)$, and together with (2.7.28) define the family of space-time rays $\{t, \mathbf{r}(t, \xi)\}$. So in the case of a moving source, to the eikonal $\phi^0(t^0)$ there corresponds the initial frequency $\omega^0(t^0) = -\partial\phi^0(t^0)/\partial t^0$ and the initial vector

$$\mathbf{k}^0(\theta, \nu, t^0) = \mathbf{l}^0(\theta, \nu)\frac{\omega^0}{c} \, n[\boldsymbol{\rho}(t^0)] ,$$

where $\mathbf{l}^0(\theta, \nu)$ is the unit vector corresponding to θ and ν

$$\mathbf{l}^0(\theta, \nu) = \mathbf{e}_z\cos\theta + \sin\theta(\mathbf{e}_x\cos\nu + \mathbf{e}_y\sin\nu) .$$

56

By way of example, Fig.2.16 depicts the space-time paths on the z, t plane for two situations. Those on Fig.2.16a correspond to a field u^0 of variable frequency $\omega^0(t^0)$ given at $z = 0$. The medium is assumed to be homogeneous but dispersive, therefore the rays possess various tilt $dz/dt = g(\omega^0)$. Figure 2.16b shows the system of z, t rays for the monochromatic field radiated by a source moving along $z = z_s(t)$.

The family of world lines $\{t, r(t, \xi)\}$, along with the frequency $\omega(t, \xi)$ and the wave vector $k(t, \xi)$, form in the extended space $\{t, r; \omega, k\}$ a family of phase trajectories (a Lagrangian manifold) described by the parameters ξ_1, ξ_2, ξ_3. The trajectories of this family do not cross, whereas in the 4-D space $\{t, r\}$ many world lines may pass through a given world point.

2.7.7 Eikonal and Wave Amplitude

On the space-time ray $\{t, r(t)\}$ the derivative of the eikonal $\phi(t, r)$ is

$$\frac{d\phi}{dt} = \frac{\partial\phi}{\partial t} + \frac{\partial\phi}{\partial r}\frac{dr}{dt} = -\omega + k\cdot g \, ,$$

which upon integration yields

$$\phi(t) = \phi^0(t^0) + \int_{t^0}^{t} (k\cdot g - \omega)dt \, , \tag{2.7.33}$$

where $\phi(t) \equiv \phi[t, r(t)]$. For an arbitrary parameter on the ray we have

$$\frac{d\phi}{d\varsigma} = \frac{\partial\phi}{\partial t}\frac{dt}{d\varsigma} + \frac{\partial\phi}{\partial r}\frac{dr}{d\varsigma} = -\omega\frac{\partial\mathcal{H}}{\partial\omega} + k\frac{\partial\mathcal{H}}{\partial k} \, , \qquad \text{whence}$$

$$\phi(\varsigma) = \phi^0 + \int_{\varsigma^0}^{\varsigma}\left(\omega\frac{\partial\mathcal{H}}{\partial\omega} + k\frac{\partial\mathcal{H}}{\partial k}\right)d\varsigma \, . \tag{2.7.34}$$

Considering that ω is constant along the world line $\{t, r(t)\}$ and using arc length σ as a parameter, we can recast (2.7.33) into

$$\phi(\sigma) = \sigma^0 + \frac{\omega}{c}\int_{t^0}^{t}(ng - c)dt = \phi^0 + \omega\int_{\sigma^0}^{\sigma}\left(\frac{1}{v_{ph}} - \frac{1}{g}\right)d\sigma \, . \tag{2.7.35}$$

The wave packet group delay time $\Delta t_g = t - t^0$ is to be derived from (2.7.27).

In a medium without dispersion, where $g = c/n$, the second term in (2.7.33 or 35) vanishes, which means that the eikonal is conserved along the world line $\{t, \mathbf{r}(t)\}$

$$\phi(t) = \text{const.} = \phi^0 \ . \tag{2.7.36}$$

The group delay in this case appears to be independent of frequency

$$\Delta t_g = \frac{1}{c} \int_{\sigma^0}^{\sigma} n(\mathbf{r})d\sigma \ . \tag{2.7.37}$$

As with monochromatic waves, the transport equations can be converted to an ordinary differential equation. Noting that $\nabla A^2 \nabla \phi + A^2 \Delta \sigma = \nabla(kA^2)$ and that $k = g[\partial(\omega^2 \epsilon)/\partial \omega]/2c^2$, we write (2.7.18) as an energy conservation law[26]

$$\partial w/\partial t + \text{div } \mathbf{S} = 0 \ , \qquad \text{where the quantities} \tag{2.7.38}$$

$$w = \frac{A^2}{2} \frac{\partial(\omega^2 \epsilon)}{\omega \partial \omega} = \frac{c}{g} nA^2 \ , \quad \mathbf{S} = w\mathbf{g} = cnA^2\mathbf{l} \tag{2.7.39}$$

have, respectively, the meaning of energy density and energy flux (the Poynting vector for this problem). The vector $\mathbf{S}$ is, within a multiplier, the vector on intensity $\mathbf{I} = p A^2 = nA^2\mathbf{l}$, with which we dealt in Sect.2.3, namely, $\mathbf{S} = \mathbf{I}c$.

On recognizing that

$$\text{div}\mathbf{S} = w\,\text{div}\mathbf{g} + \mathbf{g}\,\nabla w$$

and keeping in mind that on the world line $\{t, \mathbf{r}(t)\}$

$$\frac{dw}{dt} = \frac{\partial w}{\partial t} + \nabla w \cdot \frac{d\mathbf{r}}{dt} = \frac{\partial w}{\partial t} + \mathbf{g} \cdot \nabla w \ ,$$

we write the conservation equation as

$$\frac{dw}{dt} + w\,\text{div}\mathbf{g} = 0 \ . \tag{2.7.40}$$

But in accordance with the Liouville formula

$$\text{div}\mathbf{g} = \frac{d}{dt}\ln j(t, \xi) \ , \tag{2.7.41}$$

---

[26] The transfer equation (2.7.38) can be recast into a 4D form by contracting it into the 4D divergence equal to zero: $\partial S_\alpha / \partial x_\alpha = 0$, where $\alpha = 1,2,3,4$; and $S_4 = cw$.

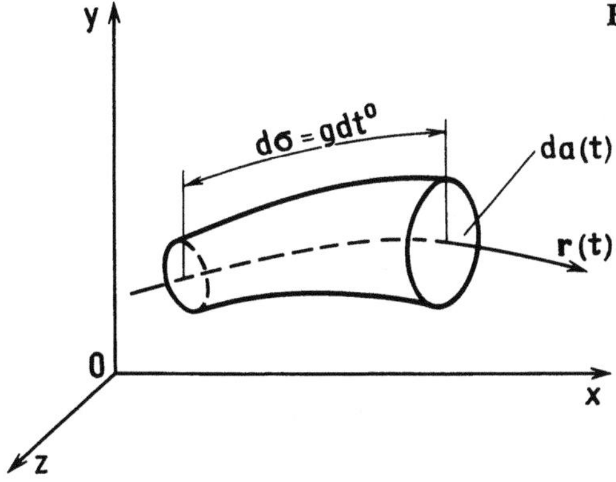

Fig.2.17. Wave packet volume dV

where $j(t,\xi)$ is the Jacobian of the transformation from the ray coordinates $\xi = (\xi_1, \xi_2, \xi_3)$ to the space coordinates $r = (x_1, x_2, x_3)$:

$$j(t,\xi) = \det\left[\frac{\partial x_i(t,\xi)}{\partial \xi_m}\right] = \left(\frac{\partial r}{\partial \xi_1} \times \frac{\partial r}{\partial \xi_2}\right) \cdot \frac{\partial r}{\partial \xi_3} \ . \tag{2.7.42}$$

Therefore, from (2.7.40) we obtain the following law for the energy density

$$w(t) = w^0(t^0) \frac{j(t^0,\xi)}{j(t,\xi)} \ , \tag{2.7.43}$$

from which we can deduce the wave amplitude

$$A(t) = A^0(t^0) \sqrt{\frac{n^0 g \, j(t^0,\xi)}{n g^0 \, j(t,\xi)}} \ . \tag{2.7.44}$$

According to (2.7.44,45) the quantities $w(t)$ and $A(t)$ are connected with their intial values at the time $t^0 = t - \Delta t_g$ retarded from the current time instant by the group-delay interval. From this, as well as directly from (2.7.40), it follows that in the space-time geometrical-optics approximation the wave packets progress along world lines $\{t, r(t)\}$, and in a coordinate space their motion is determined by the group velocity g.

The ratio of Jacobians in (2.7.40,41) describes the variation of the wave packet volume $dV(t) = j(t,\xi)d\xi_1 d\xi_2 d\xi_3$. This volume is proportional to both the cross section of the ray tube, da, and the packet length $d\sigma = g dt^0$: $dV = g \, da \, dt^0$ (Fig.2.17). Thus, when the initial conditions are supplied in the form of (2.7.29),

$$j(t,\xi) = \left[\frac{\partial r}{\partial \xi}, \frac{\partial r}{\partial \eta}, \frac{\partial r}{\partial t^0}\right] = \left(\frac{\partial r}{\partial \xi} \times \frac{\partial r}{\partial \eta}\right) \cdot \frac{\partial r}{\partial \sigma} \frac{d\sigma}{dt^0} = -g\mathscr{D}(\sigma) \ , \tag{2.7.45a}$$

so that

$$\frac{j(t^0,\xi)}{j(t,\xi)} = \frac{j^0}{j} = \frac{g^0 \mathscr{D}(\sigma^0)}{g \mathscr{D}(\sigma)} = \frac{g^0 \, da^0}{g \, da} \ . \tag{2.7.45b}$$

For the monochromatic initial field

$$u(\xi,\eta,t^0) = A^0(\xi,\eta)\exp\!\left[-i\omega t^0 + i\,\frac{\omega}{c}\psi^0(\xi,\eta)\right] ,$$

specified on a stationary Q, $r = r^0(\xi,\eta)$, all the relationships of space-time geometrical otpics reduce to the respective equations of Sects.2.1-3. Particularly from (2.7.36,37) we obtain for the phase $\phi(\sigma)$ the formula

$$\phi = -\omega t^0 + \frac{\omega}{c}\psi^0 + \omega \int_{\sigma^0}^{\sigma}\left(\frac{1}{v_{ph}} - \frac{1}{g}\right)d\sigma =$$

$$-\omega t + \frac{\omega}{c}\left[\psi^0 + \int_{\sigma^0}^{\sigma} n\,d\sigma\right] ,$$

which agrees with (2.2.15), while (2.7.44) converts to (2.3.18) since in accord with (2.7.45,24) $nj/g = -n\,\mathscr{D}(\sigma)$.

2.7.8 Space-Time Caustics

The resultant field associated with a ray family $r = R(t,\xi)$ can be expressed parametrically

$$u(t,\xi) = A(t,\xi)\exp[i\phi(t,\xi)] . \tag{2.7.46}$$

An explicit dependence of the field u on t and r can be found by eliminating the ray coordiantes ξ between (2.7.46) and the ray-family equation $r = R(t,\xi)$. If several space-time rays pass through the world point of interest, i.e., when several sets of ray coordinates are to be handled, the total field will be the sum of the ray fields (2.7.46):

$$u(t,r) = \sum_{\nu} A(t,\xi_\nu)\exp[i\phi(t,\xi_\nu)]_{\xi_\nu=\xi_\nu(t,r)} . \tag{2.7.47}$$

In multiray situations there certainly form regions where two or more rays merge and the Jacobian j vanishes. This corresponds to the world point $\{t,r\}$ situated on a space-time caustic whose position is determined by the equations

$$r = R(t,\xi) , \quad j(t,\xi) = 0 .$$

On moving closer to a space-time caustic field the amplitude rises without bound and the ray approximation is no longer valid. A specific case of such a caustic is shown in Fig.2.16a.

In their properties such caustics are very similar to spatial caustics and, in particular, they could be classified with a nomenclature analogous to that presented in Sect.2.4. Some features of these caustics will be discussed in Sect.3.9 with specific examples of illustration from [2.9].

2.7.9 Propagation of Field Discontinuities in Nondispersive Media

In nondispersive media one may count on the validity of geometrical optics not only for quasi-monochromatic high-frequency pulses (Sects.2.7.2–8), but also for aperiodic disturbances and field discontinuities. To be more precise, we need to describe the field in the vicinity of a wave front where the field is defined mainly by its high frequency components, which brings this approach closer to the theory discussed in Sects.2.7.2–8. The space-time geometrical optics describing the propagation of aperiodic disturbances and field discontinuities was the focus of many studies [2.16, 85–89].

For a particular case of initial conditions specified on a stationary surface, the fundamental results of geometrical optics can be readily reproduced with the help of the spectral representation. To demonstrate it, let

$$u(\omega, \mathbf{r})e^{-i\omega t} = f_0(\omega)e^{-i\omega t} \sum_{m=0}^{\infty} \frac{A_m(\mathbf{r})}{(ik_0)^m} e^{ik_0 \psi(\mathbf{r})} \qquad (2.7.48)$$

be a monochromatic ray field with the initial condition

$$u^0(\omega, \mathbf{r}^0)e^{-i\omega t} = f_0(\omega)e^{-i\omega t} \sum_{m=0}^{\infty} \frac{A_m^0(\mathbf{r}^0)}{(ik_0)^m} e^{ik_0 \psi^0 (\mathbf{r}^0)}$$

given on the surface Q: $\mathbf{r} = \mathbf{r}^0(\xi, \eta)$. As compared with Sects.2.1–3 we have restored here the dependence on time by introducing the factor $\exp(-i\omega t)$ and added the multiplier $f_0(\omega)$, which is the spectral amplitude of a radiated pulse $f_0(t)$

$$f_0(t) = \frac{1}{2\pi} \int_{-\infty}^{\infty} f_0(\omega)e^{-i\omega t} d\omega . \qquad (2.7.49)$$

For the nonstationary field $u(t, \mathbf{r})$ the superposition of the monochromatic waves (2.7.48) leads to the equation[27]

---

[27] We consider here that $A_m(\mathbf{r})$ and $\psi(\mathbf{r})$ are independent of frequency.

$$u(t,r) = \frac{1}{2\pi} \int_{-\infty}^{+\infty} \left(\sum_{m=0}^{\infty} \frac{A_m(r)}{(i\omega/c)^m} \, e^{i(\omega/c)\psi(r)} \right) f_0(\omega) e^{-i\omega t} \, d\omega$$

$$= \sum_{m=0}^{\infty} \tilde{A}_m(r) f_m[t - \psi(r)/c] \, , \tag{2.7.50}$$

where $\tilde{A}_m(r) = (-c)^m A_m(r)$ and

$$f_m(t) = \frac{1}{2\pi} \int_{-\infty}^{\infty} \frac{f_0(\omega) e^{-i\omega t}}{(-i\omega)^m} \, d\omega \, . \tag{2.7.51}$$

From (2.7.51) the sequence of functions $f_0(t)$, $f_1(t)$, ... is seen to satisfy the condition

$$d^m f_m(t)/dt^m = f_0(t) \quad \text{or} \quad df_m(t)/dt = f_{m-1}(t) \, . \tag{2.7.52}$$

The representation of a field by an expansion of the type (2.7.50) enables one to study the behavior of signals of arbitrary form, notably the propagation of short pulses and field discontinuities.[28] As with monochromatic waves, the main part of the field is described by the leading term of the ray expansion (2.7.50)

$$u(t,r) \simeq A_0(r) f_0[t - \psi(r)/c] \, , \tag{2.7.53}$$

which takes care of the undistorted pulse shape along the propagation route while the subsequent terms of (2.7.50) describe the distortions which the pulse shape suffers.

The functions $f_m(t)$ become smoother as m increases. For instance, if a discontinuity of the field is described by the unit step function

$$f_0(t) = \begin{cases} 0 & \text{for } t < 0 \\ 1 & \text{for } t \geq 0 \end{cases},$$

then the functions

$$f_m(t) = \begin{cases} 0 & \text{for } t < 0 \\ t^m/m! & \text{for } t \geq 0 \end{cases}$$

are continuous at $t = 0$ along with their derivatives to order (m-1). In the

---

[28] With $f_0(t) = \exp(-i\omega t)$ and $f_m(t) = (-i\omega)^m \exp(-i\omega t)$ we again have a monochromatic field (Sects.2.1-3).

neighborhood of the field discontinuity on the phase front $t-\psi(\mathbf{r})/c = 0$ all the functions $f_m(t-\psi/c)$ are small compared with f_0; therefore we may content ourselves with the zero-order approximation (2.7.53).

The domain of application of geometricl optics is limited by the fact that the inequality (2.1.4), $\mu \simeq 1/kL \simeq c/\omega L \simeq 1$, fails for the low-frequency components of $f_0(\omega)$. Therefore (2.7.53) describes only the behavior of the high-frequency, rapidly changing part of the field for which the characteristic frequencies $\omega > c/L$ or the characteristic times $\tau < L/c$, whereas the low-frequency, slowly-varying part ($\omega < c/L$, $\tau > L/c$) is subject to marked distortions. The space-time geometrical-optics method can thus describe well the field structure in the vicinity of discontinuities, but can incur substantial error at any time $\Delta t > L/c$ beyond a discontinuity.

In our deducing (2.7.50) from (2.7.48) we assumed the quantities $A_m(\mathbf{r})$ and $\psi(\mathbf{r})$ to be known from the theory for monochromatic field (Sects.2.1-3). These quantities, though, can be found directly by substituting (2.7.50) in the wave equation without dispersion

$$\Box u \equiv \Delta u - \frac{n^2(\mathbf{r})}{c^2}\frac{\partial^2 u}{\partial t^2} = 0 \qquad (2.7.54)$$

and equating to zero the coefficients of the functions $f_m(t-\psi/c)$. Performing these operations leads us to the geometrical-optics equations for monochromatic waves, namely, to the eikonal equation (2.1.11) in ψ and to the transport equation (2.1.14) in $\tilde{A}_m = (-c)^m A_m$. Thus the functions $\tilde{A}_m(\mathbf{r})$ and $\psi(\mathbf{r})$ in the ray series (2.7.50) can indeed be derived by the formulas of Sects.2.1-3.

A more general approach, feasible to describe the fields excited by a moving source or a moving interface, relies on the field representation [2.78] in the form

$$u(t,\mathbf{r}) = \sum_{m=0}^{\infty} A_m(t,\mathbf{r})f_m[\phi(t,\mathbf{r})] , \qquad (2.7.55)$$

where the functions $f_m(t)$ are deemed known and satisfy (2.7.52) for $m \geq 1$. Substituting the ray expansion (2.7.55) into (2.7.54) and equating the coefficients of like functions $f_m(t)$ leads us to the eikonal equation

$$(\nabla\phi)^2 - \frac{n^2(\mathbf{r})}{c^2}\left(\frac{\partial\phi}{\partial t}\right)^2 = 0 \qquad (2.7.56)$$

and the transport equations ($m = 0,1,2,...$)

$$2\left(\nabla A_m \nabla\phi - \frac{n^2}{c^2}\frac{\partial A_m}{\partial t}\frac{\partial\phi}{\partial t}\right) + A_m \Box\phi = -\Box A_{m-1} . \qquad (2.7.57)$$

If we denote $\mathbf{k} = \nabla\phi$ and $\omega = -\partial\phi/\partial t$ then it would be a straightforward matter to show that (2.7.56,57) coincide with the equations of geometrical optics for monochromatic waves (2.7.17,18) if we neglect dispersion in the latter. This has the implication that the formulas of Sects.2.7.2-8 are valid in deriving the desired functions $\phi(t,\mathbf{r})$ and $A_m(t,\mathbf{r})$, $m \geq 0$. The ray series (2.7.50) is, certainly, a specific case of expansion (2.7.55). The mathematical substantiation of such expansions was considered in [2.73, 82, 85, 91].

The method demonstrated was used in handling problems of moving source irradiation [2.82, 92, 93] and wave reflection from moving interfaces [2.8, 82, 94], and in other situations (extensive bibliographies can be found in [2.8, 8, 65, 87, 89, 95]).

2.8 Separation of Variables in the Eikonal Equation

2.8.1 The Complete Integral of the Eikonal Equation

Under certain conditions it proves to be more convenient to solve the eikonal equation (2.1.11) directly rather than the equation of rays, (2.2.19). The ray paths can then be readily found from the eikonal. This approach can at times be preferable to the method of characteristics, as, for example, in choosing a two- or three-dimensional model for the medium of physical problem concerned. The primary method of handling the eikonal equation explicitly is by *separation of variables.*[29]

In essence this method suggests to seek the eikonal as a sum in which each term depends only upon a single variable q_j

$$\psi = \psi_1(q_1) + \psi_2(q_2) + \psi_3(q_3) \; . \tag{2.8.1}$$

When the eikonal equation (2.1.11) alllows such a separation, the substitution (2.8.1) transforms it into three ordinary differential equations of the form

$$\mathcal{H}_j(q_j, d\psi_j/dq_j; \alpha_1, \alpha_2) = 0 \; , \tag{2.8.2}$$

where $j = 1,2,3$; and α_1 and α_2 are separation constants.

Equations (2.8.2) can be solved by quadrature if they are resolvable for $d\psi_j/dq_j$.

The function ψ found in this way should contain two independent separation constants α_1 and α_2, so $\psi = \tilde{\psi}(q_1, q_2, q_3; \alpha_1, \alpha_2)$. Considering besides that ψ enters the eikonal equation $(\nabla\psi)^2 = n^2$ as a partial derivative only, the solution of (2.6.1) may be augmented by an arbitrary additive constant α_3, after which

$$\psi = \tilde{\psi}(q_1, q_2, q_3; \alpha_1, \alpha_2) + \alpha_3 = \psi(q_1, q_2, q_3; \alpha_1, \alpha_2, \alpha_3) \; . \tag{2.8.3}$$

---

[29] Separation of variables is the general technique of solving the equations of the Hamilton-Jacobi variety [2.12, 14-17, 96, 97].

This solution involves as many independent constants α_i as it has independent variables q_i, and is called the complete integral of the eikonal equation. The complete integral meets the condition [2.16, 17]

$$\det\left[\frac{\partial^2 \psi}{\partial q_i \partial \alpha_j}\right] = 0 \ , \quad i,j = 1,2,3 \ , \tag{2.8.4}$$

which is equivalent to requiring that the α_j be independent. Separation of variables is a powerful technique for computing the complete integral (2.8.3) under conditions to be discussed in (Sects. 2.82–4).

2.8.2 Separation of Variables in Two Dimensions (Cartesian Coordinates)

In Cartesian coordinates the variables in the eikonal equation

$$\left(\frac{\partial \psi}{\partial x}\right) + \left(\frac{\partial \psi}{\partial y}\right)^2 = n^2(x, y) \tag{2.8.5}$$

may be separated to give a solution in the form $\psi = \psi_1(x) + \psi_2(y)$, only subject to the condition

$$n^2(x, y) = \epsilon_1(x) + \epsilon_2(y) \ , \tag{2.8.6}$$

where $\epsilon_1(x)$ and $\epsilon_2(y)$ are arbitrary functions.
From (2.8.5, 6) it follows that

$$[\psi_1{}'(x)]^2 - \epsilon_1(x) = - [\psi_2{}'(y)]^2 + \epsilon_2(y) \ . \tag{2.8.7}$$

This equality must be satisfied identically in x and y, implying that both its parts equal some arbitrary constant α. Then

$$\psi_1{}'(x) = \pm\sqrt{\epsilon_1(x) + \alpha} \ ,$$

$$\psi_2{}'(x) = \pm\sqrt{\epsilon_2(y) - \alpha} \ .$$

By integration this we arrive at the following expression for the complete integral of the eikonal equation

$$\psi(x, y) = \psi_1(x) + \psi_2(y)$$

$$= \pm \int_{x^0}^{x} \sqrt{\epsilon_1(x) + \alpha} \ dx \pm \int_{y^0}^{y} \sqrt{\epsilon_2(y) - \alpha} \ dy + \alpha_0 \ . \tag{2.8.9}$$

The signs and unknown constants α, α_0, x^0, and y^0 are to be chosen subject to the initial conditions for the eikonal; that is, they are decided by an actual problem statement.

2.8.3 Separation of Variables in Two Dimensions (Curvilinear Orthogonal Coordinates)

We recall that any curvilinear orthogonal coordinates q_1 and q_2 are related to the Cartesian coordinates x and y [2.94,95]:

$$q_1 + iq_2 = f(x+iy) , \qquad (2.8.10a)$$

or, using the shorthands $w = q_1+iq_2$, $z = x+iy$,

$$w = f(z) , \qquad (2.8.10b)$$

where f(z) is an arbitrary analytic function. In its region of regularity this function maps the x, y plane conformally onto the q_1, q_2 plane:

$$q_1 = q_1(x, y) = \mathrm{Re}\{f(x + iy)\} ,$$

$$q_2 = q_2(x, y) = \mathrm{Im}\{f(x + iy)\} . \qquad (2.8.10c)$$

In the q_1, q_2 coordinates the squared arc length is

$$d\sigma^2 = dx^2 + dy^2 = h_1{}^2 dq_1{}^2 + h_2{}^2 dq_2{}^2 = h^2(dq_1{}^2 + dq_2{}^2) . \qquad (2.8.11)$$

The equality of the Lame coefficients $h_1 = h_2 = h$ and the orthogonality of q_1 and q_2 ($\nabla q_1 \cdot \nabla q_2 = 0$) follow from (2.8.10) and the Cauchy-Riemann equations for an analytic f(z). So, with the nomenclature of (2.8.10) for the Lame coefficients we have [2.98]

$$h = |dw/dz|^{-1} = |dz/dw| . \qquad (2.8.12)$$

Express the eikonal equation in the curvilinear coordinates q_1, q_2

$$(\nabla\psi)^2 \equiv h^{-2}\left[\left(\frac{\partial\psi}{\partial q_1}\right)^2 + \left(\frac{\partial\psi}{\partial q_2}\right)^2 \right] = n^2 \qquad \text{or} \qquad (2.8.13a)$$

$$\left(\frac{\partial\psi}{\partial q_1}\right)^2 + \left(\frac{\partial\psi}{\partial q_2}\right)^2 = h^2 n^2 . \qquad (2.8.13b)$$

The last equation is similar to that of eikonal (2.8.5) in Cartesian coordinates. Hence, now separation of variables in (2.8.13b) is possible under the condition that

$$h^2 n^2 = \epsilon_1(q_1) + \epsilon_2(q_2) , \qquad \text{or} \qquad (2.8.14a)$$

$$n^2 = h^{-2}[\epsilon_1(q_1) + \epsilon_2(q_2)] \equiv n^2(q_1, q_2) , \qquad (2.8.14b)$$

where $\epsilon_1(q_1)$ and $\epsilon_2(q_2)$ are arbitrary functions of q_1 and q_2.

Once this condition is met, the complete integral of (2.8.13) has a form analogous to (2.8.9):

$$\psi = \pm \int_{q_1^0}^{q_1} \sqrt{\epsilon_1(q_1) + \alpha} \, dq_1 \pm \int_{q_2^0}^{q_2} \sqrt{\epsilon_2(q_2) - \alpha} \, dq_2 + \alpha_0 , \qquad (2.8.15)$$

where α, α_0 are the separation constants, and $q_{1,2}{}^0$ are the integration constants.

The condition for representing $n^2 h^2 = n^2 |dz/dw|^2$ in the form of (2.8.14a) is equivalent to

$$\frac{\partial^2}{\partial q_1 \partial q_2} \left(n^2 \left| \frac{dz}{dw} \right|^2 \right) = 0 . \qquad (2.8.16)$$

This equation defines plane systems of orthogonal curvilinear coordinates, which facilitate separation of variables in the eikonal equation (2.8.13). But it also imposes constraints on the feasible functions for the refractive index.

It should be noted that the conditions (2.8.14, 16) are identical to those for separation of variables in the Helmholtz equation (2.1.1), which were treated in depth in the literature [2.98] especially for the case of a homogeneous medium.

The solutions of (2.8.16) in a *homogeneous* medium (n = constant) have been examined in [2.98] to show that all the plane coordinate systems – polar, parabolic, elliptic, and Cartesian – are suitable for the purpose. These systems, however, exhaust the list of possibilities capable of separating variables in the two-dimensional eikonal equation for a homogeneous medium.

An *inhomogeneous* medium presents much wider possibilities for separation of variables in the two-dimensional eikonal equation. Indeed, choosing the refractive index $n(r)$ as in (2.8.14) one may achieve separation for any plane orthogonal coordinate system. For example, in addition to the aforementioned coordinate systems feasible for a homogeneous medium, the equation is separable in the hyperbolic and bipolar coordinate systems [2.98].

Note that instead of the coordinates q_1 and q_2 defined by the conformal mapping (2.8.10), we may employ more general orthogonal coordinates introduced as

$$\xi = \xi(q_1) , \quad \eta = \eta(q_2) \quad \text{or} \quad q_1 = q_1(\xi) , \quad q_2 = q_2(\eta) . \qquad (2.8.17)$$

The Lame coefficients of these coordinates are

$$h_\xi = h \frac{dq_1(\xi)}{d\xi} , \quad h_\eta = h \frac{dq_2(\eta)}{d\eta} , \tag{2.8.18}$$

where h is determined by (2.8.12). It is obvious that generally $h_\xi \neq h_\eta$.

In the two-dimensional eikonal equation written with the coordinates ξ, η

$$(\nabla\psi)^2 \equiv \frac{1}{h_\xi^2}\left(\frac{\partial\psi}{\partial\xi}\right)^2 + \frac{1}{h_\eta^2}\left(\frac{\partial\psi}{\partial\eta}\right)^2 = n^2 \tag{2.8.19}$$

the sufficient condition of separabilitycan be written in the form

$$n^2 = \frac{1}{h_\xi^2}\epsilon_1(\xi) + 1(h_\eta^2)\epsilon_2(\eta) , \tag{2.8.20}$$

where $\epsilon_1(\xi)$ and $\epsilon_2(\eta)$ are arbitrary functions. The resultant complete integral of (2.8.19) is

$$\psi = \pm \int_{\xi_0}^{\xi} \sqrt{\epsilon_1(\xi) + \alpha} \, d\xi \pm \int_{\eta^0}^{\eta} \sqrt{\epsilon_2(\eta) - \alpha} \, d\eta + \alpha_0 . \tag{2.8.21}$$

2.8.4 Separation of Variables in Three-Dimensional Space

The separability of Hamilton-Jacobi equation in three-dimensional space has been many demonstrated in the literature, but a sufficiently general separability conditions is still dominant.[30] We shall confine ourselves to the list of the most useful separating coordinates for the three-dimensional eikonal equation

$$\sum_{j=1}^{3} \frac{1}{h_j^2(q_1,q_2,q_3)} \left(\frac{\partial\psi}{\partial q_j}\right)^2 = n^2(q_1,q_2,q_3) , \tag{2.8.22}$$

where h_j are the Lame coefficients of the orthogonal coordinates q_j (j = 1, 2, 3), and we shall evaluate the form of the respective functions $n(q_1,q_2,q_3)$ governing the refractive index profile.

(i) *Cartesian coordinates x,y,z.* The complete integral of the eikonal equation

---

[30] This problem has been examined by Steckel, in particular; Steckel's conditions are considered in [2.96].

$$\left(\frac{\partial \psi}{\partial x}\right)^2 + \left(\frac{\partial \psi}{\partial y}\right)^2 + \left(\frac{\partial \psi}{\partial z}\right)^2 = \epsilon_1(x) + e_2(y) + e_3(z) \tag{2.8.23}$$

has a form similar to (2.8.9):

$$\psi(x,y,z) = \pm \int_{x^0}^{x} \sqrt{\epsilon_1(x) + \alpha_1}\, dx \pm \int_{y_0}^{y} \sqrt{\epsilon_2(y) + \alpha_2}\, dy$$

$$\pm \int_{z^0}^{z} \sqrt{\epsilon_3(z) - \alpha_1 - \alpha_2}\, dz + \alpha_3 \ . \tag{2.8.24}$$

(ii) *Cylindrical coordinates* z,r,ϕ. Now the eikonal equation

$$\left(\frac{\partial \psi}{\partial r}\right)^2 + \frac{1}{r^2}\left(\frac{\partial \psi}{\partial \phi}\right)^2 + \left(\frac{\partial \psi}{\partial z}\right)^2 = \epsilon_1(r) + \frac{1}{r^2}\epsilon_2(\phi) + \epsilon_3(z) \tag{2.8.25}$$

has the solution

$$\psi = \pm \int_{r^0}^{r} \sqrt{\epsilon_1(r) - \alpha_1 - \alpha_2/r^2}\, dr \pm \int_{\phi^0}^{\phi} \sqrt{\epsilon_2(\phi) + \alpha_2}\, d\phi$$

$$\pm \int_{z^0}^{z} \sqrt{\epsilon_3(z) + \alpha_1}\, dz + \alpha_3 \ . \tag{2.8.26}$$

(iii) *Spherical coordinates* r,θ,ϕ. The eikonal equation

$$\left(\frac{\partial \psi}{\partial r}\right)^2 + \frac{1}{r^2}\left(\frac{\partial \psi}{\partial \theta}\right)^2 + \frac{1}{r^2 \sin^2\theta}\left(\frac{\partial \psi}{\partial \phi}\right)^2$$

$$= \epsilon_1(r) + \frac{1}{r^2}\epsilon_2(\theta) + \frac{1}{r^2 \sin^2\theta}\epsilon_3(\phi) \tag{2.8.27}$$

yields this complete integral

$$\psi = \pm \int_{r^0}^{0} \sqrt{\epsilon_1(r) - \alpha_1/r^2}\, dr \pm \int_{\phi^0}^{\phi} \sqrt{\epsilon_3(\phi) + \alpha_2}\, d\phi$$

$$\pm \int_{\theta 0}^{\theta} \sqrt{\epsilon_2(\theta) + \alpha_1 - \alpha_2/\sin^2\theta}\, d\theta + \alpha_3 \; . \qquad (2.8.28)$$

The variables in parabolic, spheroidal, ellipsoidal, toroidal, and bi-spherical coordinates separate in similar way.

When separation can be achieved in several coordinate systems - as in the case with a homogeneous medium and a relevant systems of coordinates - the preference of a particular coordinate system is decided by additional physical aspects and the geometry of the problem in question.

2.8.5 Incomplete Separation of Variables

So far we have considered complete separation of variables. More frequently, though, we should have to content ourselves with partial separation that reduces the eikonal equation to a partial differential equation with a lesser number of independent variables [2.12, 14, 15]. Partial separability is feasible, for example, if one of the coordinates does not enter explicitly in the eikonal equation. In mechanics this coordinate is termed *cyclic* as normally it is an angular coordinate (another name is *implicit* or *ignored* variable). If q_1 is such a coordinate, then in the eikonal it can be separated [2.12, 14, 15]:

$$\psi(q_1, q_2, q_3) = \alpha_1 q_1 + \tilde{\psi}(q_2, q_3) \; . \qquad (2.8.29)$$

Clearly, if all the coordinates but one are cyclic, the equation at hand separates completely.

Cyclic variables often arise in the theory of wave propagation; essentially any one- (two-) dimensional inhomogeneous medium entails two (one) cyclic coordinates in the three-dimensional wave problem. Thus, in the plane-stratified medium of $n^2 = \epsilon_3(z)$, the x, y coordinates are cyclic; hence from (2.8.29) we have

$$\psi = \tilde{\alpha}_3 + \tilde{\alpha}_1 x + \tilde{\alpha}_2 y \pm \int_{z0}^{z} \sqrt{\epsilon_3(z) - \tilde{\alpha}_1^2 - \tilde{\alpha}_2^2}\, dz \; , \qquad (2.8.30)$$

which certainly agrees with the general formula for the complete integral (2.8.24) at $\epsilon_1(x) = \epsilon_2(y) = 0$.

Another example of a cyclic coordinate is time t in the space-time eikonal equation (2.7.17) for an inhomogeneous stationary medium. Separating t in the eikonal $\phi(t, \mathbf{r})$ yields

$$\phi(t, \mathbf{r}) = -\omega t \tilde{\phi}(\omega, \mathbf{r}) = -\omega t + (\omega/c)\psi(\omega, \mathbf{r}) \; ,$$

where ω takes on the role of a separation constant. This result obviously agrees with the approach to the solution of the nonstationary problem in the frequency domain (Sect.2.7).

2.8.6 The Complete Integral of Eikonal and Rays Equations

Separation of variables leads to the solution by quadratures not only to the eikonal equation but also to the equations of rays (2.2.19). The ray paths result from differentiating the eikonal equation which follows from the familiar Jacobi theorem [2.16,81]. We state this theorem for the eikonal equation $(\nabla\psi)^2 = n^2$.

Once the complete integral

$$\psi = \tilde{\psi}(q_1,q_2,q_3;\alpha_1,\alpha_2) + \alpha_3 \qquad (2.8.31)$$

of this equation is known, the solutions to the ray equations (2.2.19) are given by the equations

$$\frac{\partial}{\partial\alpha_1}\tilde{\psi}(q_1,q_2,q_3;\alpha_1,\alpha_2) = \beta_1 \; ,$$

$$\frac{\partial}{\partial\alpha_2}\tilde{\psi}(q_1,q_2,q_3;\alpha_1,\alpha_2) = \beta_2 \; , \qquad (2.8.32)$$

where β_1 and β_2 as well as α_1 and α_2 are arbitrary constants. The components of the momentum $\mathbf{p}$ defining the ray direction in a curvilinear coordinate system are to be found as

$$\tilde{p}_i = \frac{1}{h_i}\frac{\partial\psi}{\partial q_i} \; , \quad i = 1,2,3 \qquad (2.8.33)$$

with $\tilde{p}_1{}^2+\tilde{p}_2{}^2+\tilde{p}_3{}^2 = n^2$. The unit vector $\mathbf{l} = \mathbf{p}/n = \nabla\psi/n$ tangent to the ray is

$$\mathbf{l} = \frac{\mathbf{p}}{n} = \frac{\nabla\psi}{n} = \frac{1}{n}\sum\frac{\mathbf{e}_i}{h_i}\frac{\partial\psi}{\partial q_i} = \sum \mathbf{e}_i\cos\delta_i \; , \qquad (2.8.34)$$

where $\mathbf{e}_i$ are the unit vectors tangent to the curvilinear coordinates q_1, and $\cos\delta_i$ are the direction cosines in this system.

These statements are corroborated by direct verification of the ray equations (2.2.12) by differentiating (2.8.32,33) which has been done in the general form in [2.16,17,100].

If (2.8.32) can be resolved[31] for q_2 and q_3, then the ray path equations in curvilinear coordinates can be written in the explicit form

---

[31] That equations (2.8.32) are independent and, hence, resolvable for q_2 and q_3 follows from the Jacobian (2.8.4) being nonzero.

$$q_2 = q_2(q_1; \alpha_1, \alpha_2, \beta_1, \beta_2) \ ,$$

$$q_3 = q_3(q_1; \alpha_1, \alpha_2, \beta_1, \beta_2) \ . \tag{2.8.35}$$

These equations make clear the physical sense of the constants α_1, α_2, β_1, and β_2: they are determined from the initial conditions for rays, specifying the coordinates from which the rays emanate and the directions of the rays as they leave an initial surface. These constants define a *family of rays*; they are invariable for a given ray and take on other values on coming to other rays of the family. In two-dimensional space the equation of a ray is given by

$$\frac{\partial}{\partial \alpha} \psi(q_1, q_2, \alpha, \alpha_0) = \beta \ , \tag{2.8.36}$$

where $\psi(q_1, q_2, \alpha, \alpha_0)$ is the complete integral (2.8.15) of the two-dimensional eikonal equation in cuvilinear coordinates q_1, q_2. From (2.8.15, 36) we find for a two-dimensional inhomogeneous medium subject to (2.8.14) the equation for a ray family

$$\pm \int_{q_1^0}^{q_1} \frac{dq_1}{\sqrt{\epsilon_1(q_1) + \alpha}} \pm \int_{q_2^0}^{q_2} \frac{dq_2}{\sqrt{\epsilon_2(q_2) - \alpha}} = 2\beta \ . \tag{2.8.37}$$

The signs and all unknown constants in (2.8.37) are defined by the problem statement.

With the relations deduced, we are readily able to find the equations of rays in any curvilinear coordinates separating variables in the eikonal equations (the expressions for complete integrals ψ are given in Sects. 2.8.3-4).

This method proved useful in solving many propagation problems in the geometrical-optics approximation: wave refraction in a two-dimensional inhomogeneous medium [2.101, 102], horizontal gradients in radio wave propagation in the ionosphere [2.1039-105] and in wave propagation in acoustic waveguides [2.25, 39], wave propagation in smooth waveguides [2.106], and radio wave reflection from inhomogenous plasma formations [2, 107-115]. A more subtle case of separation in the eikonal equation for an anisotropic medium is handled in [2.116, 117]. We refer the reader to Chap.3 which illustrates applications of the separation method to various inhomogeneous media.

To conclude, note that besides the methods of characteristics and separation of variables the eikonal equations can be approached by the known solution techniques of first-order partial differential equations [2.16, 17, 97], for example, by the Lagrange-Sharpy technique [2.17] prescribing the construction of a second equation being an involution of the original one. In what follows we shall also examine the method of nonadditive (multiplicative) separation of variables (Sect.3.6), numerical solution techniques

(Sect.3.10), and the perturbation method, which are also helpful in evaluating the wave eikonal.

2.9 Perturbation Techniques
for Geometrical-Optics Equations

2.9.1 The Perturbation Method for the Eikonal

The exact analytic solutions to the equations of geometrical optics are available only for a few reflective index profiles such as those examined in Sect.2.8. Under the circumstances the perturbation method might be of advantage. It is applicable when the variables cannot be separated and designed to seek corrections to the eikonal, rays and amplitude caused by small perturbations

$$\nu(\mathbf{r}) = n^2(\mathbf{r}) - \bar{n}^2(\mathbf{r}) \equiv \epsilon(\mathbf{r}) - \bar{\epsilon}(\mathbf{r}) , \quad \max|\nu| \equiv \nu_m << \bar{\epsilon} , \tag{2.9.1}$$

where $\bar{\epsilon}(\mathbf{r}) = \bar{n}^2(\mathbf{r})$ is the squared undisturbed refractive index. In geometrical optics this method was used to calculate aberrations in optical systems [2.13, 30-32, 119-123] to solve statistical problems [2.124-127] and to describe refraction of radio waves in the atmosphere [2.101, 102, 128-130]. Recently, it was also employed in considering refraction at localized inhomogeneous inhomogeneities [2.131] and to account for horizontal gradients in a stratified medium [2.103, 132-134]. It is worth noting that in the theory of aberrations in optical systems it is the deviations of mirror and lens surfaces from a prescribed shape that serve as perturbations, not the medium inhomogeneities [2.13, 30-32, 119-123].

Consider first how the perturbation theory works for the eikonal. In step with [2.127] we shall seek the solution of the eikonal equation

$$(\nabla\psi)^2 = \epsilon(\mathbf{r}) + \nu(\mathbf{r}) \tag{2.9.2}$$

as the series

$$\psi(\mathbf{r}) = \psi_0(\mathbf{r}) + \psi_1(\mathbf{r}) + \psi_2(\mathbf{r}) + ... \tag{2.9.3}$$

in the small parameter ν_m. Substituting (2.9.2) into (2.9.3) and equating similar-order terms leads to

$$(\nabla\psi_0)^2 = \bar{\epsilon}(\mathbf{r}) , \tag{2.9.4}$$

$$2(\nabla\psi_0 \cdot \nabla\psi_1) = \nu(\mathbf{r}) , \tag{2.9.5}$$

$$2(\nabla\psi_0 \cdot \nabla\psi_2) = - (\nabla\psi_1)^2 \tag{2.9.6}$$

and so forth.

Assuming the solution to the unperturbed equation (2.9.4) to be known and recognizing that

$$\nabla\psi_0 = \mathbf{p} = d\mathbf{r}_0/d\tau ,$$

where $\mathbf{r}_0(\tau)$ is the unperturbed ray, we put down the first approximation (2.95) in the form

$$2(\nabla\psi_0 \cdot \nabla\psi_1) = 2d\psi_1/d\tau = \nu , \qquad (2.9.7)$$

whence the first-order correction

$$\psi_1 = \frac{1}{2} \int_{\tau^0}^{\tau} \nu d\tau = \frac{1}{2} \int_{\sigma^0}^{\sigma} (\nu/\sqrt{\epsilon})\, d\sigma . \qquad (2.9.8)$$

The integration in (2.9.8) is taken along the unperturbed ray. In a similar fashion from (2.9.5) there follow the second-order corrections:

$$\psi_2 = -\frac{1}{2} \int_{\tau^0}^{\tau} (\nabla\psi_1)^2 \, d\tau = -\frac{1}{2} \int_{\sigma^0}^{\sigma} \frac{(\nabla\psi_1)^2}{\sqrt{\epsilon}} \, d\sigma . \qquad (2.9.9)$$

The domain of application for the above equations is bounded by the distances at which the lateral deviation of rays from the unperturbed position is small compared with the characteristic transverse dimension of the perturbation, $l_\perp$. This limitation is lifted in a modification of the theory taking account of the shift of rays.

2.9.2 The Perturbation Method for Rays

This method is constructed taking the perturbation theory for a unit mass motion in mechanics as a model [2.134, 136]. We shall utilize here the simplest version of the theory following the procedure outlined in [2.132].

Introducing perturbations into the ray equation (2.2.19) we get

$$\ddot{\mathbf{r}} \equiv \frac{d^2\mathbf{r}}{d\tau^2} = \frac{1}{2}\nabla\bar{\epsilon}(\mathbf{r}) + \frac{1}{2}\nabla\nu(\mathbf{r}) . \qquad (2.9.10)$$

The solution to this equation will be sought as the series

$$\mathbf{r}(\tau) = \mathbf{r}_0(\tau) + \mathbf{r}_1(\tau) + \mathbf{r}_2(\tau) + \dots , \qquad (2.9.11)$$

where $\mathbf{r}_0(\tau)$ is the solution of the unperturbed equation

$$\ddot{\mathbf{r}}_0 = \nabla\bar{\epsilon}(\mathbf{r}_0)/2 . \qquad (2.9.12)$$

We substitute (2.9.11) in (2.9.9) and expand $\bar{\epsilon}(\mathbf{r})$ and $\nu(\mathbf{r})$ in the Taylor series in powers of the difference $\mathbf{r}-\mathbf{r}_0 = \mathbf{r}_1+\mathbf{r}_2+...$. Upon equating the terms of similar order of smallness on the right and left sides we arrive at the following iterative set of equations:

$$\ddot{\mathbf{r}}_1 - \frac{1}{2}(\mathbf{r}_1\nabla)\cdot\nabla\bar{\epsilon}(\mathbf{r}_0) = \frac{1}{2}\nabla\nu(\mathbf{r}_0) \equiv \mathbf{F}_1 \ ,$$

$$\ddot{\mathbf{r}}_2 - \frac{1}{2}(\mathbf{r}_2\nabla)\cdot\nabla\bar{\epsilon}(\mathbf{r}_0) = \frac{1}{4}(\mathbf{r}_1\nabla)^2\nabla\bar{\epsilon}(\mathbf{r}_0) + \frac{1}{2}(\mathbf{r}_1\nabla)\cdot\nabla\nu(\mathbf{r}_0) \equiv \mathbf{F}_2 \ ,$$

$$\text{--} \quad (2.9.13)$$

$$\ddot{\mathbf{r}}_n - \frac{1}{2}(\mathbf{r}_n\nabla)\cdot\nabla\bar{\epsilon}(\mathbf{r}_0) = \mathbf{F}_n \ .$$

The right-hand sides $\mathbf{F}_n$ of (2.9.13) depend only upon the preceding approximation and therefore may be viewed as known functions of τ. Once the corrections $\mathbf{r}_n(\tau)$ have been determined, the eikonal ψ is immediately found by integrating the permittivity $\epsilon = \bar{\epsilon}+\nu$ along the perturbed ray (2.9.11). The accuracy of ψ obviously improves as the number of terms retained in (2.9.11) increases. Constraining ourselves to the two first terms in (2.9.11), we obtain

$$\psi = \psi^0 + \int_{\tau^0}^{\tau}\Big(\bar{\epsilon}[\mathbf{r}_0(\tau') + \mathbf{r}_1(\tau')] + \nu[\mathbf{r}_0(\tau') + \mathbf{r}_1(\tau')]\Big)\,\mathrm{d}\tau'. \qquad (2.9.14)$$

The first-order correction (2.9.8) results here if we neglect $\mathbf{r}_1$. As we have noted in Sect.2.9.1, $\mathbf{r}_1$ may be dropped if $|\mathbf{r}_1| \ll \mathbf{l}_\perp$.

The determination of the amplitude along perturbed rays thus is reduced to computing the Jacobian $\mathscr{D}(\tau)$ accurate to an adopted approximation. So, to a first approximation

$$\mathscr{D}(\tau) = \left[\frac{\partial(\mathbf{r}_0 + \mathbf{r}_1)}{\partial\xi}\times\frac{\partial(\mathbf{r}_0 + \mathbf{r}_1)}{\partial\eta}\right]\cdot\frac{\partial(\mathbf{r}_0 + \mathbf{r}_1)}{\partial\tau} \ . \qquad (2.9.15)$$

Obviously, the error of thus derived amplitude sharply incrases near to caustics where $\mathscr{D}_0(\tau) = 0$.

2.9.3 Perturbations in Homogeneous Media

Obtain first the solution to (2.9.13) for $\bar{\epsilon}$ = constant, that is, when the unperturbed medium is homogeneous and the rays are straight lines:

$$\psi = \psi^0 + \bar{\epsilon}\tau \ , \quad \mathbf{r}_0(\tau) = \mathbf{r}^0 + \mathbf{p}^0\tau \ . \qquad (2.9.16)$$

At constant $\bar{\epsilon}$ the left-hand sides of (2.9.13) simplify considerably as $\nabla\bar{\epsilon} = 0$, then the corrections $\mathbf{r}_n$ can be found simply by repeated integration of the right-hand sides.

Consider the first-order correction which at constant $\bar{\epsilon}$ satisfies the equation

$$\ddot{\mathbf{r}}_1 = \nabla\nu[\mathbf{r}_0(\tau)]/2 = \mathbf{F}_1(\tau) \ .$$

Onefold integration yields

$$\mathbf{p}_1 \equiv \dot{\mathbf{r}} = \mathbf{p}_1^0 + \int_{\tau^0}^{\tau} \mathbf{F}_1(\tau')d\tau' \ . \tag{2.9.17}$$

Taking another integral (we assume for simplicity $\tau^0 = 0$) we get

$$\mathbf{r}_1 = \mathbf{r}_1^0 + \tau\mathbf{p}_1^0 + \int^{\tau} d\tau' \int_0^{\tau'} \mathbf{F}_1(\tau'')d\tau''$$

$$= \mathbf{r}_1^0 + \tau\mathbf{p}_1^0 + \int_0^{\tau} (\tau - \tau')\mathbf{F}_1(\tau')d\tau' \ . \tag{2.9.18}$$

The initial values $\mathbf{r}_1^0$ and $\mathbf{p}_1^0$ are to be chosen here depending on how the problem is stated. If tracing a particular ray is what matters, then $\mathbf{r}_1^0$ should be set equal to zero, and $\mathbf{p}_1^0$ equal to $\nu^0\mathbf{N}/2p_N^0$, where $\mathbf{N}$ is the unit normal to the initial surface Q, the value ν^0 is related to the point where the ray leaves the surface, and p_N^0 is the component of the unperturbed momentum in $\mathbf{N}$.

When the objective is a ray passing through a given point r, then working within the framework of the first approxiamtion of perturbation theory we are to find aprameters ξ and η of the ray primary point and the "length" of the ray τ such that[32]

$$\mathbf{r}_0(\xi,\eta,\tau) + \mathbf{r}_1(\xi,\eta,\tau) = \mathbf{r} \ . \tag{2.9.19}$$

This equation is a specific case of (2.99) and can have several roots, implying that several rays pass through r. If only one ray passed through r in the absence of perturbations, whereas there are more than one when perturbations have set in, the implication is that the perturbation has brought about caustics.

It is important that these caustics can be brought to light already in the *first order* of the perturbation theory for rays whereas the perturbation

---

[32] Accounting for subsequent approximations gives rise to the terms $\mathbf{r}_2$, $\mathbf{r}_3$, etc. in (2.9.19).

theory for the eikonal (Sect.2.9.1) is principally incapable of describing the appearance of caustics. The point is that with an analytic perturbation $\nu(\mathbf{r})$ the perturbation theory for the eikonal yields to any order an analytic dependance of ψ on the coordinates, whereas at caustics the phase is *nonanalytic*.

2.9.4 Perturbations in Nonhomogeneous Media

In order to solve (2.9.13) for the general case of $\bar{\epsilon}$ being an arbitrary function of coordinates, we recur to the known property of the equations of calculus of variations [2.135, 136] that the derivatives of the unperturbed trajectory $r_0(\tau) = r_0(\alpha_k, \tau)$ with respect to the free parameters α_k, that is, the vector functions $\rho_k(\tau) = \partial r_0(\tau)/\partial \alpha_k$ taken together form the fundamental solution of the set of linear homogeneous equations

$$\ddot{\rho}_k - \frac{1}{2}(\rho_k \nabla) \cdot \nabla \bar{\epsilon}(r_0) = 0 \ . \tag{2.9.20}$$

This statement is verified simply by differentiating (2.9.12) with respect to α_k.

In the general case the ray path is defined by six parameters α_k ($k = 1, ..., 6$) which may be realized as, say, three components of the initial vector $\mathbf{r}_0^0 \equiv \mathbf{r}_0(0)$ and three components of the initial momentum $\mathbf{p}_0^0 \equiv \mathbf{p}_0(0)$, though any other feasible set may also suit.

The left-hand sides in (2.9.13) are the same as in (2.9.20), but their right-hand sides $\mathbf{F}_n$ are other than zero. Since (2.9.13) are linear, a natural way toward their solution would be by variation of their constants. We shall perform it for the equation of the first approximation of (2.9.6). To do so we represent $\mathbf{r}_1$ as a linear combination of the fundamental solutions $\rho_k(\tau)$:

$$\mathbf{r}_1(\tau) = \sum_{k=1}^{6} C_k(\tau)\rho_k(\tau) = \sum_{k=1}^{6} C_k(\tau) \frac{\partial r_0(\tau)}{\partial \alpha_k} \ , \tag{2.9.21}$$

where $C_k(\tau)$ are the coefficients to be defined. On performing the standard manipulations in variation of the constants we obtain for C_k the system of two equations in vector form:

$$\sum_{k=1}^{6} \dot{C}_k \rho_k(\tau) = 0 \ , \quad \sum_{k=1}^{6} \dot{C}_k(\tau)\dot{\rho}_k(\tau) = \mathbf{F}_1 \equiv \frac{1}{2}\nabla\nu(\mathbf{r}_0) \ , \tag{2.9.22a}$$

or, which is the same, the set of six scalar equations:

$$\sum_{k=1}^{6} \dot{C}_k(\tau)\rho_{kj}(\tau) = 0 \ , \qquad \sum_{k=1}^{6} \dot{C}_k(\tau)\rho_{kj}(\tau) = F_{1j} \ , \qquad (2.9.22b)$$

where $j = 1,2,3$; and ρ_{kj} and F_{1j} are the Cartesian coordinates of ρ_k and F_1.

Since ρ_k are linearly independent, the determinant of the set (2.9.22) is other than zero so that this equation can always be solved for $\dot{C}k(\tau)$ whereupon $C_k(\tau)$ may be found by quadratures of the form

$$C_k(\tau) = C_k(0) + \int_0^\tau \sum_{j=1}^{3} Q_{kj}(\tau')F_{1j}(\tau')d\tau' \ , \qquad (2.9.23)$$

with $C_k(0)$ to be determined from the initial conditions. Hence the search for the first-order correction has been reduced to the evaluation of the integrals of type (2.9.23). Higher-order corrections r_n can be found in a similar fashion. Some particular cases (for example, plane-stratified media) enjoy simpler equations than (2.9.23) because ρ_k take on an especially simple form [2.128].

If we write the vector $\mathbf{p} = \dot{\mathbf{r}}$ tangent to the ray as the series

$$\mathbf{p} = \mathbf{p}_0 + \mathbf{p}_1 + \mathbf{p}_2 + \ldots = \dot{\mathbf{r}} + \dot{\mathbf{r}}_1 + \dot{\mathbf{r}}_2 \ldots \ ,$$

then in accord with (2.9.21) the first-order correction is

$$\mathbf{p}_1 = \frac{d}{d\tau} \sum_{k=1}^{6} C_k\rho_k = \sum_{k=1}^{6} (\dot{C}_k\rho_k + C_k\dot{\rho}_k) \ .$$

But by virtue of (2.9.22) $\Sigma C_k\rho_k = 0$ so that

$$\mathbf{p}_1 = \sum_{k=1}^{6} C_k(\tau)\dot{\rho}(\tau) \ . \qquad (2.9.24)$$

Obviously, the correction $\mathbf{p}_1$ is defined by the same coefficients $C_k(\tau)$ that enter (2.9.21). It can be used to express in its terms the correction $\mathbf{l}_1$ of the unperturbed unit vector $\mathbf{l}_0$

$$\mathbf{l}_1 = [\mathbf{p}_1 - \mathbf{l}_0(\mathbf{l}_0 \cdot \mathbf{p}_1)]/\sqrt{\bar{\epsilon}} = \mathbf{p}_\perp / \sqrt{\bar{\epsilon}} \ .$$

The quantity $\mathbf{l}_1$ defines the angles of incidence of the perturbed ray.

A perturbation theory can also be constructed for space-time rays (Sect.4.6.6 and references cited in [2.9]. A perturbation technique using two-dimensional expansion is proposed in [2.133,134].

2.10 Applicability of Geometrical Optics

2.10.1 Existent Estimators of Method's Errors

Inequality (2.1.4) which, when satisfied, enables us to assume the wave being almost plane and the medium almost one-dimensional, is a *necessary* but not sufficient condition for the method of geometrical optics to be applied [saying "geometrical-optics field" we imply here and below only the zero approximation field, $u_0(r)$]. The sufficient conditions of applicability should in one or another way consider the accumulating errors due to the fact that the zero approximation field $u_0 = A_0 \exp(ik_0\psi)$ is not an exact solution of (2.1).[33]

If we attempted to formulate a sufficient condition of the method's applicability and estimate its error

$$\gamma \simeq |u_e - u_0|/|u_0| \, , \tag{2.10.1}$$

we would soon find out that the available analytic possibilities are very limited, as in the majority of situations the exact solution, u_e, is unavailable. The known rigorous solutions are able only dto indicate the value of γ for some typical conditions. Similar indications can be derived by comparing geometrical-optics fields with the known approximate (but more general than for rays) solutions of the wave problem.

More definite estimates of γ could be inferred by assuming that the difference $|u_e - u_0|$ is of the order of the first dropped term A_1/ik_0 in the ray series (2.1.13):

$$\gamma \simeq \left| \frac{A_1}{k_0 A_0} \right| = \left| \frac{1}{2k_0 A_0 \sqrt{\mathscr{D}(\tau)}} \int_{\tau_0}^{\tau} \Delta A_0(\tau') \sqrt{\mathscr{D}(\tau')} \, d\tau' \right| \tag{2.10.2}$$

[we have used here (2.3.34)]. Nevertheless, this estimate would be difficult to use because of the computational difficulties that one would encounter in handling its integral.

In what follows we wish to formulate a sufficient condition for the geometrical-optics method to be applicable, working on heuristic grounds [2.49,74], i.e., using only the elementary concepts of Huygens and Fresnel on the interference of secondary waves, which we have already discussed in connection with the Fermat principle in Sect.2.2.

---

[33] The difference $\tilde{u} \equiv u - u_0$ satisfies the equation $\Delta\tilde{u} + n_0^2\tilde{u} = -\exp(ik_0\psi)\Delta\mathscr{N}_0$.

2.10.2 Fresnel Zones and Fresnel Volume of Rays in Inhomogeneous Media

To determine the Fresnel zones in an inhomogeneous medium we invoke the Green's function which relates the field u(r) at the point of observation r to the values of the field $u^0(r)$ and its normal derivative at points r′ belonging to an arbitrary (but smooth) surface Q [2.19, 139], viz.,

$$u(r) = \frac{1}{4\pi} \int \left[\frac{\partial u^0(r')}{\partial N'} g(r,r') - u^0(r') \frac{\partial g(r,r')}{\partial N'} \right] dS' \,, \qquad (2.10.3)$$

where $g(r,r')$ is the Green's function of the wave problem, $\partial/\partial N'$ is the derivative in the direction of the external normal to the closed surface Q, and dS is the elementary area of the surface. Green's formula (2.10.3) represents a rigorous formulation of the Huygens principle for a scalar wave field in an inhomogeneous medium.

In order to employ the method of stationary phase to compute (2.103), we represent the pirmary wave $u^0(r')$ and the Green's function $g(r,r')$ in geometrical-optics form

$$u^0(r') = A^0(r')\exp[ik_0\psi^0(r')] \,,$$

$$g(r,r') = G(r,r')\exp[ik_0\psi_g(r,r')] \,,$$

where A^0 and G are the amplitudes slowly varying over the wavelength. In computing the derivatives in N′, these amplitudes may be deemed virtually constant so that

$$\frac{\partial u^0(r')}{\partial N'} \sim ik_0 A^0(r') \frac{\partial \psi^0(r')}{\partial N'} \exp[ik_0\psi^0(r')] \,,$$

$$\frac{\partial g(r,r')}{\partial N'} \sim ik_0 G(r,r') \frac{\partial \psi_g(r,r')}{\partial N'} \exp[ik_0\psi_g(r,r')] \,.$$

As a result, Green's formula (2.10.3) takes the form

$$u(r) \simeq \frac{ik_0}{4\pi} \int A^0(r')G(r,r')\exp[ik_0\psi^0(r')+\psi_g(r,r')] \frac{\partial}{\partial N'}[\psi^0(r')+\psi_g(r,r')]dS' \,.$$

$$(2.10.4)$$

If we denote by ∇' the operator of differentiation with respect to r′ and by $\nabla'_Q = \nabla'-N'(N'\cdot\nabla')$ the operator of differentiation over the surface Q, then the equation

$$\nabla'_Q[\psi^0(r') + \psi_g(r,r')] = 0 \qquad (2.10.5)$$

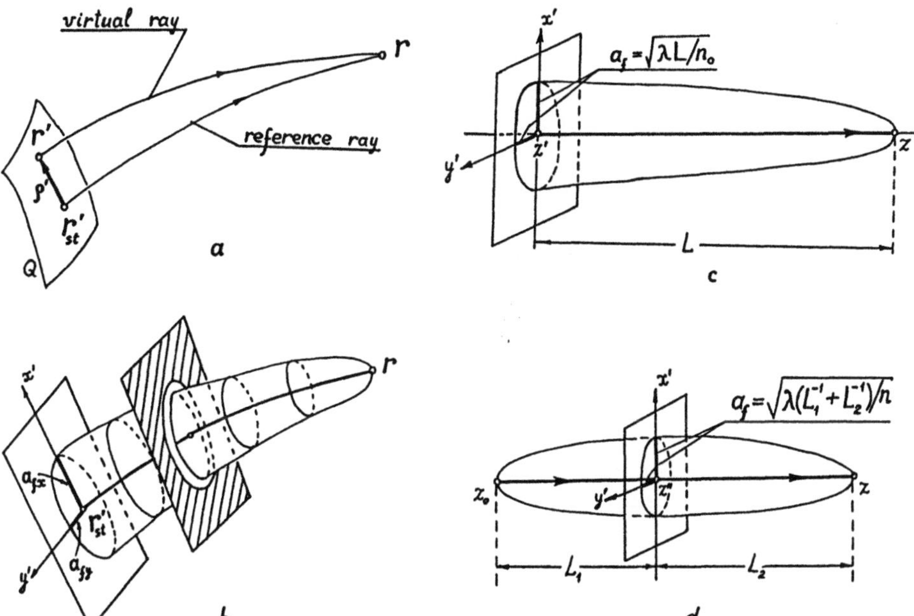

Fig.2.18. (a) The reference ($r_{(st)}' \rightarrow r$) and virtual ($r' \rightarrow r$) rays corresponding to the Huygens-Fresnel principle. The Fresnel volume, or the region of localization, of the ray $r_{st}' \rightarrow r$ (b) for an arbitary phase front in an inhomogeneous medium, (c) for a plane wave in a homogeneous medium, and (d) for the field due to a point source in a homogeneous medium

singles out on this surface a stationary point r_{st}' which becomes the origin of the ray $r_{st}' \rightarrow r$ passing through the point of observation r (Fig.2.18a). We call this the *reference* (basic) ray to contrast it with the multitude of virtual rays $r' \rightarrow r$ which correspond to the secondary Huygens sources on the surface Q and which also satisfy (2.2.12).

The difference between the *virtual* eikonal $\psi_{virt}(r,r') = \psi^0(r')+\psi_g(r,r')$ and the *reference* eikonal $\psi_{ref}(r) = \psi^0(r_{st}')+\psi_g(r,r_{st}')$ will be denoted by $\tilde{\psi}(r')$, namely,

$$\tilde{\psi}(r') = \psi_{virt}(r,r') - \psi_{ref}(r)$$

$$= \psi^0(r') + \psi_g(r,r') - \psi^0(r_{st}') - \psi_g(r,r_{st}') . \tag{2.10.6}$$

This quantity is obviously dependent on the observation-point radius-vector r, too, but for brevity we shall not indicate this dependence explicitly.

It is known from the times of Fresnel that the major contribution to such integrals as in (2.10.4) is due to a small neighborhood of the stationary point where the virtual eikonal ψ_{virt} differs from its extreme value $\psi_{ref}(r)$ by at most $\lambda_0/2$, that is, where $|\tilde{\psi}(r')| < \lambda_0/2$. The corresponding phase $\Psi = k_0\tilde{\psi}$ does not exceed π in this neighborhood, which justifies referring to this domain as the π neighborhood of the stationary point.

This π neighborhood actually forms the field at the point of observation and is known as the first Fresnel zone in wave theory. Higher Fresnel zones correspond to values of $|\tilde{\psi}| \leq m\lambda_0/2$ or $|\tilde{\Psi}| \leq m\pi$, where m = 1,2, etc. For a uniform primary field and smooth variations of medium properties, the secondary wavelets of higher Fresnel zones cancel out each other, and the resultant field u(r) is defined by the closest neighborhood of the reference ray, that is, by the first Fresnel zone. This qualitative argument may be considered a sort of the physical basis of the stationary phase method.

If we denote by M the product of all slowly varying factors in (2.10.4), namely,

$$M(r,r') = \frac{ik_0}{4\pi}A^0(r')G(r,r')\frac{\partial}{\partial N}[\psi^0(r') - \psi_g(r,r')] ,\qquad (2.10.7)$$

and make use of the notation of (2.10.6), then the Green's formula (2.10.4) takes on the form [2.140]

$$u(r') = \exp[ik_0\psi(r)]\int_Q M(r,r')\exp[ik_0\tilde{\psi}(r')]dS' .\qquad (2.10.8)$$

We have here written $\psi(r)$ instead of $\psi_{ref}(r)$ to emphasize that the quantity $\psi_{ref}(r) = \psi^0(r'_{st})+\psi_g(r,r'_{st})$ is the eikonal of the primary wave arriving at the point of observation r.

In what follows we replace the surface Q in the neighborhood of the stationary point by a plane perpendicular to the ray (this simplifying assumption does not imply any loss of generality for our consideration) and expand $\tilde{\psi}(r')$ into a power series of the radius vector $\rho' = r-r'_{st}$ lying in this plane. The leading term of this expansion is zero in view of the definition of (2.10.6), and the linear term vanishes because of the stationarity condition (2.10.5). Therefore, the expansion begins with the quadratic terms

$$\tilde{\psi}(r') = \frac{1}{2}(\rho\nabla'_Q)^2\tilde{\psi}(r')|_{r'=r'_{st}} + ...$$

$$= \frac{1}{2}\left[\tilde{\psi}_{xx}x'^2 + 2\tilde{\psi}_{xy}x'y' + \tilde{\psi}_{yy}y'^2\right] + ... ,\qquad (2.10.9)$$

where x' and y' are the components of $\rho = (x',y')$ in the plane Q (Fig. 2.18b), and

$$\tilde{\psi}_{xx} = \frac{\partial^2\tilde{\psi}}{\partial x'^2} , \quad \tilde{\psi}_{xy} = \frac{\partial^2\tilde{\psi}}{\partial x'y'} , \quad \tilde{\psi}_{yy} = \frac{\partial^2\tilde{\psi}}{\partial y'^2}$$

are the second-order derivatives taken at the stationary point r'_{st}. If we take the variable σ' to characterize the position of the stationary point on the reference ray so that $\mathbf{r}_{st} = \mathbf{r}_{st}(\sigma')$, then the derivatives $\tilde{\psi}_{xx}$, $\tilde{\psi}_{yy}$, $\tilde{\psi}_{xy}$ will be functions of σ'.

In applying the method of stationary phase to (2.10.8), we make use of the expansion (2.10.9) and move M with the value at the stationary point $\mathbf{r}'_{st}$ outside the integral, viz.,

$$u(\mathbf{r}) = \frac{2\pi}{k_0}\exp[ik_0\psi(\mathbf{r}) + i\pi\beta/4]M(\mathbf{r},\mathbf{r}_{st}')\left[\tilde{\psi}_{xx}\tilde{\psi}_{yy} - \psi_{xy}^2\right]^{-1/2} , \qquad (2.10.10)$$

where β is the signature of the matrix of second derivatives, i.e., the number of positive eigenevalues minus the number of negative eigenvalues.

It is worth noting that $M(\mathbf{r},\mathbf{r}'_{st})$ is somewhat simpler than the previous function $M(\mathbf{r},\mathbf{r}')$. This is because the eikonals $\psi^0(\mathbf{r}')$ and $\psi_g(\mathbf{r},\mathbf{r}')$ have $|\nabla'\psi^0| = |\nabla'\psi_g|$ at all points $\mathbf{r}'$ as both ψ^0 and ψ_g satisfy the eikonal equation, and at the stationary point of Q they have $|\nabla'_Q\psi^0| = |\nabla'_Q\psi_g|$ as follows from the stationarity condition (2.10.6). Consequently, at the stationary point the derivatives $\partial\psi^0/\partial N'$ and $\partial\psi_g/\partial N'$ also have identical absolute values, but they differ in sign: on moving from $\mathbf{r}'_{st}$ to $\mathbf{r}$, the eikonal of the primary field $\psi^0(\mathbf{r}')$ increased whereas the eikonal of the Green's function $\psi_g(\mathbf{r},\mathbf{r}')$ decreases. Therefore, at the stationary point

$$\frac{\partial}{\partial N'}(\psi^0 - \psi_g)_{st} = 2\left[\frac{\partial\psi^0}{\partial N'}\right]_{st} ,$$

so that

$$M(\mathbf{r},\mathbf{r}'_{st}) = \frac{ik_0}{2}A^0(\mathbf{r}'_{st})G(\mathbf{r},\mathbf{r}'_{st})\left[\frac{\partial\psi^0}{\partial n'}\right]_{st} . \qquad (2.10.11)$$

By direct differentiation one can verify that the field (2.10.10) satisfies the equations of geometrical optics. Thus, the geometrical optics-approximation evolves from the exact integral representation (2.10.3) by approximate computation of this integral with the method of stationary phase. Accordingly, the conditions of applicability of the method of stationary phase are simultaneously the conditions of applicability of ray theory. This argument will be used later in formulating the range of validity of geometrial optics, and at this point we introduce the concept of the Fresnel volume of a ray.

We revisit (2.10.9) and choose, in the Q plane, the coordiantes x' and y' such that the mixed derivative vanishes to render the second-derivative matrix diagonal. Then

$$\tilde{\psi}(\mathbf{r}') = \tfrac{1}{2}\left[\tilde{\psi}_{xx}x'^2 + \tilde{\psi}_{yy}y'^2\right] + \dots . \qquad (2.10.12)$$

For the elliptic type of stationary point, the derivatives $\tilde{\psi}_{xx}$ and $\tilde{\psi}_{yy}$ have the same sign, that is, $\tilde{\psi}_{xx}\tilde{\psi}_{yy} > 0$. In this case the curves of constant

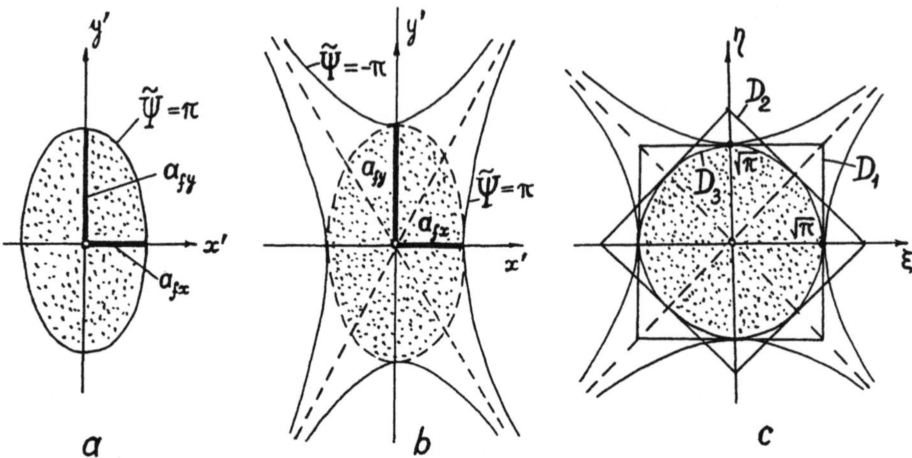

Fig.2.19. Shape of the first Fresnel zone for the saddle point of (a) elliptic and (b) hyperbolic type. Integration over the squares D_1 and D_2 and over the circle D_3 in the ξ,η plane indicates that the effective region of formation (shown by stippling) is of finite size nonwithstanding the infinite tails

phase $\tilde{\psi}(\mathbf{r}') = $ const. trace ellipses in the x'y' plane. The first Fresnel zone $|\tilde{\psi}| - \lambda_0/2 = 0$ is associated with an ellipse (Fig.2.19a) defined as

$$\frac{1}{2}|\tilde{\psi}_{xx}|x'^2 + |\tilde{\psi}_{yy}|y'^2 - \frac{\lambda_0}{2} = 0 \; . \tag{2.10.13}$$

The semiminor and semimajor axes of the Fresnel ellipse are, respectively,

$$a_{fx} = \sqrt{\frac{2\pi}{k_0|\tilde{\psi}_{xx}|}} = \sqrt{\frac{\lambda_0}{|\tilde{\psi}_{xx}|}} \; , \quad a_{fy} = \sqrt{\frac{2\pi}{k_0|\tilde{\psi}_{yy}|}} = \sqrt{\frac{\lambda_0}{|\tilde{\psi}_{yy}|}} \; . \tag{2.10.14}$$

The contribution of the first Fresnel zone (2.10.13) is known to be twice as large as the resultant field from the entire surface Q. This enables us to think of the first Fresnel zone as the region responsible for the formation of the field at the point of observation. Integration of all Fresnel zones through which the reference ray traverses will be called the *Fresnel volume* of the ray (Fig.2.18b). This name [2.74] seems more suitable to us than the terms "three-dimensional Fresnel zone" [2.137], "spatial Fresnel zone" [2.134], "region essential for diffraction" [2.139], and similar therms that have been applied to various specific cases to identify a spatial domain essential for the field forming at a given point.

Equation (2.10.13), in which the derivatives $\tilde{\psi}_{xx} = \tilde{\psi}_{xx}(\sigma')$ and $\tilde{\psi}_{yy} = \tilde{\psi}_{yy}(\sigma')$ depend on the coordinate σ' measured along the reference ray, may be viewed as the equation defining the boundary of the Fresnel volume

$F(x', y', \sigma') = 0$. We illustrate this statement by taking as examples two familiar problems.

Let a plane wave $u = \exp(ik_0 n_0 z)$ propagate in a uniform medium of refractive index n_0. The equation of the Fresnel volume for straight rays, corresponding to the plane wave, may be obtained from (2.10.13). Letting $\psi^0 = n_0 z'$ and $\psi_g(r,r') = n_0|r-r'|$, and treating z' as a parameter on the reference ray passing through $r = (0,0,z)$ we obtain

$$\tilde{\psi}_{xx}(z') = \tilde{\psi}_{yy}(z') = |z - z'|^{-1} ,$$

and (2.10.13) becomes

$$\frac{n_0}{2|z - z'|}(x'^2 + y'^2) \equiv \frac{n_0 \rho'^2}{2|z - z'|} = \frac{\lambda_0}{2} .$$

Thus the Fresnel volume of the ray is bound by a paraboloid of revolution (Fig.2.18c). The radius of a cross section through this paraboloid by a plane at the distance $L = |z-z'|$ from the observation point is

$$a_f = \sqrt{\lambda_0 L/n_0} . \tag{2.10.15}$$

For a spherical wave $u = \exp(ik_0 n_0|r-r'|)/|r-r'|$ (the source is at r_0), the Fresnel volume is known to shape as an ellipsoid of revolution. An approximate equation for the boundary of this volume can readily be derived by computing derivatics of the eikonal

$$\tilde{\psi}(r') = n_0(|r_0 - r| + |r' - r| - |r - r_0|) .$$

Assume that the z-axis is along the straight line from the source to the point of observation, and employ z' as the ray parameter σ'. Then for points on the z axis we get

$$\tilde{\psi}_{xx}(z') = \tilde{\psi}_{yy}(z') = n_0(|z_0 - z'|^{-1} + |z - z'|^{-1}) = n_0(L_1^{-1} + L_2^{-1}) ,$$

where $L_1 = |z-z'|$ and $L_2 = |z_0-z'|$ are the distances of the source and observation point from the plane $z' = $ const (Fig.2.18d). As a result, the equation for the boundary of the Fresnel volume becomes

$$\frac{n_0}{2}\left[\frac{1}{L_1} + \frac{1}{L_2}\right]\rho'^2 = \frac{\lambda_0}{2} ,$$

whence for the Fresnel radius a_f in the z' plane we obtain

$$a_f(z') = \sqrt{\frac{\lambda_0}{n_0}\left[\frac{1}{L_1} + \frac{1}{L_2}\right]^{-1}} . \tag{2.10.16}$$

In an inhomogeneous medium the reference ray leading to the point of observation bends, as does the Fresnel volume of the ray (Fig.2.18b). Below we have specific examples of Fresnel-volume construction.

The case of a saddle-like phase function $\tilde{\psi}(\varrho')$ with the derivatives $\tilde{\psi}_{xx}$ and $\tilde{\psi}_{yy}$ in (2.10.12) being of opposite signs, $\tilde{\psi}_{xx}\tilde{\psi}_{yy} < 0$, calls for a separate analysis. Let for definiteness $\tilde{\psi}_{xx} > 0$ and $\tilde{\psi}_{yy} < 0$, then the equation for the first Fresnel zone takes on the form

$$|\tilde{\psi}| - \frac{1}{2}\lambda_0 \simeq \frac{1}{2}|\tilde{\psi}_{xx}|x'^2 - |\tilde{\psi}_{yy}|y'^2 - \frac{1}{2}\lambda_0 = 0 . \qquad (2.10.17a)$$

Whereas for an elliptic stationary point the Fresnel zone is an ellipse (Fig.2.19a), for a saddle point it is confined by hyperbolas (Fig.2.19b). The equations for these hyperbolas assume a simple form in the variables $\xi = x'(k_0|\tilde{\psi}_{xx}|/2)^{1/2}$ and $\eta = y'(k_0|\tilde{\psi}_{yy}|/2)^{1/2}$, namely, (2.10.17) can be rewritten as

$$|\tilde{\Psi}| - \pi = |\xi^2-\eta^2| - \pi = 0 . \qquad (2.10.17b)$$

Therefore, the first Fresnel zone $|\tilde{\Psi}| = |\xi^2-\eta^2|$ takes a cross-like shape extending infinitely along the diagonals $\eta = \pm\xi$ (Fig.2.19b). However, the region of actual formation of the field if finite. This can be verified by comparing the integral

$$U_j = \iint_{D_j} d\xi d\eta \exp[i(\xi^2 - \eta^2)] ,$$

taken over a finite area D_j in the ξ,η plane, with the value U_∞ of this integral but taken within infinite limits

$$U_\infty = \iint_{-\infty}^{+\infty} d\xi d\eta \exp[i(\xi^2 - \eta^2)] = \left|\int_{-\infty}^{\infty} \exp(i\xi^2)d\xi\right|^2 = \pi .$$

On integration over the square $|\xi| \le p$, $|\eta| \le p$ (D_1 in Fig.2.19c) we obtain

$$U_1(p) = \int_{-p}^{+p} d\xi \int_{-p}^{+p} d\eta \exp[i(\xi^2-\eta^2)] = 4|F(p)|^2 ,$$

where

$$F(p) = \int_0^p \exp(it^2)dt$$

is the complex Fresnel integral. Analysis of U_1 as a function of the square size p indicates that at $p = \sqrt{\pi}$ the relative error of $U_1(\sqrt{\pi})$ with respect to U_∞ amounts to 3%.

If we tilt the square of side 2p by 45° (D_2 in Fig.2.19c), then the respective integral will be

$$U_2(p) = 2Si(2p^2) \qquad \text{where}$$

$$Si(t) = \int_0^t \frac{\sin t}{t}\, dt \ .$$

For $p = \sqrt{\pi}$, $u_2(\sqrt{\pi})$ differs from U_∞ merely by 2.2%.

Finally, the integral $U_3(p)$ taken over the circle $\xi^2 + \eta^2 \leq p$ (D_3 in Fig.2.19c) is expressed through the Bessel and Struve functions. For $p = \sqrt{\pi}$, $U_3(\sqrt{\pi})$ differs from U_∞ by less than 8%. Similar estimates may be justified for the Gaussian window as well.

A more detailed account of this numerical analysis is given in [2.149, 150].

Thus the integration over the three various shapes in the ξ,η plane indicates that nonwithstanding the inifite extension of the first Fresnel zone $|\tilde{\psi}| \leq \pi$, the field is actually formed by a relatively small neighborhood around the stationary point $\xi^2 + \eta^2 \leq \pi$. In the variable x',y' it is described by the ellipse $(x'/a_{fx})^2 + (y'/a_{fy})^2 = 1$ with the semiaxes (2.10.14) shown in Fig.2.19b by stippling. Formally this is the same region as in the case of the elliptical stationary point (Fig.2.19a), therefore we take (2.10.13) to define the Fresnel volume boundary no matter what the signs of $\tilde{\psi}_{xx}$ and $\tilde{\psi}_{yy}$ are in (2.10.12). We arrive at the same conclusion in attempts to localize the ray by means of the Gaussian window [2.149, 150].

If the second derivatives $\tilde{\psi}_{xx}$ and $\tilde{\psi}_{yy}$ vanish, the eikonal expansion (2.10.12) should be augmented by higher terms represented as one of the typical normal forms or a sum of such forms. By equating the absolute value of each of these forms to $\lambda_0/2$ the Fresnel-volume boundary can be determined for this complicated situation as well [2.149].

2.10.3 The Physical Meaning of the Ray

When the initial conditions are smooth and the properties of the medium vary smoothly, the secondary wavelets emanating from the higher wave zones cancel out each other and the resultant field is defined by the nearest neighborhood of the ray, namely, by its Fresnel volume.

If we are going to treat the ray as a physical object, the Fresnel volume should be thought of as a *region where the ray is localized*. Indeed, if we wish to single out a given ray by directing a wave on an aperture in a screen (Fig.2.18b), the dimension of this aperture must be larger than the cross section of the Fresnel volume [2.74, 141, 149, 150]. If we narrowed the aperture closer than the first wave zone, it would obstruct the field on this ray.

Therefore, we should distinguish between the mathematical ray, which is an infinitely thin line satisfying the equation of rays, and the physical ray, which does have a *finite thickness* defined by the Fresnel volume.

2.10.4 Heuristic Criteria on Geometrical-Optics Applicability

Working from the viewpoint that the Fresnel volume defines the region in space which forms the field at a given point, we formulate the following two criteria of the applicability of the geometrical-optics method:

(i) *The parameters of the medium and the parameters of the wave - its amplitude and phase gradient - must not vary significantly over the cross section of the Fresnel volume.*

(ii) *Fresnel volumes of rays belonging to one and the same congruence and reaching one and the same point must not penetrate into one another significantly.*

According to the first we shall have

$$a_f \left| \nabla_\perp A/A \right| \ll 1 \tag{2.10.18}$$

and similar inequalities for the components of the momentum $p = \nabla\psi$ and for the refractive index n:

$$a_f \left| \nabla_\perp p_i/p \right| \ll 1 \; , \quad a_f \left| \nabla_\perp n/n \right| \ll 1 \; . \tag{2.10.19}$$

Here a_f is the largest cross section of the Fresnel volume, and $\nabla_\perp = \nabla - 1(1\nabla)$ is the operator of differentiation perpendicular to the ray. From (2.10.18) there follow definite constraints on the rate of change of the Jacobian $\mathscr{D}$ and on the value of both principal radii of curvature of the phase front, $R_{1,2}$

$$a_f \ll \left| R_{1,2} \right| \; . \tag{2.10.20}$$

It is worth noting that within a small distance from the initial surface Q (Fig.2.18b) $a_f \simeq \lambda$ [see, e.g., (2.10.15) for $z \simeq \lambda$]; therefore (2.10.1) coincide with the necessary applicability condition (2.1.3).

The second criterion may be formalized by

$$\delta S_f \ll S_f \; , \tag{2.10.21a}$$

where δS_f is the common area of the first Fresnel zone of rays reaching a given observation point have in common. In order that the Fresnel volumes of adjacent rays should not overlap significantly, the phase difference along these rays should not exceed π, i.e.,

$$\left| \Psi_1 - \Psi_2 \right| = k_0 \left| \psi_1 - \psi_2 \right| \geq \pi \quad \text{or} \quad \left| \psi_1 - \psi_2 \right| \geq \lambda_{0/2} \; . \tag{2.10.21b}$$

This criterion provides that the contributions of the same secondary wavelets leaving the region where the first Fresnel zones of two close rays overlap should not be accounted twice in the resultant field. Criterion II can be shown to be a corollary of criterion I, but criterion II is more vivid and is more convenient for assessing the region where the ray method is applicable when close to caustics and, hence, is valuable as it stands.

The conditions (2.10.19-21) have arisen in one form or another in many works, starting with those of Fresnel. In some cases they allow for rigorous corroboration. The novel point might be in that we emphasize these criteria as uniformly valid and sufficent - in all specific cases we know that criteria I and II agree completely with other methods of evaluation of geometrical-optics validity limits (Chap.3 and [2.74, 149]).

2.10.5 Applicability Conditions for Space-Time Geometrical Optics

Let us $u^0(t', z')$ be the field of a unidimensional primary wave in the plane z' at time t', and $\phi^0(t', z')$ be the phase of this wave. In a nondispersive, stationary and homogeneous medium the wave travels without distortion of its shape, this is, the primary disturbance propagates along space-time rays $z-z'=c(t-t')$ so that $u(t, z) = u^0(t', z-c(t-t'))$.

In a dispersive medium the pulse profile suffers distortions which can be described as resulting from space-time diffraction. Conditions under which such a dispersion-induced diffraction may be neglected are convenient to formulate with the aid of the time Fresnel zone that can be introduced by analogy with the spatial problem.

To find the field at point (t, z) we introduce the phase of the Green's function $\phi_g(t, z; t', z')$ corresponding to the virtual space-time ray (t', z') – (t, z). In a homogeneous medium, the equation of this ray is

$$z - z' = g(\omega')(t - t') \qquad (2.10.22)$$

where $g(\omega') = [dk(\omega')/d\omega']^{-1}$ is the group velocity. This equation may be viewed as the equation for the frequency ω' dependant on t, z, t', and z'. The virtual ray $(t', z') \rightarrow (t, z)$ is associated with the phase

$$\phi_{\text{virt}}(t, z; t', z') = \phi^0(t', z') + \phi_g(t, z; t', z') \; . \qquad (2.10.23)$$

The stationary value of this phase

$$\phi_{\text{ref}}(t, z) = \phi_{\text{virt}}(t, z; t'_{st}, z) = \phi^0(t'_{st}, z'f) + \phi_g(t, z; t'_{st}, z') \qquad (2.10.24)$$

relates to the time instant $t' = t_{st}'$ defined from the stationarity condition

$$\frac{\partial \phi_{\text{virt}}(t, z; t', z')}{\partial t'} = 0 \; . \qquad (2.10.25)$$

The stationary value of the phase (2.10.24) is obtained at the reference ray $(t'_{st}, z') \rightarrow (t, z)$ (Fig.2.20). In applying the Huygens-Fresnel concept to the problem at hand we recognize that at point (t, z) the field is formed

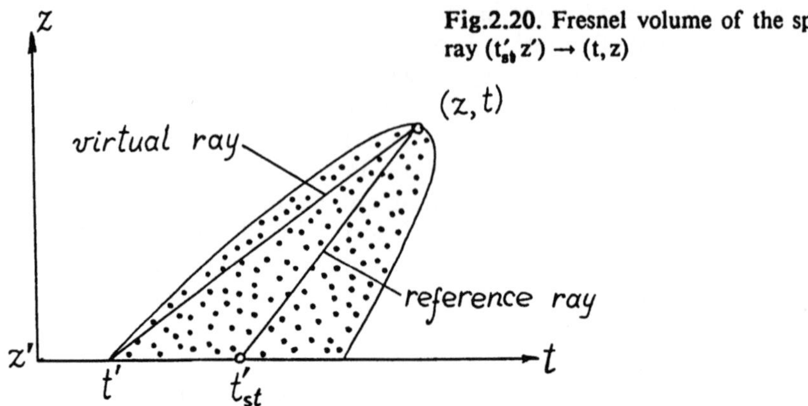

Fig.2.20. Fresnel volume of the space-time ray $(t'_{st}, z') \rightarrow (t, z)$

predominantly by the π neighborhood of the stationar point t'_{st}, i.e., by the region where the phase (2.10.23) on a virtual ray differs from the reference phase (2.10.24) by less than π, viz.,

$$|\tilde{\phi}| = |\phi_{\text{virt}}(t, z; t', z') - \phi_{\text{ref}}(t, z)| \leq \pi . \tag{2.10.26}$$

The interval $\tau_f = |t' - t'_{st}|$ in which this inequality holds may be treated as the temporal radius of the first Fresnel zone, or, in short, as the Fresnel interval. The family of all first Fresnel zones belonging to the given reference ray is assumed to be the Fresnel volume of the space-time ray. In Fig.2.20, this volume is indicated by stippling.

By expanding the pahse difference $\tilde{\phi}$ in a series in powers of $t - t'_{st}$ (this expansion begins with a quadratic term), in place of (2.10.26) we obtain the condition

$$\frac{1}{2} \left| \frac{\partial^2 \tilde{\phi}}{\partial t'^2} \right| (t' - t'_{st})^2 \leq \pi ,$$

from which it follows the expression for Fresnel interval

$$\tau_f = \sqrt{2\pi |\partial^2 \tilde{\phi} / \partial t'^2|^{-1}} . \tag{2.10.27}$$

By way of a simple example of τ_f computation, we consider a homogeneous medium with the dispersion law $k = k(\omega)$. The primary field will be represented by a harmonic wave of frequency ω_0

$$u^0(t', z') = \exp\{i[k(\omega_0)z' - \omega_0 t']\} ,$$

so that

$$\phi^0(t', z') = k(\omega_0)z' - \omega_0 t' . \tag{2.10.28}$$

90

The virtual space-time ray (2.10.22) is associated with the phase of the Green's function

$$\phi_g(t,z;t',z') = k(\omega')(z-z') - \omega'(t-t') ,$$

so that

$$\phi_{virt}(t,z;t',z') = k(\omega_0)z' - \omega_0 t' + k(\omega')(z-z') - \omega'(t-t') .$$

The frequency ω' and the corresponding group velocity $g(\omega')$ must be chosen such the space-time ray (2.10.22) traces from the point (t',z') to the point (t,z). This implies that the quantity ω' depends both on the initial (t',z') and on the final (t,z) coordinates of the ray, i.e. $\omega' = \omega'(t',z';t,z)$.

In this case the stationarity condition (2.10.25) takes on the form

$$\frac{\partial \phi_{virt}}{\partial t'} = \left[\frac{dk(\omega')}{d\omega'}(z-z') - (t-t') \right] \frac{\partial \omega'}{\partial t'} + (\omega'-\omega_0) = 0 . \tag{2.10.29}$$

This condition is satisfied by the reference ray

$$z - z' = g(\omega_0)(t-t_{st}') , \quad g(\omega_0) = [dk(\omega_0)/d\omega_0]^{-1}$$

corrsponding to the primary wave frequency $\omega_{st}' = \tilde{\omega}_0$.

By virtue of (2.10.29) the second derivative of $\tilde{\phi}$ at the stationary point is

$$\frac{d^2 \tilde{\phi}}{dt'^2} = \frac{\partial^2 \phi_{virt}}{\partial t'^2} = \left(\frac{\partial \omega'}{dt'} \right)^2 k''(\omega_0)(z-z') + 2 \frac{\partial \omega'}{\partial t'} ,$$

where $k''(\omega) = \partial^2 k(\omega)/\partial \omega^2$. The derivative $\partial \omega'/\partial t'$ can be determined from the virtual-ray equation (2.10.22) rewritten in the form

$$\frac{1}{g(\omega')} = \frac{dk(\omega')}{d\omega'} = \frac{t - t'}{z - z'} .$$

Differentiation this equation with respect to t' yields

$$k''(\omega')\partial \omega'/\partial t' = - (z-z')^{-1} \qquad \text{whence}$$

$$\frac{\partial \omega'}{\partial t'} = - \frac{1}{k''(\omega')(z - z')} ,$$

consequently

$$\frac{\partial^2 \tilde{\phi}}{\partial t'^2} = - \frac{1}{k''(\omega)(z-z')} .$$

Substituting this expression into (2.10.27) gives for the Fresnel interval

$$\tau_f = \sqrt{2\pi |k''(\omega)| |z-z'|} \ . \tag{2.10.30}$$

In nondispersive media, $k'' = 0$ and the Fresnel redius vanishes implying undistorted propagation of the light pulse.

The formula (2.10.29) for the constant-frequency field resembles structurally the expression $a_f = \sqrt{\lambda |z-z'|}$ for the spatial Fresnel radius corresponding to a plane wave.

The wave $u^0(t',z') = \exp[i\phi^0(t',z')]$ of variable frequency $\omega^0 = \partial\phi^0/\partial t'$ is similar to the spatial wave with distorted phase front. Computations for this case turn out to be more complicated than for a harmonic wave. By way of example we give the expression for the Fresnel interval τ_f in the plane $z' = $ const. for the case of a honogeneous dispersive medium, viz.,

$$\tau_f = \left(\left|2\pi k''(\omega'_{st})\right|\right)^{1/2} \left|(z-z')^{-1} - (z_c-z')^{-1}\right|^{-1/2} \ , \tag{2.10.31}$$

where the quantity

$$z_c = z' - \left[k''(\omega')\frac{\partial\omega'}{\partial t'}\right]^{-1} \quad \text{at} \quad t'=t'_{st}$$

corresponds to a space–time caustic.

In the absence of frequency modulation of the primary field, i.e., at $\partial\omega'/\partial t = 0$, the distance z_c-z' becomes infinite, and (2.10.31) transforms into (2.10.30). Expression (2.10.31) resembles (2.10.16) for the Fresnel radius of a spherical wave, where in both cases $L_1 = z-z'$ is the distance from the plane $z' = $ const to the point of observation. The quantity $z'-z_c$ characterizing the distance from the plane $z' = $ const. to the caustic z_c in (2.10.31) is an analog of the distance to the source $L_2 = z'-z_0$ in (2.10.16).

This consideration agrees well with the results derived in a pulse-shape analysis of pulses propagating in a dispersive medium with both the space–time parabolic equation [2.142, 143] and Fourier transformation [2.86].

For the problem at hand, the applicability conditions of space–time geometrical optics reduce to the requirement that the wave parameters vary only insignificantly within Fresnel time intervals:

$$\tau_f \left|\frac{1}{A}\frac{\partial A}{\partial t}\right| \ll 1 \ , \quad \tau_f \left|\frac{1}{\omega}\frac{\partial\omega}{\partial t}\right| \ll 1 \ . \tag{2.10.32}$$

In addition, we should require that the Fresnel volumes of rays should not intersect. Similar validity conditions can be stated for the general case [2.74].

Besides formulating the applicability conditions for geometrical optics, the above stated criteria can elucidate a number allied, essentially diffractional, problems of value.

2.10.6 Heuristic Accuracy Estimates of Geometrical Optics

Small parameters of the type

$$\nu \simeq a_f^2 |\nabla_\perp A/A|^2 \ll 1 \qquad (2.10.33)$$

appearing in (2.10.18, 19) might be employed as a heuristic measure of accuracy of the zero approximation field u_0:

$$\gamma_{\text{heur}} \simeq \nu . \qquad (2.10.34)$$

This assumption qualitatively agrees, particularly, with the estimate (2.10.2). The relation (2.10.33) should, of course, be viewed as a rough one. It can only suggest where the error is certainly small and where it is large. It should be convenient to trace the boundary between regions where the approximation works well and where it does not at $\gamma_{\text{heur}} \simeq 1$. This condition is equivalent to changing in the above inequalities, which define the applicability of geometrical otpics, the symbols $\ll$ to $\le$; the legitimacy of this substitution is corroborated by multiple computations in [2.74, 149] (see also examples in Chap. 3).

2.10.7 Estimating the Width of a Caustic Zone

The width of a caustic zone where the ray approximation fails (Fig. 2.8) can be estimated either from the condition $\gamma_{\text{heur}} \simeq 1$, or with the help of criterion II. We derive the estimates for a simple caustic. When the observation point r is far from the caustic, the Fresnel volumes of two rays leaving the source and reaching r are widely spaced apart (Fig. 2.21a). These volumes overlap markedly when one of the rays appears entirely inside the Fresnel volume of the other (Fig. 2.21b). This situation is described by the condition opposite to (2.10.21a)

$$k_0 \, |\psi_1 - \psi_2| < \pi , \quad \text{or} \quad |\psi_1 - \psi_2| < \lambda_0/2 . \qquad (2.10.35)$$

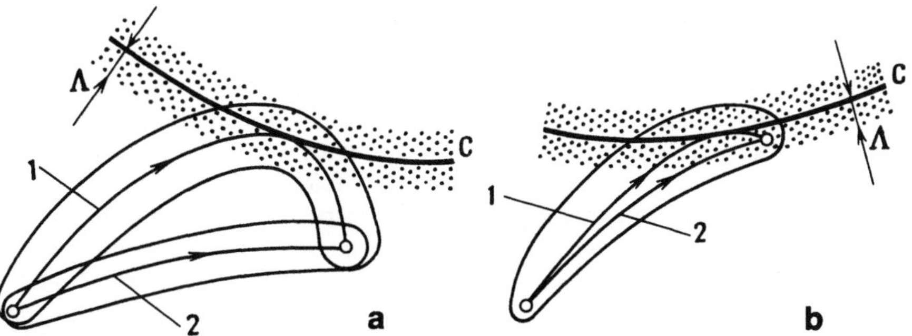

Fig. 2.21a,b. Fresnel volumes of two rays (*1* and *2*) near a simple caustic: (a) the observation point is far from the caustic and the Fresnel volumes are separated in space, (b) the observation point is within the caustic zone and the Fresnel volumes overlap markedly

A simple geometrical argument indicates [2.144, 145] that the difference of the eikonal near a simple caustic is

$$\delta\psi = |\psi_1 - \psi_2| \simeq \frac{4}{3}\beta^{1/2}|l_N|^{3/2} , \quad \beta = 2n_c|K_{rel}| , \tag{2.10.36}$$

where l_N is the normal distance to the caustic, n_c the refractive index at the caustic, $K_{rel} = |K_c - K_{ray}\cos\delta|$ the quantity defining the relative curvature of the ray and caustic, K_c the curvature of the normal section of the caustic cut by the plane tangent to the ray, K_{ray} the curvature of the ray at its point of tangency, and δ the angle between the normal to the caustic and the principal normal to the ray (the normal to the caustic is assumed to point to the lighted region - Fig.2.7a).

Substituting (2.10.36) into (2.10.35) we get for the width of the caustic zone l_c the relationship

$$\frac{4}{3}k_0\beta^{1/2}l_c^{3/2} \simeq \pi , \quad \text{whence}$$

$$l_c \simeq \left(\frac{3\pi}{4k_0}\right)^{2/3}\beta^{-1/3} = 1.77\Lambda , \quad \Lambda \equiv k_0^{-2/3}\beta^{-1/3} . \tag{2.10.37}$$

By similar computations it is possible, in principle, to estimate the width of a caustic zone in more complex situations, for example, at the vertex of a caustic cusp or in the vicinity of a caustic pocket. The corresponding computations are presented in Sect.3.1 and in [2.74, 149].

Compare now the estimate (2.10.37) with what wave theory yields and which is able to compute the field at a caustic. According to the local asymptotic formulas [2.13, 16, 144, 148], *Maslov* [2.20], and the uniform asymptotic Kravtsov-Ludwig theory [2.19, 67, 146, 147] ,the wave field at a simple caustic is described by the Airy function $Ai(\xi)$. Away from the caustic, for $|\varsigma| \gg 1$, the WKB asymptotics becomes valid for the function and the caustic asymptotics converts into the geometrical-optical one. Consequently, the caustic zone width may be estimated as $|\varsigma| < |\varsigma_a|$, where ς_a is the value of the argument starting with which for $Ai(\xi)$ its asymptotic is valid. The value ς_a can easily be obtained from Fig.2.22 illustrating the Airy function and its WKB asymptotics. As can be seen, the fitting yields $\varsigma_a \simeq -0.75$ accurate to within 10% and $\varsigma_a \simeq -0.98$ accurate to 5%.[34] For practical reasons one often uses $\varsigma_a = -1.02$ corresponding to the first maximum of the Airy function, which warrants an accuracy of fit around 4%.

It has been demonstrated in [2.144-146] that in the vicinity of a caustic the Airy function argument is $\varsigma = -l_N/\Lambda$, where $\Lambda = (k_0^2\beta)^{-1/3}$ is the characteristic dimension of field variation near the caustic. In the simplest case of a uniform medium of the refractive index $n = 1$ where the rays are

---

[34] Evidently, at the zeros of $Ai(\xi)$ the relative approximation error rises sharply.

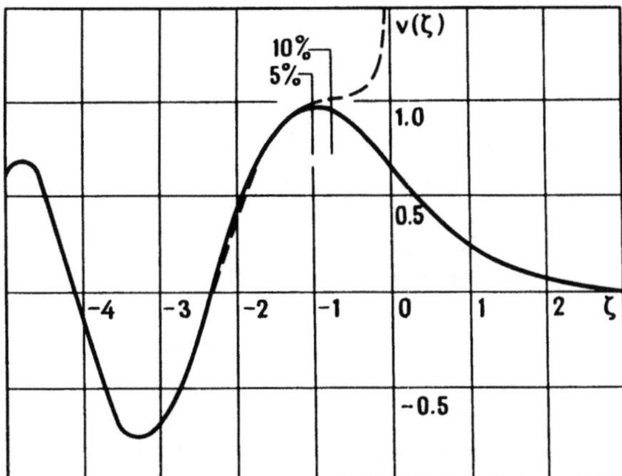

Fig.2.22. Behavior of the exact solution, Airy's function, (solid line) and the ray-optical solution (dashed line) near a caustic. The leaders point to the matching areas with accuracy better than 5% and 10%, respectively

straight lines ($K_{ray} = 0$) and $\beta = 2/R_c$ ($R_c = 1/K_c$ is the caustic's radius of curvature), the dimension Λ, according to [2.62], will be $\Lambda = (R_c/2k_0^2)^{1/3}$.

By virtue of (2.10.23), to the width of caustic zone l_c there corresponds $\zeta_c = -1.77$ comparable in value with the aforementioned values $\zeta_a = -0.75$, -0.98, and -1.02. Note in passing that the value $\zeta_c = -1.77$ lies approximately midway between the first maximum ($\zeta = -1.02$) and the first zero ($\zeta = -2.34$) of the Airy function.

A minor modification of the above argument leads us to a fairly accurate estimate of the distances to the first maximum and the first zero of the interference field near a caustic. To demonstrate, considering the phase retardation on a caustic $\delta\Psi = -\pi/2$, the condition that the waves incident on and reflected from a caustic be in antiphase assumes, in contrast to (2.10.35), the following form

$$k_0|\psi_1 - \psi_2| = 3\pi/2 .$$

By virtue of (2.10.22), this equation is satisfied by the distance $l_N = 2.32\Lambda$, which almost coincides with that to the first zero of the Airy function, $l_N = 2.34\Lambda$.

On the other hand, if two ray fields added are in phase, then, upon accounting for the caustic phase shift,

$$k_0|\psi_1 - \psi_2| = \pi/2 .$$

This equation leads us immediately to the distance $l_N = 1.11\Lambda$, very close to that of the first maximum of the Airy function, $l_N = 1.02\Lambda$.

All the distances in our analysis have been measured in terms of the characteristic dimension $\Lambda = (k_0^2\beta)^{-1/3}$ which henceforth will be adopted as an estimate of the caustic zone width $l_c \simeq \Lambda$.

2.10.8 Indistinguishability of Rays in the Caustic Zone

Inequality (2.10.35) furnishes a condition of the physical indistinguishability of rays, telling us that for $\Lambda < l_N < l_c \simeq 1.77\Lambda$ no physical facilities are able to measure separately the parameters of rays. Specifically, the rays cannot be distinguished by an opaque screen with an aperture because at $|\psi_1 - \psi_2| < \lambda_0/2$ the Fresnel volumes of two rays practically merge, as in Fig.2.21b. The rays here are devoid of their physical meaning (Sect.2.10.3), though they continue playing the role of the geometrical skeleton of the wave field [2.5, 19, 55, 67, 147, 149].

Another ray-separating technique - by the direction of arrival - would also fail. Toward a caustic, the angle between the rays diminsihes and in order to resolve them one should have to narrow the directional pattern of a receiving antenna, θ_a, say by increasing its length l_a, since $\theta_a \simeq \lambda/l_a = \lambda_0/n_c l_a$. Then from the restriction on angular resolution $\theta_a < \theta$ there follows the restriction on l_a, namely, $l_a > \lambda_0/\theta_a n_c$. Obviously, the antenna's length cannot be greater than the width of the caustic zone, $l_c \simeq \Lambda$, that is, $l_a \leq \Lambda$ (otherwise the field within the antenna would be markedly nonuniform, and it could no longer be considered as a superposition of two local plane waves). This results in the minimum resolvable angle $\theta_{min} \simeq \lambda_0/n_c\Lambda$. Computations indicate [2.74, 149] that it exactly corresponds to the boundary of the caustic zone.

2.10.9 Observability of Caustics

In order to observe a caustic it is natural to require that the field at the caustic be at least several times that beyond the caustic zone. The degree of focusing is characterized by the focusing index determined relative to the far field away from the caustic [2.37] or relative to the field in a homogeneous medium [2.25].

In case of two adjacent caustics, we can, obviously, observe both provided that they are separated farther than the total width of the caustic zone $\Lambda_1 + \Lambda_2$. Otherwise we face a new physical formation, and separate caustics should be thought of only as mathematical objects which cannot be observed separately (we have already touched upon this topic in Sect.2.4 in connection with the selection of physically justifiable models for actual caustics). This distinction between a real and an unobservable (imaginary) caustic is illustrated in Fig.2.23, showing two identical pockets on a simple caustic for various widths of the caustic zone.[35] We may speak of an actual pocket when the caustic zone is narrow (Fig.2.23a), whereas when it is wide (Fig.2.23b) no caustic pocket can be observed by physical devices.

Thus the caustic becomes a real physical object if surrounded by the *caustic volume* similar to how the ray acquires the physical content when considered along with its Fresnel volume (Sect.2.10.3). Then the question of the observability of individual branches of caustics is reduced to estimating the degree of overlapping of the caustic volumes.

---

[35] For a simple caustic, say, according to (2.10.37), the width depends upon the wavelength and the radius of curvature of the caustic.

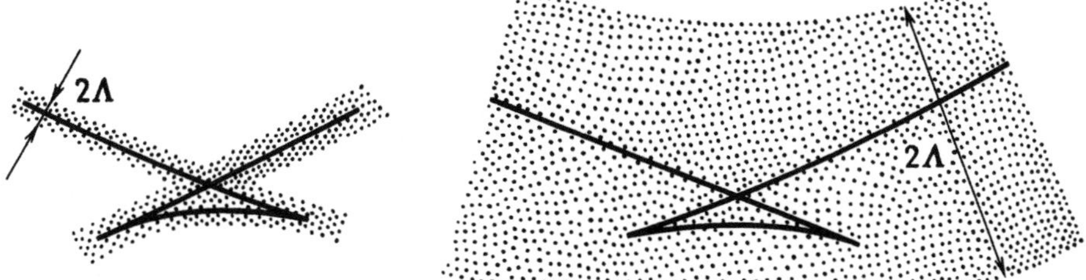

Fig.2.23a,b. Swallowtail-type caustic surrounded by the caustic zone (indicated by a dotted area): **(a)** a situation with discernible caustic branches, **(b)** a situation with the caustic branches physically indiscernible

It goes without saying that the observability (contrast) of casutics is related not only to the focusing of the field on the caustic but also to the formation of the caustic shadow region. The shadow region, when it appears, is even easier to track than the field focusing. The situation is exactly the same with the resolution of distinct caustic branches. For example, if two caustic branches draw closer to one another and their caustic volumes overlap, then the caustics are indiscernible from the illuminated region, but in the shadow region the intensity drops sharply from the field level at the caustic, pointing out to its presence. This situation is observed, for instance, in open resonators [2.71, 72] (see also Sect. 3.7).

2.10.10 Field Estimations Beyond the Validity Region of Geometrical Optics

In a number of situations the ray method is able to provide estimates of the field in the region where the method is no longer valid, which, though rough, are correct in the order of magnitude. In order to obtain such an estimate, one may invoke either the values of geometrical-optical field at the applicability limit (i.e., where $\gamma_{\mathrm{heur}} \simeq 1$), or the energy conservation principle by assuming that the energy flow, determined when nearing the applicability limit, spreads more or less uniformly over the region where the method fails.

We illustrate the suggested approach by estimating the field in the vicinity of a simple caustic. Let $l^0 = l^0(\Lambda)$ be the width of the ray tube that is Λ wide near the caustic, as shown in Fig.2.24. Assuming the energy is conserved in a ray of finite size, we have

$$(A^0)^2 l^0 n^0 \simeq |u_c|^2 \Lambda n_c , \qquad (2.10.38)$$

where A^0 and n^0 are the initial values of the amplitude and the refractive index, respectively, and u_c and n_c are the values of the field and the refractive index at the caustic. From (2.10.38) we have estimate

$$|u_c| \simeq A^0 (l^0 n^0 / \Lambda n_c)^{1/2} . \qquad (2.10.39)$$

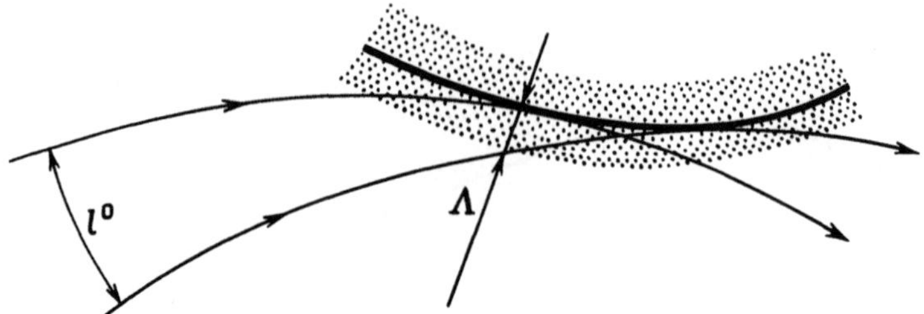

Fig.2.24. Illustration of the field estimation (2.10.21,25) on a caustic by the principle of conservation of energy applied to the ray tube of finite section

The same order of magnitude for $|u_c|$ is predicted by the asymptotic formulas for the field near a simple caustic that are based on the Airy function [2.74, 149]. Thus *Kravtsov* and *Orlov* [2.74] have indicated that the field at the first - that is, the closest to the caustic - interference maximum at $l_N = 1.02 \Lambda$ is only 1.34 times that of the energy-flux estimate (2.10.39)

$$|u_{1,\max}| = 1.34 A^0 (l^0 n^0 / \Lambda n_c)^{1/2} .$$

The energy estimates can also be employed to assess the field intensity around an ideal focus, at the vertex of a caustic cusp, in the region of a space-time focus, etc. [2.55, 74, 149].

2.10.11 Field-Focusing Indices at Caustics

Computations indicate [2.70] that in (2.10.24) $l^0 \propto \sqrt{\Lambda}$ so that $|u_c| \simeq \Lambda^{-1/4} \simeq k_0^{1/6}$ or, if we have recource to the small parameter $\mu = 1/kL$, $|u_c| \simeq A^0 \mu^{-1/6}$. This estimate is a specific case of the general dependence for the field at a caustic:

$$|u_c| \simeq A^0 \mu^{-\sigma_f} . \tag{2.10.40}$$

We will call the exponent σ_f the focusing index.[36] It takes on the smallest value at a simple caustic, $\sigma_f = 1/6$. Its largest value corresponds to an ideal focus: in a three-dimensional problem $\sigma_f = 1$; in a two-dimensional problem $\sigma_f = 1/2$. The values of σ_f for their caustics can be derived from energy estimations of the type (2.10.38), but they can also be inferred from the analysis of integral field representations. Table 2.1 of Sect.2.4.3 lists the values of σ_f borrowed from [2.29].

Formulas of the type (2.10.40) can be used to deduce approximate, or a priori, estimates of the field at caustics, taking every time L to be a characteristic length of the problem.

---

[36] In [2.29,52], σ_f is called the "singularity index".

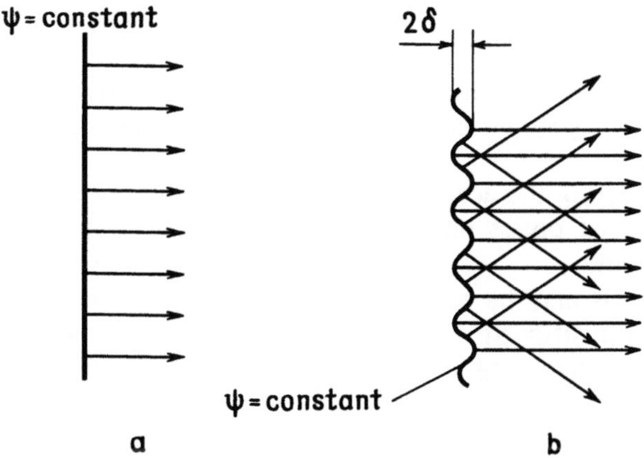

Fig.2.25a,b. Phase front of a plane wave: (a) prior to perturbation, (b) after a perturbation causing the appearance of "new rays". For $\delta \ll \lambda$ the ray-optical description of these rays breaks down and they should be deemed fictitious physically

2.10.12 Stability with Respect to Small Perturbations

The gist of the problem can be disclosed with a simple example, already discussed in Sect.2.4. Consider a plane wave to which there correspond a parallel bundle of rays (Fig.2.25a). If we subject the original phase front to weak periodic perturbations, the ray pattern changes drastically (Fig.2.25b). The geometrical-optical field changes markedly, too. On the other hand, it is obvious that if the perturbing amplitude δ is small compared with the wavelength, the actual wave field remains very nearly the same (it incorporates only small disturbances of order $\delta/\lambda \ll 1$).

The considered example indicates that the geometrical-optical field is *unstable* (or, more correctly, *very sensitive*) with respect to small perturbations of initial conditions (and medium parameters). The point is that the conditions of the ray-method applicability (2.10.19) fail for weak, small-sized perturbations. Notably, the "new" rays that appeared in Fig.2.24b as a result of the plane, phase-front perturbation turn out to be fictitious physically, since their surrounding Fresnel volumes contain many inhomogeneities for $\delta \ll \lambda$, rendering the geometrial-optics method inapplicable because (2.10.19) no longer holds.

2.10.13 Wave-Pattern Analysis in General

Such an analysis is of value when one needs a rapid and general (rough) assessment of the structure of a shortwave field, either to furnish engineering estimates, or to provide a precomputational forecast.

As we have seen above, ordinary geometrical-optics augmented by the notion of Fresnel volume becomes an analytic tool for high-frequency fields, capable of yielding beyound their qualitative structure some quantitative estimates as well, even where geometrical optics is invalid. Such an analysis involves:

(i) Defining the geometrical-optical field (evaluation of rays, phase fronts, caustics, amplitudes, etc.).

(ii) Determining the Fresnel volumes of rays and the domains of applicability and invalidity of geometrical optics by means of heuristic criteria.

(iii) Heuristically estimating the accuracy of the field within the applicability domain of geometrial optics and the estiamtes of the amplitude beyond the applicability limit.

If necessary, the analysis may be extended to estimating exponentially small scattered fields [2.74] mentioned in Sect.2.5. Some examples of wave-pattern analysis in general, involving Fresnel volumes of rays, are presented in Chap.3.

3. Applications of the Ray Methods

This chapter is devoted to the most important and frequently encountered problems of wave physics allowing for ray treatment. We begin with the simplest case of homogeneous media and proceed to rays and ray fields with various inhomogeneities present: interfaces, layered media, localized inhomogeneities, etc. Ample space has here been reserved for the analysis of complex-shaped caustics. Separate sections are devoted to the space-time analysis of pulse propagation in dispersive media. They are intended to illustrate the multitude of problems encountered in modern applied physics.

3.1 Waves in Homogeneous Media

3.1.1 Rays and the Eikonal

From the ray equation (2.2.12) it follows that in a medium of constant refractive index n the momentum along a ray remains unchanged, $\mathbf{p} = \mathbf{p}^0(\xi,\eta)$, and $\tau = \sigma/n$ so that the rays are straight lines

$$\mathbf{r} = \mathbf{r}^0 + \tau\mathbf{p}^0 = \mathbf{r}^0 + (\sigma/n)\mathbf{p}^0 = \mathbf{r}^0 + \sigma\mathbf{l}^0 = \mathbf{r}(\xi,\eta,\sigma) \ . \tag{3.1.1}$$

Here $\mathbf{r}^0 = \mathbf{r}^0(\xi,\eta)$ is the initial point on the ray, $\mathbf{l}^0 = \mathbf{p}^0/n = \mathbf{l}^0(\xi,\eta)$ the unit vector of ray direction at point $\mathbf{r}^0$, and $\sigma = |\mathbf{r}-\mathbf{r}^0|$ the distance along the ray.

The wave eikonal (2.2.13) in a homogeneous medium becomes

$$\psi = \psi^0(\xi,\eta) + \tau n^2 = \psi^0(\xi,\eta) + \sigma n \ , \tag{3.1.2}$$

where $\psi^0 = \psi^0(\xi,\eta)$ is the initial value of the eikonal at $\sigma = 0$ (at point $\mathbf{r}^0$).

Equation (3.1.1) defines a family (congruence) of straight lines normal to the wave front $\psi = $ constant, that is, a *normal congruence of straight lines*. The orthogonality condition for the congruence has the form [3.1]

$$\mathbf{r}^0_\xi \cdot \mathbf{l}^0_\eta = \mathbf{r}^0_\eta \cdot \mathbf{l}^0_\xi \ , \tag{3.1.3}$$

where $\mathbf{r}^0_\xi = \partial\mathbf{r}^0/\partial\xi$, $\mathbf{l}^0_\xi = \partial\mathbf{l}^0/\partial\xi$, etc. It implies, particularly, that in contrast to the spatial (3-D) case, a plane (2-D) congruence of straight lines ($\mathbf{l}_\eta=0$, $\mathbf{r}^0_\eta=0$) is always orthogonal.

3.1.2 The Wave Amplitude

Using the ray family equation (3.1.1), we determine the Jacobian $\mathscr{D}(\sigma)$:

$$\mathscr{D}(\sigma) = \left[\frac{\partial \mathbf{R}}{\partial \xi} \times \frac{\partial \mathbf{R}}{\partial \eta}\right] \cdot \frac{\partial \mathbf{R}}{\partial \sigma} = [\mathbf{l}^0 \times (\mathbf{r}_\xi^0 + \sigma \mathbf{l}_\xi^0)] \cdot (\mathbf{r}_\eta^0 + \sigma \mathbf{l}_\eta^0)$$

$$= a\sigma^2 + b\sigma + c = a(\sigma - \sigma_1)(\sigma - \sigma_2) \, , \tag{3.1.4}$$

where $\sigma_{1,2}$ denote the roots of the equation $\mathscr{D}(\sigma) = 0$:

$$\sigma_{1,2} = (-b \pm \sqrt{b^2 - 4ac})/2a \, , \tag{3.1.5}$$

and the notation employed is

$$a = (\mathbf{l}^0 \times \mathbf{l}_\xi^0) \cdot \mathbf{l}_\eta^0 \, ,$$

$$b = (\mathbf{l}^0 \times \mathbf{l}_\xi^0) \cdot \mathbf{r}_\eta^0 - (\mathbf{l}^0 \times \mathbf{l}_\eta^0) \cdot \mathbf{r}_\xi^0 \, , \tag{3.1.6}$$

$$c = (\mathbf{l}^0 \times \mathbf{r}_\xi^0) \cdot \mathbf{r}_\eta^0 \, .$$

Then, for the divergence of the ray tube we have

$$\mathscr{J} = \frac{\mathscr{D}(\sigma)}{\mathscr{D}(0)} = \frac{(\sigma - \sigma_1)(\sigma - \sigma_2)}{\sigma_1 \sigma_2}$$

$$= \frac{\sigma^2}{\sigma_1 \sigma_2} - (1/\sigma_1 + 1/\sigma_2)\sigma + 1 \, , \tag{3.1.7}$$

where $1/\sigma_1 \sigma_2 = a/c$, $1/\sigma_1 + 1/\sigma_2 = -b/c$. The field amplitude A is related to the divergence $\mathscr{J}$ by (2.3.9)

$$A = \frac{A^0(\xi,\eta)}{\sqrt{\mathscr{J}}} = A^0(\xi,\eta) \sqrt{\frac{\sigma_1 \sigma_2}{(\sigma - \sigma_1)(\sigma - \sigma_2)}} \, . \tag{3.1.8}$$

Hence, to a zero approximation of geometrical optics, the wave field in a homogeneous medium varies according to

$$u = A e^{ik_0 \psi} = u^0(\xi,\eta) \sqrt{\frac{\sigma_1 \sigma_2}{(\sigma - \sigma_1)(\sigma - \sigma_2)}} \, e^{ik_0 n \sigma} \, , \tag{3.1.9}$$

where $u^0(\xi,\eta) = A^0(\xi,\eta) \exp[ik_0 \psi^0(\xi,\eta)]$ is the wave field on the initial surface $\mathbf{r} = \mathbf{r}^0(\xi,\eta)$.

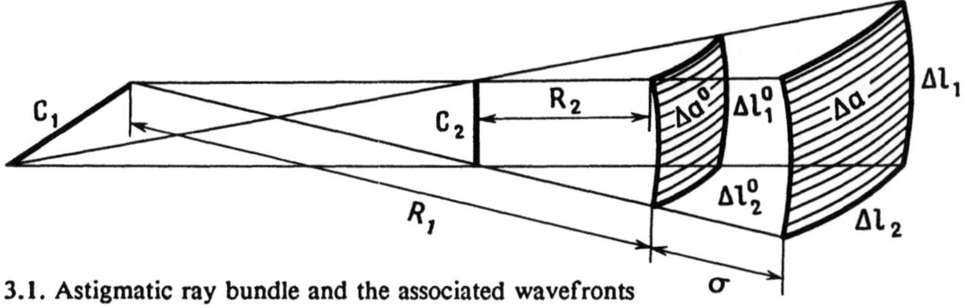

3.1. Astigmatic ray bundle and the associated wavefronts

In order to disclose the geometrical meaning of σ_1 and σ_2 in (3.1.8-9), we consider a case when the initial surface $\mathbf{r} = \mathbf{r}^0(\xi,\eta)$ is a wave front. Adopting the lines of curvature[1] of the front as coordinates ξ and η, we have by the equations of O. Rodrigues [3.2]:

$$\mathbf{r}_\xi^0 = R_1 \mathbf{l}_\xi^0 ,$$

$$\mathbf{r}_\eta^0 = R_2 \mathbf{l}_\eta^0 , \tag{3.1.10}$$

where $R_{1,2}$ are the principal radii of curvature of the front, and $\mathbf{l}^0$ is the normal to the front. Substituting (3.1.10) into (3.1.5) and observing that $b = -a(r_1+r_2)$ and $c = aR_1R_2$, we obtain

$$\sigma_1 = R_1 \quad \text{and} \quad \sigma_2 = R_2 . \tag{3.1.11}$$

Thus the quantities $\sigma_{1,2}$ are equal to the *principal radii of curvature* of the initial wave front (at point $\mathbf{r}^0$). When $R_1 \neq R_2$, the ray bundle (3.1.1) is said to be *astigmatic* (Fig.3.1), in contrast to a *homocentric* or *stigmatic* bundle when $R_1 = R_2$.

By virtue of (3.1.1, 8) we have for the field amplitude

$$A = A^0 \sqrt{\frac{R_1 R_2}{(R_1 - \sigma)(R_2 - \sigma)}}$$

$$= A^0(K/\tilde{K})^{-1/2} = A^0(1 - 2H\sigma + K\sigma^2)^{-1/2} , \tag{3.1.12}$$

where $K^0 = 1/R_1^0 R_2^0$ and $\tilde{K} = 1/\tilde{R}_1 \tilde{R}_2$ are the total (Gaussian) curvatures of the initial ($\sigma = 0$) and instantaneous wave fronts, $\tilde{R}_{1,2} = R_{1,2}^0 - \sigma$ are the principal radii of curvature of the front, and $H^0 = (1/R_1^0 + 1/R_2^0)/2$ is the mean curvature of the initial front.

---

[1] We recall that a line of curvature of a surface is the curve at every point of which the tangent belongs to the plane of the principal normal section of the surface at the point. Two lines of curvature can be drawn through every nonsingular point of a surface, the lines being always orthogonal [3.2].

We could have readily obtained (3.1.12) on geometrical grounds by calculating the cross section of an astigmatic ray tube (Fig.3.1) in an initial point (Δa^0) and instantaneous points (Δa)

$$\Delta a^0 = \Delta l_1^0 \Delta l_2^0 \quad \text{and} \quad \Delta a = \Delta l_1 \Delta l_2 \; , \tag{3.1.13a}$$

and recognizing that

$$\Delta l_{1,2} = \Delta l_{1,2}^0 \frac{|R_{1,2}| + \sigma}{|R_{1,2}|} = \Delta l_{1,2}^0 \frac{R_{1,2} - \sigma}{R_{1,2}} \tag{3.1.13b}$$

(for the diverging front in Fig.3.1, $R_{1,2} = -|R_{1,2}| < 0$). In a two-dimensional problem, say, in the x, z plane, we have $R_2 \rightarrow \infty$ and from (3.1.12) it follows that ($R_1 = R_2$)

$$A = A^0 \left[\frac{R}{R - \sigma} \right]^{1/2} = A^0 \left(1 - \frac{\sigma}{R} \right)^{-1/2} \; , \tag{3.1.14}$$

where the radius of curvature is given by ($a \equiv 0$)

$$R = -\frac{c}{b} = -\frac{(l^0 \times r_\xi^0) \cdot e_y}{(l^0 \times l_\xi^0) \cdot e_y} = l_z^0 \left(l_x^0 \frac{\partial z^0}{\partial \xi} - l_z^0 \frac{\partial x^0}{\partial \xi} \right) \left(\frac{\partial l_x^0}{\partial \xi} \right)^{-1} \; .$$

It is a straightforward exercise to verify that for an axially symmetric three-dimensional problem (z is the axis of symmetry)

$$A = A^0 \left[\frac{\rho}{r} \frac{R}{R - \sigma} \right]^{1/2} = A^0 \left[\frac{r}{\rho} \left(1 - \frac{\sigma}{R} \right) \right]^{-1/2} \; , \tag{3.1.15}$$

where ρ and r are the cylindrical coordinates of the exit point and the observation point, respectively, and the radius of curvature R is determined in a meridional cross section of the front.

3.1.3 Caustics

By virtue of the condition $\mathscr{D}(\sigma_{1,2}) = 0$ the quantities σ_1 and σ_2 are equal to the *distances to the caustics* measured along the ray.[2] Each ray in the three-dimensional case cannot touch the caustics more than *twice*[3] (at $\sigma =$

---

[2] Strictly speaking, this statement holds only for real-valued σ_1 and σ_2, which can be readily verified with the condition (3.1.3) on the normality of a congruence of straight lines [Ref.3.1, Chap.14].

[3] This fact has been konwn from the earliest development of optics (Morolicus, 1575) [Ref.3.1, Chap.14].

σ_1 and $\sigma = \sigma_2$), each time its phase being retarded by $-\pi/2$ (Sect.2.4). For a homocentric bundle rays ($R_1 = R_2$) forming a focal point, the caustic phase shift equals $-\pi$.

From (3.1.11) it follows that caustic points on a ray lie at the centers of curvature of the initial wave front; the caustics of an astigmatic bundle (C_1 and C_2 in Fig.3.1) are at right angles to the central ray [3.3]. The caustics that correspond to a family of rays "in general" are the *loci of the centers of curvature* of the wave front and form two cavities, each of which is consistent with one family of lines of curvature. The position of the caustics with respect to the initial front is specified by the surface type of the front. Specifically, for elliptic points ($R_1 R_2 > 0$) on the front, they form two *real* ($R_1 > 0$, $R_2 > 0$) or *imaginary* ($R_1 < 0$, $R_2 < 0$) caustics placed on one side of the front [3.3]. If the front is axially symmetric, then according to (3.1.15) one of the caustics is a surface of revolution while the other degenerates into a focal line (the axis of symmetry) and is referred to as an *axial caustic*. In the two-dimensional (plane) problem, the caustic is an evolute and the fronts are involutes [3.2, 4].

The vector parametric equations of caustics follow from (3.1.1) at $\sigma = \sigma_{1,2}$ as

$$\mathbf{r} = \mathbf{r}^0(\xi,\eta) + \sigma_{1,2}\mathbf{l}^0(\xi,\eta) = \mathbf{r}^0(\xi,\eta) + R_{1,2}\mathbf{l}^0(\xi,\eta) \equiv \mathbf{F}_{1,2}(\xi,\eta) \, , \quad (3.1.16)$$

where $R_{1,2} = \sigma_{1,2} = \sigma_{1,2}(\xi,\eta)$ are found from (3.1.5). In a homogeneous medium, caustics may assume quite different configurations (see Sect.2.4 and illustrations in [3.1, 3-7]). From an analysis of (3.1.16) one may infer that cusps on a caustic correspond to the extrema of the radius of curvature of the front; when the radius is at a minimum, the caustic cusp points toward the initial wave front; when at a maximum, away from the front. To the parabolic points ($R_1 R_2 \to \infty$) of the front (to the inflection points in two-dimensional space) there correspond the branches of caustics (3.1.16) asymptotically approaching infinity. The noted features of a caustic in a two-dimensioanl problem follow from the properties of the evolute.

Rays in two-dimensional space can be constructed especially easily by the shape of the caustic because here the fronts are involutes of the caustic. In three-dimensional space, in addition, one needs to give, at the caustic surface, the directional field of rays leaving the caustic.

3.1.4 The Plane Phase-Amplitude Screen

Consider the problem with initial conditions for the wave field given on a plane,[4] say $z = 0$. This plane can be viewed as a *phase-amplitude screen* modulating the plane wave $\exp(ik_0 z)$ by the complex transmission coefficient $A^0(\xi,\eta)$, i.e.,

$$u^0(\xi,\eta) = A^0(\xi,\eta)\exp[ik_0\psi^0(\xi,\eta)] \, ,$$

---

[4] Some problems become more amenable with their initial conditions imposed at the caustic [3.7].

where ξ, η are Cartesian coordinates in the plane of the screen.

Behind the screen ($z > 0$) the ray equations (3.1.1) have the form (we assume below that $n = 1$)

$$x = \xi + p_x^0 \sigma, \quad y = \eta + p_y^0 \sigma, \quad z = p_z^0 \sigma, \qquad (3.1.17)$$

where $p_{x,y,z}^0$ are given by (2.3.31) for $\epsilon^0 = 1$. Equations (3.1.17) indicate that for $z > 0$ the rays exist if either $p_z^0 > 0$ or $(p_x^0)^2 + (p_y^0)^2 < 1$. Otherwise there exists no real-valued outgoing component of momentum p_z^0 perpendicular to the screen, and the wave $u^0 = A^0 \exp(ik_0\psi^0)$ travels at a slower phase speed over the screen (a surface wave [3.8]).

A straightforward computation shows that

$$\mathscr{D}(\sigma) = \frac{\partial(x,y,z)}{\partial(\xi,\eta,\sigma)} = [a_0 \sigma^2 + b_0 \sigma + (p_z^0)^2]/(p_z^0), \qquad \text{whence} \qquad (3.1.18)$$

$$u(x,y,z) = \sum_{\nu} A^0(\xi_\nu, \eta_\nu) \sqrt{(p_2^0)^2 /[a_0 \sigma^2 + b_0 \sigma + (p_z^0)^2]}$$

$$\cdot \exp\{ik_0[\psi^0(\xi_\nu, \eta_\nu) + \sigma]\} \qquad \text{with} \qquad (3.1.19)$$

$$a_0 = \psi_{\xi\xi}^0 \psi_{\eta\eta}^0 - (\psi_{\xi\eta}^0)^2,$$

$$b_0 = \psi_{\xi\xi}^0[1 - (\psi_\eta^0)^2] + \psi_{\eta\eta}^0[1 - (\psi_\xi^0)^2] + 2\psi_\xi^0 \psi_\eta^0 \psi_{\xi\eta}^0,$$

$$\sigma = \sqrt{(x-\xi)^2 + (y - \eta)^2 + z^2};$$

ξ_ν, η_ν are the ray coordinates of the point $\{x, y, z\}$ defined from the set of ray equations (3.1.17), and the summation is taken over all rays passing through the observation point.

In a two-dimensional problem (on the x, z plane) with $p_y^0 = 0$, (3.1.9) yields

$$u(x,y) = \sum_{\xi_\nu} A^0(\xi)(1 - \sigma/R)^{-1/2} \exp\{ik_0[\psi^0(\xi) + \sigma]\}, \qquad (3.1.20)$$

where

$$R = [1 - (\psi_\xi^0)^2]/(\psi_{\xi\xi}^0), \quad \sigma = \sqrt{(x-\xi)^2 + z^2}, \qquad (3.1.21)$$

and $\xi_\nu = \xi_\nu(x,z)$ are the roots of the ray equations (3.1.17) for $p_y^0 = 0$, which can be alternatively written as

$$x = \xi + z\,\psi^0_\xi \Big/ \sqrt{1 - (\psi^0_\xi)^2} \; . \tag{3.1.22}$$

By analogy with (3.1.20) and (3.1.15) we may easily derive for an axially symmetric eikonal ψ^0 for which $\psi^0 = \psi^0(\rho)$, where $\rho = \sqrt{\xi^2 + \eta^2}$ is the radius from $z = 0$,

$$A(z, r, \phi) = A^0(\rho, \phi)\sqrt{\rho R / r(R - \sigma)} \; , \tag{3.1.23}$$

where R is obtained from (3.1.21) by substituting ρ for ξ.

The caustics for the rays (3.1.22) are given by

$$x = \xi - \psi^0_\xi [1 - (\psi^0)^2] / \psi^0_{\xi\xi} \equiv x_c(\xi) \; ,$$

$$z = - [1 - (\psi^0_\xi)^2]^{3/2} / \psi^0_{\xi\xi} \equiv z_c(\xi) \; , \tag{3.1.24}$$

from which it follows that for $z > 0$ a real caustic can form only for $\psi^0_{\xi\xi} < 0$ (the respective part of the screen provides the *focusing* effect). The caustic cusps (3.1.24) are defined subject to $dx_c/d\xi = 0$ (or $dz_c/d\xi = 0$) which yields

$$\psi^0_\xi \left(\psi^0_{\xi\xi\xi}[1 - (\psi^0_\xi)^2] + 3\psi^0_\xi (\psi^0_{\xi\xi})^2 \right) = 0 \; . \tag{3.1.25}$$

The asymptotes of the caustic (3.1.24) are the rays (3.1.22) leaving the screen at the points where $\psi^0_{\xi\xi} = 0$. The geometry of caustic asymptotes is discussed in [3.5.]. Caustics corresponding to the near field of plane apertures were studied in [3.9].

3.1.5 The Sinusoidal Phase Screen. An Illustrative Example

Let

$$\psi^0(\xi) = b\sin(\kappa\xi) = b\sin(2\pi\xi/h) \; , \tag{3.1.26}$$

where b is the phase modulation amplitude, and $\kappa = 2\pi/h$ is the wave number of the sinusoidal disturbance. The ray equation (3.1.22) now becomes

$$\frac{x-\xi}{z} = \frac{\kappa b\cos(\kappa\xi)}{\sqrt{1 - \kappa^2 b^2 \cos^2(\kappa\xi)}} \; . \tag{3.1.27}$$

We can determine various regions of multipath propagation by means of the caustic equations (3.1.24) which assume the form

$$\kappa x = \kappa\xi + \cot(\kappa\xi)[1 - \kappa^2 b^2 \cos^2(\kappa\xi)] \; ,$$

$$\kappa z = \frac{1}{\kappa b\,\sin(\kappa\xi)}[1 - \kappa^2 b^2 \cos^2(\kappa\xi)]^{3/2} \; . \tag{3.1.28}$$

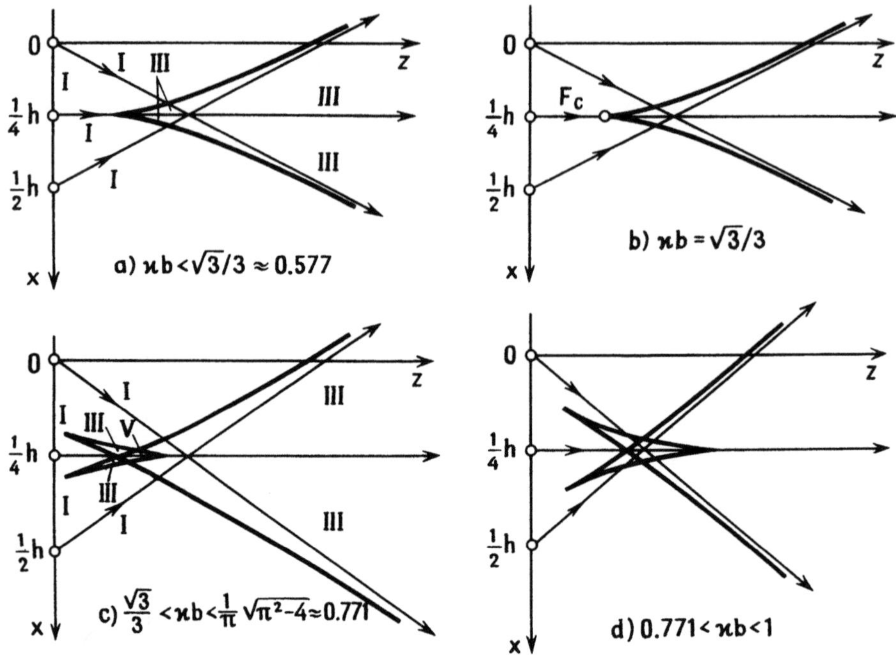

3.2. Evolution of the caustic that occurs when a plane wave passes through the sinu-soidal screen of period h ($\kappa = 2\pi/h$) and phase modulation amplitude b

The caustic (3.1.28) is a periodic sequence of curves with the period h $= 2\pi/\kappa$ whose shape is determined by the product κb. One period of the caustic is shown in Fig.3.2; Roman numerals in the figure indicate the number of rays traveling through the observation point in various regions of the x,z plane. For $\kappa b < \sqrt{1/3} \simeq 0.577$, the caustic exhibits one cusp pointed to the screen (Fig.3.2a), and for $0.577 < \kappa b < 1$ it has three different cusps (Fig.3.2c, d) deduced, according to (3.1.25), from the conditions

$$\kappa\xi = \frac{\pi}{2}m , \quad m = \pm 1, \pm 2, ...,$$

$$\sin(\kappa\xi) = \pm \frac{1}{\kappa b} \sqrt{\tfrac{1}{2}(1 - \kappa^2 b^2)} . \tag{3.1.29}$$

As $\kappa b \rightarrow \sqrt{1/3}$ the caustic pockets contract to the focus F_c (Fig.3.2b). The caustics in Fig.3.2 are typical cross sections of a butterfly-type surface (Sect.2.4). The rays leaving the screen at $\xi = mh = 2\pi m/\kappa$ (m = 0,±1,..) are the asymptotes for the caustics.

For $\kappa b > 1$ the part of the screen where $\partial\psi^0/\partial\xi > 1$ does not emit any rays. The caustic in this case has no asymptotes and touches the screen at the points where $\partial\psi^0/\partial\xi = 1$, forming a single cusp pointing outward from the screen, as in Fig.3.3a.

Figure 3.3 presents a general pattern of the caustics for $\kappa b < 0.577$ and shows by Roman numerals the number of rays in various parts of space.

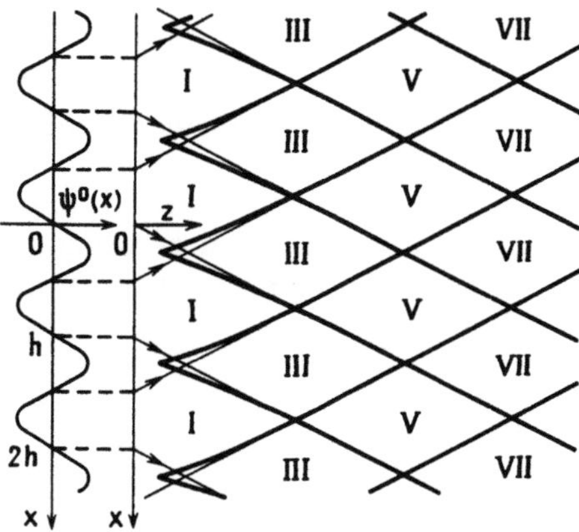

Fig.3.3. Rays increase in number (shown by Roman figures) as the point of observation moves away from the phase screen

Away from the screen, the number of rays passing through an observation point increases, thus complicating the interference structure of the wave field.

3.1.6 Applicability Conditions for Geometrical Optics

On the condition that $|R_{1,2}| \gg \lambda_0/n$ the dimensions of a Fresnel zone on the initial wave may be obtained from a formula similar to (2.10.16):

$$a_f^{1,2} = \sqrt{\frac{\lambda_0}{n} \left| \frac{1}{R_{1,2}} - \frac{1}{\sigma} \right|^{-1}} = \sqrt{\frac{\lambda_0}{n} \left| \frac{\sigma R_{1,2}}{\sigma - R_{1,2}} \right|} \, , \qquad (3.1.30)$$

where $R_{1,2}$ stands for the principal radii of curvature of the front ($R_{1,2} < 0$ for a convex diverging front). On the initial surface Q the Fresnel volume cuts out a zone measuring

$$b_f^{1,2} = a_f^{1,2}/|\cos \theta| \, , \qquad (3.1.31)$$

where θ is the angle between the unit vector l^0 along the ray and the normal **N** to the surface θ (the angle at which the ray emetrges from Q).

In accord with the criteria of geometrical-optics applicability (2.10.12, 14) we have[5]

$$\frac{a_f^{1,2}}{|\cos\theta|} \left| \frac{1}{A^0} \frac{\partial A^0}{\partial \xi_j} \right| \ll 1 \, , \quad a_f^{1,2} \ll |R_{1,2}| \, , \qquad (3.1.32)$$

---

[5] For a homogeneous medium the applicability conditions (3.1.32) may be given a rigorous substantiation by asymptotically examining the Kirchhoff-Huygens diffraction integral [3.10].

where $\xi_1 = \xi$, $\xi_2 = \eta$ are cuvilinear coordinates on Q. By substituting (3.1.30) for $a_f^{1,2}$ in (3.1.32) it is a straightforward matter to demonstrate that in the absence of caustics ($R_{1,2} < 0$) the inequality $a_f^{1,2} \ll |R_{1,2}|$ holds provided that $|R_{1,2}| \gg \lambda_0/n$.

If, on the other hand, the initial surface Q is of finite size, $d_{1,2} \gg \lambda_0/n$, then for geometrical optics to be applied we need, in addition to (3.1.32), another restriction $2b_f^{1,2} \ll d_{1,2}$. In view of (3.1.30,31) we readily obtain the following restriction on distance σ

$$\frac{\sigma}{\lambda} \ll \left[\left(\frac{2\lambda}{d_{1,2}\cos\theta} \right)^2 - \left| \frac{\lambda}{R_{1,2}} \right| \right]^{-1} , \quad \lambda = \frac{\lambda_0}{n} , \tag{3.1.33a}$$

which, in the specific case of

$$\lambda |R_{1,2}| \gg (\tfrac{1}{2} d_{1,2}\cos\theta)^2 , \qquad \text{becomes}$$

$$\lambda\sigma \ll (\tfrac{1}{2} d_{1,2}\cos\theta)^2 . \tag{3.1.33b}$$

The last condition holds for any σ when

$$\lambda^2 \ll \lambda |R_{1,2}| \ll (\tfrac{1}{2} d_{1,2}\cos\theta)^2 .$$

To exemplify, if an initial surface Q is a diverging wave front of $A^0 = $ constant, then subject to the condition

$$1 \ll |R_{1,2}|/\lambda \ll (d_{1,2}/2\lambda)^2$$

the geometrical optics approximation is valid along the entire route of wave propagation. The validity boundaries of the ray approximation for a converging front, when caustics or a focus arise, are considered in depth below in Sects. 3.1.9, 10.

3.1.7 Geometrical Optics in Far and Near Antenna Fields. Wave Beam Propagation

Consider the field of a plane, equiphase, aperture antenna with the aperture diameter $2b \gg \lambda$. The initial (at $z = 0$) field distribution is $u^0(x, y) = A^0(x, y)$; that is, the initial phase is assumed zero, $\psi^0 = 0$. We set out to investigate the applicability of geometrical optics for both near- and far-field analyses and to demonstrate in what the ray evaluations differ for these two regions.

In the vicinity of a plane, equiphase, aperture antenna of large dimension, the geometrical-optics approximation (3.1.19) describes the field of a parallel ray bundle (a collimated beam, Fig. 3.4a) with the same amplitude distribution that has been on the aperture

$$u(x, y, z) = u^0(x, y)\exp(ik_0 z) = A^0(x, y)\exp(ik_0 z) . \tag{3.1.34a}$$

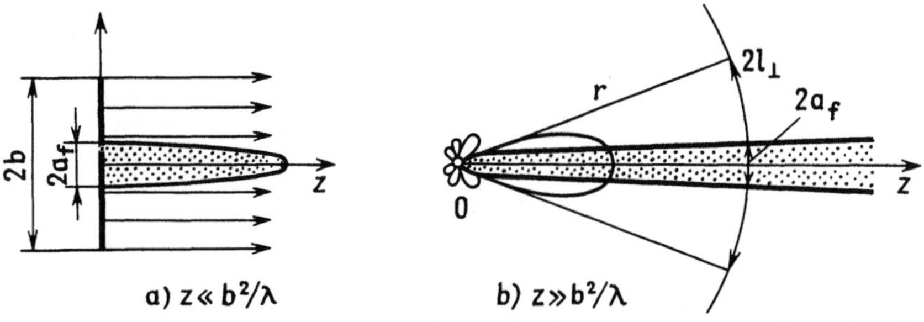

3.4. In the near (collimated beam) field (a) of the antenna, the Fresnel radius a_f is small compared to the aperture size b, whereas in the far field (b) a_f is small compared to the main lobe width of the antenna radiation pattern $l_\perp \sim \lambda\, r/b$. The geometrical optics approach is valid in both areas, but the initial field conditions in these situations are different

By virtue of (3.1.33b) the approximation is valid subject to the condition[6] $\sigma = z \ll b^2/\lambda$ (here we have allowed for $\theta = 0$ and $|R_{1,2}| \to \infty$). This condition corresponds to the *near field* of the antenna.

Far from the aperture, the field is a directional spherical wave whose main lobe width is of order λ/b, so that at a distance r from the antenna the characteristic dimension of field change becomes $l_\perp \sim \lambda r/b$ (Fig.3.4b). The Fresnel zone dimension on a sphere of radius r, for an infinitely distant observation point, is $a_f = \sqrt{\lambda r}$. Therefore, according to the criterion of geometrical-optics applicability, must be the following inequality satisfied:

$$a_f/l_\perp = \sqrt{b^2/\lambda r} \ll 1 \quad \text{or} \quad r \gg b^2/\lambda \,. \tag{3.1.34b}$$

This condition is seen to coincide with the ordinary criterion of the far field. If instead of (3.1.34b) we impose the relaxed requirement $a_f \leq l_\perp$, then we may deem that the far field begins at $r \geq b^2/\lambda$. In terms of the geometrical optics of a collimated beam, for an observer at a distance of $r = b^2/\lambda$, the aperture contains a single Fresnel zone.

Thus the geometrical-optics description is applicable both in the near (Fig.3.4a) and far (Fig.3.4b) fields of a plane, equiphase aperture; however, the difference in the initial conditions should be recognized. For the near-field distribution, initial conditions are imposed on the aperture, whereas evaluating the far-field distribution takes as initial conditions the directional spherical wave field, corresponding to the well-known radiation pattern of the antenna.

In the intermediate region $r \sim b^2/\lambda$ the field is of a more complicated character which geometrical optics is unable to describe (for the ray and quasi-ray approaches for this case, see [3.9,11]). It should be observed, though, that the situation considered reflects the features of the bundle of parallel rays (plane wave front). As we have already noted in Sect.3.1.6, for a curvilinear initial wave front (say, for a plane aperture with a nonlinear phase excitation) a unified ray description may be applied to the near, intermediate, and far antenna fields.

---

[6] The derived estimate follows directly from (2.10.9) for a_f and the requirement $a_f \ll b$.

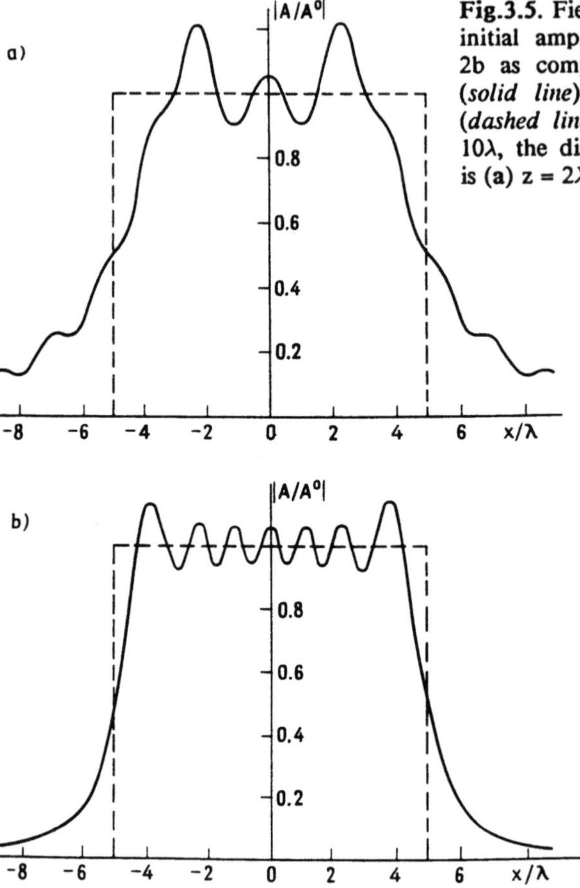

Fig.3.5. Field of a beam with a uniform initial amplitude distribution of width 2b as computed by diffraction theory (*solid line*) and by ray-optical theory (*dashed line*). The beam width is 2b = 10λ, the distance from the initial plane is (a) z = 2λ, and (b) z = 10λ

All the above arguments concerning the applicability boundaries of geometrical optics hold for finite wave beams both in their near and far fields. The ray approximation in the near field of the beam was given in (3.1.19, 20, 23).

To illustrate the accuracy of the ray approximation, Figs.3.5,6 plot amplitude distributions in two different sections z = constant for a two-dimensional beam,[7] computed both with the aid of diffraction theory [3.11, 12] (solid curves) and geometrical optics (dashed curves). Figure 3.5 refers to a beam of a uniform initial amplitude distribution, $A^0(x)$, and Fig.3.6 to a beam of a cosine distribution, $A^0(x) = A^0\cos(\pi x/2b)$, the beam width being 2b = 10λ, and the initial eikonal $\psi^0 = 0$. The bottom plots corresponds to the distance z = 2λ ($a_f \simeq 1.4\lambda$), and the top plots to z = 10λ ($a_f \simeq 3.2\lambda$).

As is apparent from Figs.3.5,6, geometrical optics yields more accurate approximations for smoothly varying distributions of amplitude because these have a smaller transverse gradient of the amplitude and, conse-

---

[7] These patterns may, of course, refer also to the near field of the plane aperture antenna.

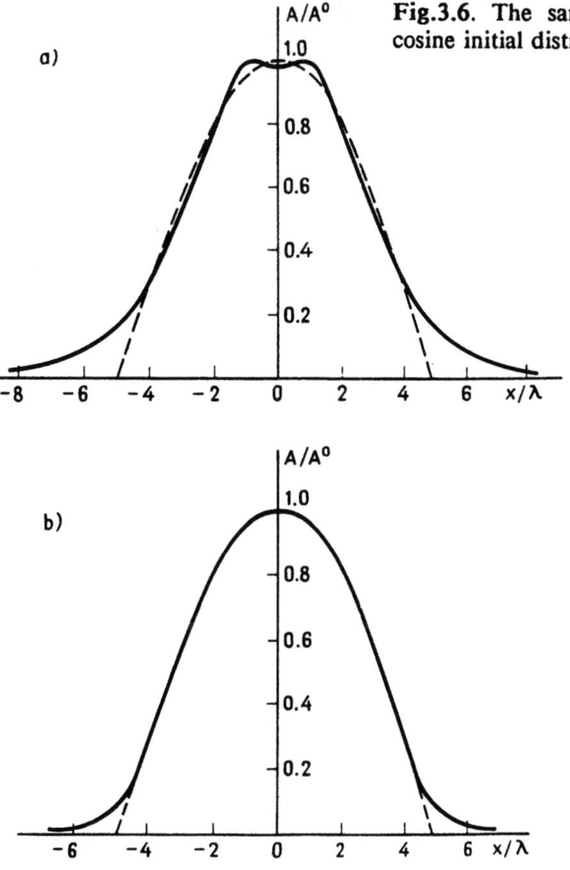

Fig.3.6. The same as in Fig.3.5 but for the cosine initial distribution $A^0(x) = A_0 \cos(\pi x/2b)$

quently, a smaller transverse diffusion. Obviously, the accuracy of geo-metrical-optics computations decreases (diffraction effects become more pronounced) as the distance z from the initial aperture increases, that is, as the observation point travels farther away from the near field to the inter-mediate field.

If the initial phase changes in the $z = 0$ plane, the field structure of the wave field differs markedly from those displayed in Figs.3.5,6. The near field of the beam now is substantially influenced by the beam diver-gence and the field focusing on caustics (3.1.24). These effects were stud-ied, for example, in [3.9, 11, 12].

3.1.8 On the Phase Center of an Antenna or a Scatterer

The far field of an antenna or a scatterer is given by

$$u(r, \theta, \phi) = \frac{e^{ik_0 r}}{r} F(\theta, \phi) e^{if(\theta, \phi)} , \qquad (3.1.35)$$

where r, θ, and ϕ are the spherical coordinates, and F and f are the given amplitude and phase functions, respectively. As the function Fe^{if} is complex, the resultant phase of the field (3.1.35) depends on θ and ϕ, i.e.,

$$\Phi = k_0 r + f(\theta, \phi) \equiv \Phi(r, \theta, \phi) \ . \tag{3.1.36}$$

As is stated in [3.8, 13, 14], the field (3.1.35) has a *phase* center situated at $r_p = \{x_p, y_p, z_p\}$ only if

$$f(\theta, \phi) = k_0(1 \cdot r_p) = k_0(x_p \cos\phi + y_p \sin\phi)\sin\theta + k_0 z_p \cos\theta \tag{3.1.37}$$

(at $f \equiv 0$, the phase center coincides with the origin $r = 0$). At $f \neq k_0(1 \cdot r_p)$ the equiphase surfaces $\Phi = $ constant are not spherical and a common (for all θ and ϕ) phase center does not exist. As we have seen above, when the equiphase surface (phase front) deviates from spherical shape, it brings about astigmatism - there appear two caustics formed by the intersection of rays (normal to the wave front) belonging to the principal normal cross sections of the front.[8] These caustics may be thought of as two *apparent (partial, local)* phase centers consistent with these sections and a given direction θ, ϕ; their poisiton is, of course, a function of θ and ϕ. In an arbitrary (not principal and normal) section, normals to the front (rays) cannot intersect and do not form any partial phase center.

Therefore, instead of a phase center for the field (3.1.35), in the general case one should consider two "phase caustics" whose geometry and position can be determined from (3.1.16). Obviously, in a vector wave field, phase caustics are different for different field components.

3.1.9 Field Near a Lens Focus

Consider a lens of an aperture 2b and a focal length F. Such a lens generates a convergent wave of a radius of curvature F which corresponds to the following initial eikonal distribution:

$$\psi^0(\rho) = -\sqrt{\rho^2 + F^2} \ ,$$

where $\rho = \sqrt{\xi^2 + \eta^2}$ is the cylindrical coordinate in the plane of the lens, $z = 0$.

Although geometrical optics is incapable of deriving the field at the focus directly, it can be used to produce estimates of the field, working on the grounds set forth in Sect.2.10.10. First, we evaluate the size of the focal spot by means of the criteria of geometrical-optics applicability (Sect.2.10. 4). To this end we shall seek the radius of the Fresnel zone a_f in the plane of the lens.

Let our observation point z be placed on the lens axis so that $z < F$. The quantity a_f can be found from (2.10.16), setting there $n = 1$, $R_2 = z$,

---

[8] We recall that the position of two principal normal cross-sections of a surface varies from point to point

and taking the quantity -F for R_1, which now plays the role of a negative radius of curvature of the converging spherical wave. This results in

$$a_f = \sqrt{\lambda z F/(F-z)} \, . \tag{3.1.38}$$

This formula also follows from the equation of the Frensel volume boundary (2.10.13). The shape which the Fresnel volume may take - among others for observation points beyond the lens focus - was examined more closely in [3.15, 16].

As $z \to F$, the radius a_f tends to infinity and the whole plane $z = 0$ becomes the region from which the focus can collect rays. When a_f increases as a result of $z \to F$, the validity criteria (2.10.18) fail, and the geometrical-optics approximation becomes invalid when a_f equals the lens radius b.

Let $\zeta = F-z$ be the distance between the observation point z and the focus. For geometrical optics to be applicable we require that $a_f \leq b$. Then we arrive that the restriction on ζ

$$\zeta \geq \lambda(F/b)^2 \equiv l_\| \, . \tag{3.1.39}$$

The expression $l_\| = \lambda F^2/b^2$ defines the longitudinal spread of the focal spot. The ray AF traveling from the rim of the lens to the focus (Fig.3.7) meets the plane $z = F-\zeta$ at point B at the distance

$$l_\perp = b l_\|/ F = \lambda F/b \tag{3.1.40}$$

from the lens axis. This quantity characterizes the transverse dimension of the focal spot.

We now evaluate the field u_F in the lens focus. The geometrical-optics approximation (3.1.20,23) relates the field amplitude A(z) at a plane z = constant to the initial field amplitude A^0 in the plane $z = 0$ by

$$A(z) = A^0 \left[\frac{F}{F-z} \right]^m = A^0 \left(\frac{F}{\zeta} \right)^m , \tag{3.1.41}$$

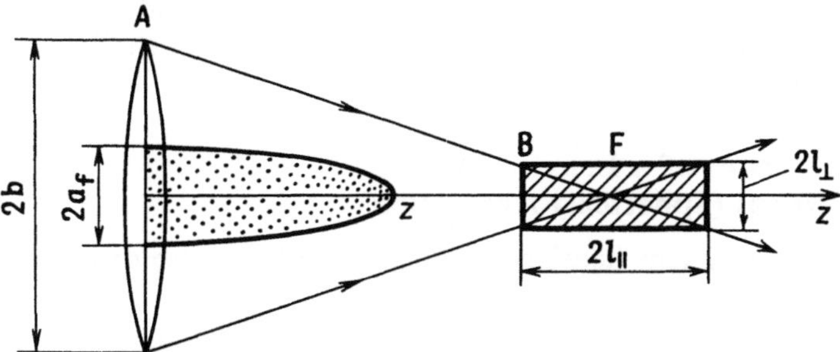

Fig.3.7. Illustration of the estimation of the transverse dimension $(l_\perp)$ and longitudinal spread $(l_\perp)$ of the focal spot (*hatched*)

where m = 1 for three-, and m = $\frac{1}{2}$ for two-dimensional problems. Putting our observation point z at the boundary of the focal spot, i.e., letting $\zeta = l_\parallel$, we get

$$A = A^0(F/l_\parallel)^m = A^0(b^2/\lambda F)^m \ . \tag{3.1.42}$$

By virtue of Sect.2.10.10, this is the value of the amplitude that may be adopted as the sought estimate of the field in the focus

$$|u_F| \sim A^0\left(\frac{b^2}{\lambda F}\right)^m = \begin{cases} A^0 b^2/\lambda F \\ A^0 b/\sqrt{\lambda F} \end{cases} \quad \begin{array}{l} \text{in 3-D} \\ \text{in 2-D} \end{array} \tag{3.1.43}$$

(the upper line refers to the three-dimensional focus, and the lower line to the two-dimensioanl one).

We could arrive at the same estimate by adopting the energy-flow approach (Sect.2.10.10). Equating the energy flows through the lens $(A^0)^2\pi b^2$ and the focal spot $|u_F|^2\pi l_\perp^2$ (we assume the energy flow in the lens focus to be uniformly distributed over the circle of radius $l_\perp$) we have for the three-dimensional problem

$$(A^0)^2\pi b^2 = |u_F|^2\pi l_\perp^2 = |u_F|^2\pi(\lambda F/b)^2 \ . \tag{3.1.44}$$

Likewise, for a two-dimensional problem (cylindrical lens), we obtain

$$(A^0)^2 = |u_F|^2 l_\perp = |u_F|^2\lambda F/b \ . \tag{3.1.45}$$

The estimates of $|u_F|$ following from (3.1.44, 45) are equivalent to (3.1.43) in both cases.

The estimates (3.1.43) agree well with the wave fields in the focus u(F) predicted by wave theory (the Kirchhoff-Huygens integral) [3.15, 16]:

$$|u(F)| = \begin{cases} A^0\pi b^2/\lambda F \\ A^0 2b/\lambda F \end{cases} \ . \tag{3.1.46}$$

We stress that the values (3.1.46) refer to the center of the focal spot (to the core of the focal point), whereas estimates (3.1.43) refer to the boundary of the focal spot.

3.1.10 Field at the Focus of a Lens with Cylindrical (Spherical) Aberration

Consider now a focusing system (lens) generating at z = 0 a phase distribution being quadratic in ρ:

$$u^0(\rho) = \exp[ik_0\psi^0(\rho)] \ , \quad \psi^0(\rho) = -\rho^2/2F \ . \tag{3.1.47}$$

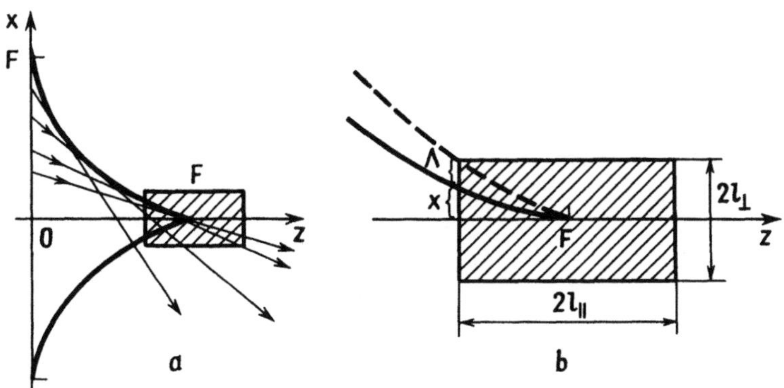

Fig.3.8. (a) Position of the focal spot for a lens with cylindrical (spherical) aberration. (b) The transverse dimension of the spot $l_\perp$ is comparable with the halfwidth of the caustic cusp x and with the width of the caustic zone Λ

With an initial phase distribution like this, the rays described by (3.1.17) form a caustic which, according to (3.1.24), has the shape of an astroid (Fig.3.8a)

$$r^{2/3} + z^{2/3} = F^{2/3} \ . \tag{3.1.48}$$

The cusp point of the caustic with the coordinates $r = 0$, $z = F$ is the focus of the lens in the paraxial approximation ($r \ll F$). Caustic (3.1.48) suggests that *spherical aberration* is present in the focusing system. In an analogous two-dimensional system (in a cylindrical lens) this leads to our saying that *cylindrical aberration* takes place.

Now we evaluate the field at the lens focus with spherical (cylindrical) aberration present, keeping the same computational procedure as in the absence of abberations (Sect.3.1.9). Let the observation point be in the lens focus. We determine the radius a_f of the first Fresnel zone immediately at the lens exit (i.e., at the $z = 0$) from (2.10.13) by letting there

$$\psi_{\text{virt}} = \sqrt{a_f^2 + F^2} - a_f^2/2F \ , \quad \psi_{\text{ref}} = F \ ,$$

which leads to

$$\sqrt{a_f^2 + F^2} - a_f^2/2F - F - \lambda/2 = 0 \ . \tag{3.1.49}$$

If $a_f \ll F$, then from (3.1.49) it follows

$$a_f = (4\lambda F^3)^{1/4} \ . \tag{3.1.50}$$

Unlike the strictly spherical wave (Sect.3.1.9), when to the lens focus there corresponds an infintely long Fresnel zone on the screen, now the Fresnel zone is of finite dimension even when the observation point is in the focus.

The central area of the lens within a radius of order a_f can be viewed as an ideal lens of the effective radius $b_{eff} \simeq a_f$. With this assumption in mind, we can estimate the width of the focal spot $2l_\perp$ in the plane $z = F$, the spread of the spot $2l_\parallel$ along the z-axis, and the magnitude of the field in the focus $|u_F|$, by substituting into (3.1.39, 40, 43) the radius $b_{eff} \simeq a_f = (4\lambda F^3)^{1/4}$ for the lens radius b. This leads to

$$l_\parallel \simeq \frac{1}{2} \sqrt{\lambda F} ,$$

$$l_\perp \simeq (\lambda^3 F/4)^{1/4} \simeq 0.7(\lambda^3 F)^{1/4} , \tag{3.1.51}$$

$$|u_F| \simeq \begin{cases} A^0(4F/\lambda)^{(1/2)} \\ A^0(4F/\lambda)^{1/4} , \end{cases} \tag{3.1.52}$$

where the upper line refers to the three-dimensional problem and the lower line to the two-dimensional one.

Estimates (3.1.51, 52) can be corroborated by comparing them with the calculations based on the Kirchhoff-Huygens integral. For example, the computations of the field at the lens focus give [3.15]

$$|u_F| = \begin{cases} \pi A^0(2F/\lambda)^{1/2} = 2.2A^0(4F/\lambda)^{1/2} \\ 1.8A^0(4F/\pi\lambda)^{1/4} = 1.35\, A^0(4F/\lambda)^{1/4} . \end{cases} \tag{3.1.53}$$

Hence the exact value of the field in the focus (3.1.53) exceeds the energy-flow estimate (3.1.52) by 2.2 times for the three-dimensional problem and by 1.35 times for the two-dimensional one.

The given estimates of the caustic-zone size and the field $|u_F|$ inside of this zone agree well with the asymptotic expressions for the field near the cusp of the caustic (3.1.48). To demonstrate this, the two-dimensional wave field is described by the Pearsey integral [3.17]

$$I(X, Y) = \int_{-\infty}^{\infty} \exp[i(Yt + Xt^2 + t^4)]dt , \tag{3.1.54a}$$

where

$$X = \tilde{x}\left[\frac{12\pi}{\lambda\delta}\right]^{1/2} ,$$

$$\tag{3.1.54b}$$

$$Y = \tilde{y}\left[\frac{192\pi^3}{\lambda^3\delta}\right]^{1/4} ,$$

and δ is the parameter determined from the equation of a caustic near the cusp:

$$\tilde{x}^3 = -9\delta\tilde{y}^2/8 \ .$$

The caustic equation (3.1.48) in two-dimensional space ($r \rightarrow x$) near the paraxial focus $z = F$ can be written in the form

$$x \simeq (2/3)^{3/2} F^{-1/2}(F - z)^{3/2} \ , \tag{3.1.55a}$$

whence

$$(z - F)^3 = \frac{27}{8}Fx^2 \ , \quad \delta = 3F \ , \tag{3.1.55b}$$

with $x = z - F$ and $y = x$.

Using these results, we can determine, for example, the dimensions of the focal spot, $l_\perp$ and $l_\parallel$. We shall look for them as the dimensions of the region where the geometrical optics approximation is invalid. Figure 3.9 illustrates how the wave field varies both along the symmetry axis of the caustic structure and perpendicular to it. Solid lines represent the results of wave theory, derived by the numerical computation of the functions $|I(X,0)|$ and $|I(0,Y)|$ [3.17], and dashed lines give the results of geometrical optics. As is readily apparent from this figure, the longitudinal (for $Y = 0$) and transverse (for $X = 0$) dimensions of the invalidity domain are defined, respectively, by the condition $|X| \leq 2$ and $|Y| \leq 4$ (the respective accuracy of fit is better than 10%). This gives further

$$(192\pi^3/\lambda^3\delta)^{1/4}|x| \leq 4 \ , \tag{3.1.56}$$

$$(12\pi/\lambda\delta)^{1/2}|z-F| \leq 2 \ .$$

Noting that $\delta = 3F$ and taking the equalities in (3.1.56) we find for $l_\perp = |x|$ and $l_\parallel = |z - F|$

$$l_\perp = \left[\frac{4}{\pi^3}\lambda^3 F\right]^{1/4} \simeq 0.6(\lambda^3 F)^{1/4} \ ,$$

$$\tag{3.1.57}$$

$$l_\parallel = \left[\frac{1}{\pi}\lambda F\right]^{1/2} \simeq 0.56\sqrt{\lambda F} \ .$$

These values are in good agreement with (3.1.51).

It is worth noting that the estimates (3.1.57) and, consequently, (3.1.51) match with the estimate of the width from the ray-optics invalidity domain over the nonsingular region of the caustic (3.1.48). Considering that the radius of curvature of the caustic (3.1.48) is $R_c = 3(xzF)^{1/3}$, we

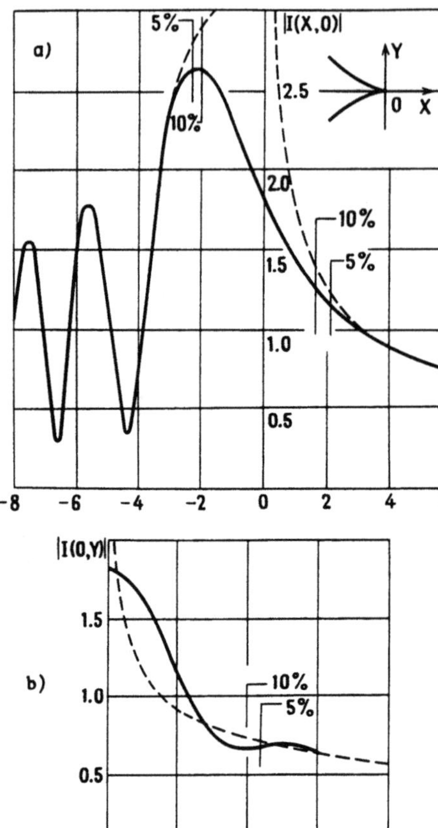

Fig.3.9. Wave field variation (a) along the X axis of symmetry of the caustic cusp and (b) in the perpendicular direction Y. The solid lines represent the numerical computations of the integral (3.1.54a) while the dashed line show the estimates by means of ray theory. The arrows point towards the regions where ray theory is in error by respectively not more than 5% and 10%

obtain near the focus, i.e., for small $F-z$, $R_c \simeq [6F(F-z)]^{1/2}$. This leads to the quantity Λ at the edge of the focal spot, i.e., at $|F-z| = l_{\parallel} = \sqrt{(\lambda F)}/2$

$$\Lambda = (R_c/2k_0^2)^{1/3} \simeq 3^{1/6}(\lambda^3 F)^{1/4}/2\pi^{2/3} \simeq 0.28(\lambda^3 F)^{1/4} \ .$$

The distance of the caustic point, where the caustic zone width has been determined, from the z-axis is obtained from (3.1.55a):

$$x \simeq (\lambda^3 F)^{1/4}/3\sqrt{3} \simeq 0.19(\lambda^3 F)^{1/4} \ .$$

As a result, for the transverse spread of the focal spot (Fig.3.8b) we find

$$l_{\perp} \simeq x + \Lambda \simeq 0.47 \, (\lambda^3 F)^{1/4} \ ,$$

which agrees, in the order of magnitude, with both (3.1.51 and 57). This example illustrates that caustic volumes even for complex caustics can be estimated by the simple relationships for the width of the caustic zone of a nonsingular caustic $\Lambda = (k_0^2 \beta)^{-1/3}$.

120

3.2 Reflection and Refraction at an Interface Between Homogeneous Media

3.2.1 Reflection Formulas

The problem of reflection and refraction of an arbitrary wave at a curvilinear interface between two homogeneous media has been solved in the general form by *Fock* [3.18,19] and considered later in [3.7,20-28]. The ray geometry in the presence of interfaces has been examined in many papers on instrumental geometrical optics [3.1,3,29-34]. In situations with mulitple interfaces, matrix computations have proved valuable for ray tracing [3.23,32]. Matrix techniques have also been employed to derive wave fields [3.23,24].

The reflection formulas of Sect.2.5, for homogeneous media (Sect.3.1) assume the form

$$u_r(r) = \Gamma(\xi,\eta) A_0^{\,i}(\xi,\eta)\, \mathscr{J}_r^{\,-1/2} \exp\{ik_0[\psi_i(\xi,\eta) + \sigma n_1]\} \ , \qquad (3.2.1a)$$

$$u_t(r) = D(\xi,\eta) A_0^{\,i}(\xi,\eta)\, \mathscr{J}_t^{\,-1/2} \exp\{\psi_i(\xi,\eta) + \sigma n_2]\} \ , \qquad (3.2.1b)$$

where ξ, η are curvilinear coordinates on the interface Q; $A_0^{\,i}$ and ψ_i are the amplitude and eikonal of the incident wave at the interface; Γ and D are the Fresnel reflection and transmission coefficients (2.5.15,16), respectively; $\mathscr{J}_r$ and $\mathscr{J}_t$ are the divergences (3.1.7) of reflected and refracted rays; and $\sigma = |r-r^0|$ is distance along the ray measured from the interface $r = r^0(\xi,\eta)$. The higher-order approximations of geometrical optics are defined by (2.5.11) and (2.3.33); the higher-approximations for a number of specific situations have been studied in [3.35-38].

3.2.2 Divergence of Reflected and Refracted Rays

A family of rays reflected from the surface $r = r^0(\xi,\eta)$ is given by

$$r = r^0(\xi,\eta) + \sigma l_1(\xi,\eta) \equiv r_1(\xi,\eta,\sigma) \ , \qquad (3.2.2)$$

where $l_1(\xi,\eta)$ is the unit vector of the reflected ray, defined by the reflection law

$$l_1 = l_0 - 2N(l_0 \cdot N) = l_0 + 2N\cos\theta \ . \qquad (3.2.3)$$

Here $l_0 = l_0(\xi,\eta)$ is the unit vector of the incident ray, defined by the eikonal of the incident wave, $l_0 = \nabla\psi_i/n_1$; $N = N(\xi,\eta)$ is the unit vector normal to the interface, directed opposite the incident wave (Fig.3.10); and θ is the angle of incidence (reflection).

In order to evaluate the divergence $\mathscr{J}_r$ or reflected rays we need to calculate the Jacobian $\mathscr{D}(\sigma) = r_{1\xi} \cdot l_{1\eta}$. Similar to (3.1.7) we may readily obtain

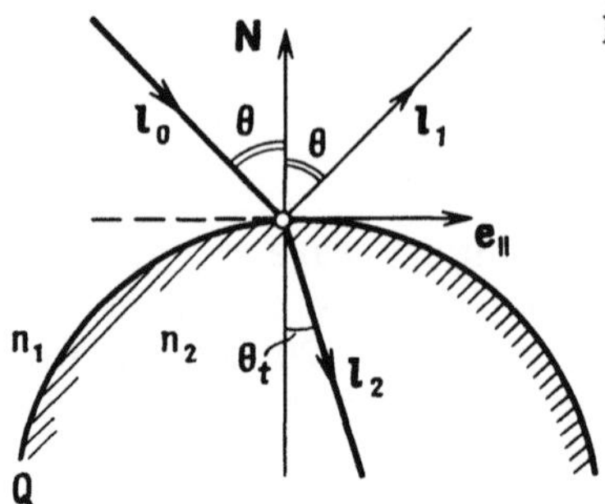

Fig.3.10. Ray unit vector set at interface reflection

$$\mathscr{J}_r = \frac{\mathscr{D}(\sigma)}{\mathscr{D}(0)} = \frac{a}{c}\sigma^2 + \frac{b}{c}\sigma + 1 \equiv K_r\sigma^2 + 2H_r\sigma + 1 \ , \qquad (3.2.4a)$$

where

$$a = (l_1 \cdot l_{1\xi}) \cdot l_{1\eta}$$

$$b = (l_1 \times r^0_\xi) \cdot l_{1\eta} + (l_1 \times l_{1\xi}) \cdot r^0_\eta$$

$$c = (l_1 \times r^0_\xi) \cdot r_\eta \ , \qquad (3.2.4b)$$

and $K_r = a/c$ and $H_r = -b/2c$ are, respectively, the total and mean curvature of the reflected front whose principal radii of curvature are

$$R_{1,2} = (-b \pm \sqrt{b^2 - 4ac})/2a = \left(H_r \pm \sqrt{H_r{}^2 - K_r}\right)/K_r \ . \qquad (3.2.5)$$

In a similar manner we can find the divergence of refracted (transmitted) rays described by the equation

$$r = r^0(\xi,\eta) + \sigma\, l_2(\xi,\eta) \ , \qquad (3.2.6)$$

where l_2 is the unit vector of the refracted ray (Fig.3.10):

$$l_2 = -\cos\theta_t\, N + \sin\theta_t\, e_{\parallel} = (n_1 l_0 - M_{21} N)/n_2 \ , \qquad (3.2.7)$$

θ_t is the refracted angle according to Snell's law (2.5.6), $e_\parallel$ is the unit vector tangent to the interface and lying in the plane of incidence, and $M_{21} = n_2\cos\theta_t - n_1\cos\theta$. Comparing the ray equations (3.2.6) and (3.2.2) suggests that to find the divergence

$$\mathscr{J}_t = K_t\sigma^2 - 2H_t\sigma + 1 \qquad (3.2.8)$$

and principal radii of curvature for the refracted front we may invoke (3.2.4,5) where l_2 should be substituted for l_1.

122

Formulas (3.2.4,5) and their counterparts for the refracted wave, (3.2.8), complete, in principle, the geometrical part of the problem - the evaluation of the divergence of reflected and refracted rays - thus paving the way for obtaining the refracted and reflected fields (3.2.1). Physical studies, though, find it more convenient to deal with the expressions in invariant form following from (3.3.2.4,5) and providing the divergences $\mathscr{J}_r$ and $\mathscr{J}_t$ via the radii of curvature of the interface and the incident wavefront. The solution of the general problem has been outlined in [3.20-24].

For the sake of simplicity, instead of an arbitrary wave[9] we consider here a spherical wave generated by a source in point $\mathbf{r}_s$. For this wave

$$\mathbf{l}^0 = [\mathbf{r}^0(\xi,\eta) - \mathbf{r}_s]/\left|\mathbf{r}^0(\xi,\eta) - \mathbf{r}_s\right| = \mathbf{R}_0/R_0 , \tag{3.2.9}$$

where R_0 is the distance from the source to the point of incidence $\mathbf{r}^0(\xi,\eta)$ of the ray. For the coordinate lines ξ = constant and η = constant on the interface, it is convenient to choose (as in Sect.3.1.2) the lines of curvature because they simplify the calculation of the coefficients a, b, and c with (3.2.3). Dropping the uncomplicated, though cumbersome, manipulations we present the resultant expressions for the complete and mean curvatures of the reflected and refracted rays, which follow from (3.2.4):

$$K_r = 4\tilde{K} - 2\cos\theta(2\tilde{H} + \tilde{\kappa}\tan^2\theta)/R_0 + 1/R_0^2 , \tag{3.2.10a}$$

$$H_r = \cos\theta(2\tilde{H} + \tilde{\kappa}\tan^2\theta) - 1/R_0 , \tag{3.2.10b}$$

$$K_t = \frac{n_1^2}{n_2^2\cos^2\theta_t}\left[\frac{1}{n_1^2}M_{21}^2\tilde{K} + \frac{\cos^2\theta}{n_1 R_0}M_{21}(2\tilde{H} + \tilde{\kappa}\tan^2\theta) + \frac{\cos^2\theta}{R_0^2}\right] ,$$

$$\tag{3.2.11a}$$

$$H_t = \frac{-1}{2n_2\cos^2\theta}\left[M_{21}\cos^2\theta_t(2\tilde{H}+\tilde{\kappa}\tan^2\theta) + \frac{1}{R_0}n_1(\cos^2\theta+\cos^2\theta_t)\right] .$$

$$\tag{3.2.11b}$$

Here $\tilde{K} = 1/a_1 a_2$ is the total (Gaussian) curvature of the interface surface at the reflection point; $\tilde{H} = (1/a_1+1/a_2)/2$ is its mean curvature, $\tilde{K}$ is the curvature of the interface Q in the normal section of the incidence plane: $\tilde{\kappa} = (\cos^2\alpha)/a_1+(\sin^2\alpha)/a_2$, α is the angle between the plane of incidence and that of the first, principal normal section; and a_1 and a_2 are the principal radii of curvature of the interface Q ($a_1 > 0$, $a_2 > 0$ when the lines of curvature on Q bend toward $\mathbf{N}$, whereas for a convex surface, as that in Fig.3.10, $a_1 < 0$ and $a_2 < 0$). Formulas (3.2.10,11) may be used for a plane incident wave by letting $R_0 \to \infty$.

---

[9] The problem of the incidence of an arbitrarily shaped wave is of special value for studies of multiple scattering on rough surfaces [3.22] and on complex-shaped bodies [3.23, 39,40].

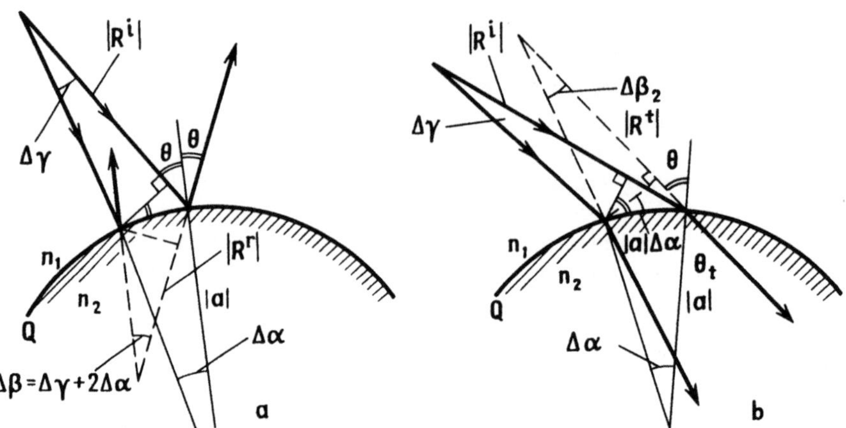

Fig.3.11. Illustrating the derivation of (3.2.12) for the parameters of (a) reflected and (b) refracted (transmitted) waves

In a two-dimensional (planar) problem, (3.2.10,11) yield as corollaries the formulas for divergences which are valid for an arbitrary incident wave

$$\mathscr{J}_r = 1 - \sigma/R^r , \tag{3.2.12a}$$

$$\mathscr{J}_t = 1 - \sigma/R^t ,$$

$$1/R^r = 1/R^i + 2/a\cos\theta , \tag{3.2.12b}$$

$$1/R^t = (n_1\cos^2\theta/R^i - M_{21}/a)n_2\cos^2\theta_t ,$$

where R^i, R^r, and R^t are the radii of curvature of the incident, reflected, and refracted wave front, respectively (for a convex front, $R < 0$; for example, for a spherical wave, $R^i = -R_0$, $R_0 > 0$). Formulas (3.2.12) easily result also from geometrical constructions shown in Fig.3.11, where we should observe that $\Delta\beta = \Delta\gamma+2\Delta\alpha$,

$$|a|\Delta\alpha = |R^i|\Delta\gamma/\cos\theta = |R^r|\Delta\beta/\cos\theta = |R^t|\Delta\beta_2/\cos\theta_t ,$$

and the angles θ and θ_t are related by the refractive law (2.5.6,3.1.57).

In close similarity to (3.1.15,3.2.12) for an axially symmetric interface and an incident front of the same symmetry, we have

$$\mathscr{J}_r = \frac{r}{\rho}(1 - \sigma/R^t) , \tag{3.2.13}$$

$$\mathscr{J}_t = r/\rho (1 - \sigma/R^r) ,$$

where r and ρ are the cylindrical coordinates of the observation and incidence points, respectively, and R^r and R^t are defined in the meridional section by (3.2.12).

124

3.2.3 Effective Scattering Surface of a Body
   in the Geometrical-Optics Approximation

Consider the reflection of a plane wave from an arbitrary interface and evaluate the field at large distances when $\sigma \gg |a_{(1,2)}|$. Noticing that by virtue of (3.2.4a, 10a) the divergence is $\mathcal{J}_r \simeq K_r \sigma^2 = 4K\sigma^2 = 4\sigma^2/a_1 a_2$, we find from (3.2.1a) the amplitude of the reflected far field

$$A_0^r \simeq \frac{1}{2\sigma}\Gamma A_0^i \sqrt{a_1 a_2} \simeq \frac{1}{2r}\Gamma A_0^i \sqrt{a_1 a_2} \; , \tag{3.2.14}$$

where r is distance from the origin. For the effective scattering surface defined as

$$\Sigma = \lim_{r \to \infty} 4\pi r^2 |A^r/A^i|^2 \; , \tag{3.2.15}$$

we obtain then the following geometrical-optical representation [3.41, 42]

$$\Sigma = \pi |a_1 a_2||\Gamma|^2 \; , \tag{3.2.16}$$

where the main radii of curvature of the interface, a_1 and a_2, and the reflection coefficient Γ are taken at the point of incidence (reflection) of the ray. For an ideal reflector ($|\Gamma| = 1$) we have $\Sigma = \pi |a_1 a_2|$. In a specific case of sphere ($a_1 = a_2 = a$) we get $\Sigma = \pi a^2$.

For a cylindrical reflector, when the effective scattering surface is given by

$$\Sigma = \lim_{r \to \infty} 2\pi r |A^r/A^i|^2 \; , \tag{3.2.17}$$

from (3.2.1, 12) we obtain $\Sigma = \pi |a| \cos\theta |\Gamma|^2$. In a particular case of back-scattering ($\theta = \pi$) from an ideal reflector we have $\Sigma = \pi a$.

3.2.4 Reflection Far Field of a Directional Point Source

This field can be evaluated in a manner similar to that for (3.2.15)

$$u_r \simeq r^{-1} \exp[ik_0(r + \psi_i)]F_r \; ,$$
$$F_r = \Gamma A_0^i K_r^{-1/2} \; . \tag{3.2.18}$$

Here K_r is given by (3.2.10a). For a directional point source at point r_s

$$A_0^i = F_0(l_0)/R_0 \; ,$$
$$\psi_i = R_0 + f_0(l_0) \; , \tag{3.2.19}$$

where F_0 and f_0 are the functions describing the amplitude and phase radiation patterns (in infinite space), respectively, and $R_0 = |r^0 - r_s|$. Then (3.2.18, 19) yield the radiation pattern of the reflected field

$$F_r(l_1) = \Gamma F_0(l_0)\left[1 - 2R_0\cos\theta(2\tilde{H} + \tilde{\kappa}\tan^2\theta) + 4R_0^2\tilde{K}\right]^{-1/2}, \qquad (3.2.20)$$

where l_1 is the unit vector (3.2.3) of the reflected ray. Evidently, the resultant field of the source equals the sum of all rays reaching the observation point (for example, the primary and reflected).

In the two-dimensional problem, using (3.2.12a) we readily obtain

$$F_r(l_1) = \Gamma F_0(l_0)[a\cos\theta/(a\cos\theta - 2R_0)]^{1/2}. \qquad (3.2.21)$$

The above stated formulas are widely used in design of effective scattering surfaces and radiation patterns for antennas, especially in case of complex-shaped bodies [3.40-43].

3.2.5 Caustics of Refracted and Reflected Rays

The ability of a curvilinear interface to focus rays is used to advantage in optical devices. The relevant geometry of reflected and refracted rays has been extensively developed in optical-instrument research [3.1, 3, 29–34]. In dealing with an interface it is covenient to distinguish between the caustics of reflected rays, called *catacaustics*, and those of refracted rays, called *diacaustics*. Like in an infinite homogeneous medium (Sect.3.1), in 3-D space a reflected front forms two catacaustics, while a refracted front two diacaustics.[10]

The equations of both catacaustics are defined from (3.2.2) for $\sigma = R_1$ and $\sigma = R_2$, where $R_{1,2}$ are given by (3.2.5). Diacaustics are described by (3.2.6) for $\sigma = R_{1,2}$, where $R_{1,2}$ are given by the expressions similar to (3.2.5). The conditions $R_{1,2} > 0$ give rise to real-valued caustics, while $R_{1,2} < 0$ to imaginary ones. In the two-dimensional problem the equations of a catacaustic and a diacaustic become

$$r = r^0(\xi) + R^r l_1 \equiv r_{cc}(\xi), \qquad (3.2.22a)$$

$$r = r^0(\xi) + R^t l_2 \equiv r_{dc}(\xi), \qquad (3.2.22b)$$

where R^r and R^t are yielded by (3.2.12). For a specific case of a plane incident wave, when $R^i \to \infty$, (3.2.22) yields for the catacaustic

$$x = x^0(\xi) + \tfrac{1}{2}a(\xi)\cos\theta\cos2\theta \equiv x_{cc}(\xi), \qquad (3.2.23a)$$

$$y = y^0(\xi) - \tfrac{1}{2}a(\xi)\cos\theta\sin(2\theta)\,\mathrm{sgn}(dy^0/dx^0) \equiv y_{cc(\xi)}, \qquad (3.2.23b)$$

---

[10] In an axially symmetric problem, one of the caustics degenerates into the axis of symmetry (Sect.3.1).

and for the diacaustic

$$x = x^0(\xi) + a(\xi)\cos^2\theta_t(n_1/M_{21} + \cos\theta) \equiv x_{dc}(\xi) , \qquad (3.2.24a)$$

$$y = y^0(\xi) - a(\xi)\cos^2\theta_t\sin\theta\,\mathrm{sgn}(dy^0/dx^0) \equiv y_{dc}(\xi) , \qquad (3.2.24b)$$

where, for convenience, it is assumed that the x axis is directed opposite to the incident wave, that is, $l_0 = -e_x$. equations (3.2.23), notably, corroborate an obvious geometrical fact that it is a concave boundary ($a > 0$) that focuses a parallel ray pencil on a caustic. The condition for a cusp to form on a catacaustic (3.2.23), consistent with the requirement $dx_{cc}/d\xi = 0$, has the form

$$3\tan\theta + \frac{d}{d\xi}a(\xi) = 0 \qquad (3.2.25a)$$

or, recognizing that $1/a = -d\theta/d\xi$,

$$\frac{d}{d\xi}(a\cos^3\theta) = 0 . \qquad (3.2.25b)$$

3.2.6 Examples of Catacaustics and Diacaustics

A point source irradiating a close interface generates only an imaginary catacaustic the equation of which, according to (3.2.24), has the form

$$(x/b_1)^{2/3} + (y/b_2)^{2/3} = 1 , \qquad (3.2.26)$$

where $b_1 = x_s n_2/n_1$ and $b_2 = b_2(1-n_2^2/n_1^2)^{-1/3}$. Diacaustic (3.2.26) is an astroid with a cusp F at $y = 0$, $x = x_F = x_s n_2/n_1$, and for $n_1 < n_2$ the cusp will point toward the interface; whereas for $n_1 > n_2$, away from it (Fig.3.12). When $n_1 > n_2$, the diacaustic comes up the interface at the points $y =$

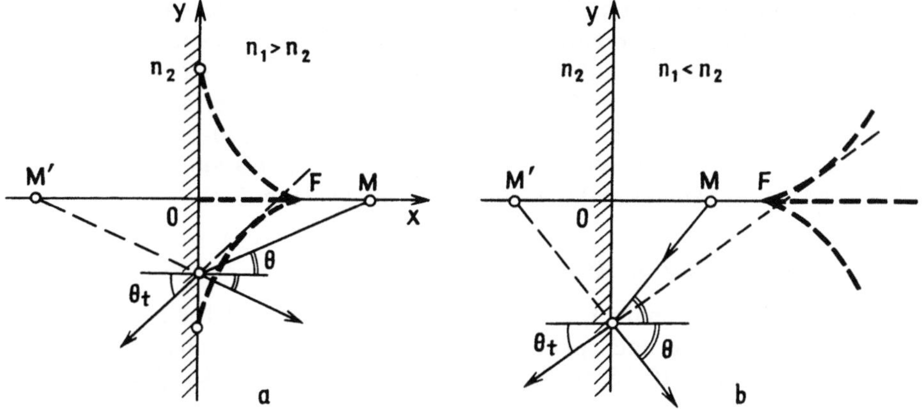

Fig.3.12. Caustics of refracted rays (diacaustics) occurring in the reflection of the point source field from the plane interface of two media with (a) $n_1 > n_2$, (b) $n_1 < n_2$

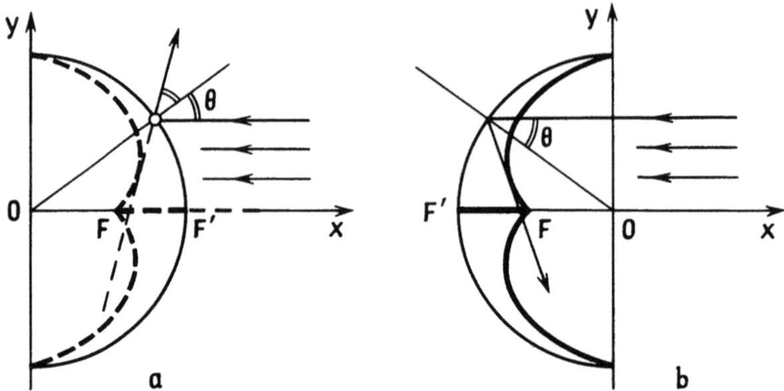

Fig.3.13. Caustics of reflected rays (catacaustics) emerging in reflection from (a) convex and (b) concave surface

$\pm b_2$, which correspond to the critical angle θ_{cr} = arcsin (n_1/n_2) (at $\theta = \theta_{cr}$ the refracted ray travels along the interface). It is well known [3.16] that for $\theta > \theta_{cr}$ the reflection is accompanied by a *lateral displacement* of rays, the amount of which is decided by the phase of the reflection coefficient Γ and which gives rise to a catacaustic. Its geometry was examined in [3.43] where an extensive discussion is given to the case of a plamsa of $\epsilon_2 \equiv n_2^2 < 0$.

For the refraction of a plane wave at a cylindrical (spherical) interface of radius a_0, the catacaustic equations (3.2.24) become

$$x = \pm (a_0/2)\cos\theta(1 + 2\sin^2\theta) \ ,$$

$$y = a_0 \sin^3\theta \ , \tag{3.2.27}$$

where the upper (lower) sign refers to a concave (convex) reflector. Caustic (3.2.27) is an epicycloid, built around a circle of radius $a_0/2$, with a cusp F on the x axis at $x = a_0/2$ (Fig.3.13a).[11] The caustic in Fig.3.13b corresponds to a cylindrical (spherical) aberration of the reflector, which must be taken into consideration in designing both optical devices [3.1,3,29-34] and one- or two-mirror antennas [3.44,45]. Caustics of refracted and reflected rays have been studied in, e.g., [3.1,3,34,46-49]. Geometry of rays and caustics arising in reflection and refraction at interfaces of shapes other than those discussed can be found in [3.1,34,35,50-52]. An important role of caustics should be noted for optical aberration theory [3.1,3, 29-34,53] and open-resonator theory [3.47,47,51] (Sect.3.7).

3.2.7 Applicability of Reflection Formulas

The applicability domain for (3.2.1) has been considered for a number of specific cases (plane and spherical interfaces, plane and spherical waves), for example, in [3.10,16,54,55]. Particularly, *Brekhovskikh* [3.16,55] has

---

[11] The segment FF' of the x-axis is an axial caustic for a spherical mirror only.

extensively studied the problem of irradiation of a point source under a plane interface surface.

The validity domain of the reflection formulas in the general case is specified by criteria I and II of Sect.2.10.4, and may be determined from (3.1.32), where A^0 should be viewed as the amplitudes of the reflected and refracted waves at the interface and $R_{1,2}$ as the principal radii of curvature of the reflected and refracted fronts.[12] Consider, for simplicity, (3.1.32) in the plane of wave incidence and assume that the principal sections of the interface and the fronts (with their radii of curvature a_1, R_1^i, R_1^r, and R_1^t) belong to this plane. According to (3.1.32) the applicability domain of the ray formulas (3.2.1) is restricted by the conditions

$$\frac{a_f}{\cos\theta}\left|\frac{1}{A_0^i}\frac{\partial A_0^i}{\partial\xi}\right| \ll 1 \;,$$

$$\frac{a_f}{\cos\theta}\left|\frac{1}{\Gamma}\frac{\partial\Gamma}{\partial\xi}\right| \ll 1 \;, \quad a_f \ll |R_1^r| \tag{3.2.28}$$

for the reflected field, and

$$\frac{\hat{a}_f}{\cos\theta_t}\left|\frac{1}{A_0^i}\frac{\partial A_0^i}{\partial\xi}\right| \ll 1 \;,$$

$$\frac{\hat{a}_f}{\cos\theta_t}\left|\frac{1}{D}\frac{\partial D}{\partial\xi}\right| \ll 1 \;, \quad \hat{a}_f \ll |R_i^t| \tag{3.2.29}$$

for the refracted field. The halfwidth of the Fresnel zone on the primary reflected (a_f) and refracted ($\hat{a}_f$) fronts is defined by (3.1.30) where R_1 is to be replaced, respectively, by R_1^r or R_1^t.

If in the absence of real caustics we require that (3.2.28,29) be satisfied along the entire ray ($\sigma \to \infty$), this will lead us to the following inequalities:

$$\sqrt{\frac{\lambda_0}{n_1}|R_1^r|}\,\left|\frac{1}{A_0^i}\frac{\partial A_0^i}{\partial\xi}\right| \ll \cos\theta,$$

$$\sqrt{\frac{\lambda_0}{n_1}|R_1^r|}\,\left|\frac{1}{\Gamma}\frac{\partial\Gamma}{\partial\xi}\right| \ll \cos\theta \;, \tag{3.2.30a}$$

---

[12] Similar to Sect.3.1, the validity criteria of geometrical-optics formulas in this case may be substantiated with the Fresnel integral technique.

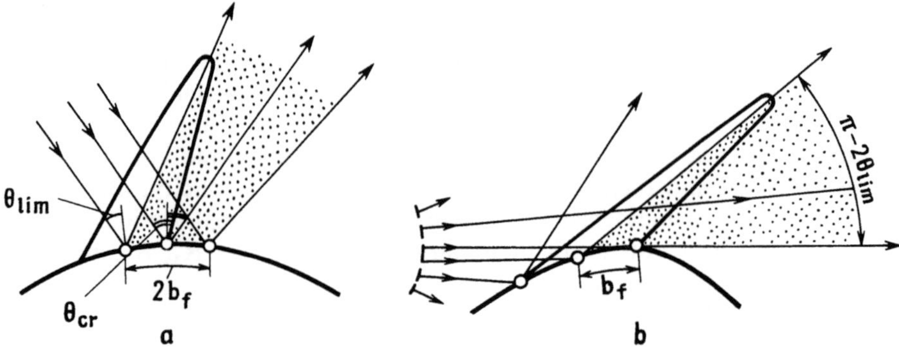

Fig.3.14. Invalidity domain of the reflection formulas (*stippled*) (**a**) in the vicinity of the critical angle and (**b**) in the neighborhood of the light-shadow boundary

$$\sqrt{\frac{\lambda_0}{n_2}|R_1^t|}\ \left|\frac{1}{A_0^i}\frac{\partial A_0^i}{\partial\xi}\right| \ll \cos\theta_t\ ,$$

(3.2.30b)

$$\sqrt{\frac{\lambda_0}{n_2}|R_1^t|}\ \left|\frac{1}{D}\frac{\partial D}{\partial\xi}\right| \ll \cos\theta_t\ ,$$

$$|R_1^r| \gg \frac{\lambda_0}{n_1}\ ,\quad |R_1^t| \gg \frac{\lambda_0}{n_2}\ ,$$

(3.2.30c)

where R_1^r and R_1^t are given by (3.2.12) of the two-dimensional problem. Conditions (3.2.30) have simple physical meaning. They require that the amplitude of the incident field A_0^i, the reflection coefficient Γ, and the transmission coefficient D vary sufficiently slowly along the interface, and the curvature of the reflected and refracted fronts be small (Sect.2.10).

From (3.2.30) it follows, in particular, that the validity conditions for the reflection formulas (3.2.1) fail near the rays tangent to a curvilinear interface ($\theta \simeq \pi/2$, $|R_1^r| \simeq 0$), close to the angle of total internal reflection [$\theta \simeq \theta_{cr1} = \arcsin(n_2/n_1)$, $n_2 < n_1$, $|D| \simeq 0$, $|\partial\Gamma/\partial\xi| \to \infty$], close to the Brewster angle of total refraction[13] ($\theta \simeq \theta_{cr2}$, $|\Gamma| \simeq 0$), and in the vicinity of zeros of the radiation pattern for the primary field ($|A_0^i| \simeq 0$). Following the procedure outlined in Sect.2.10 we can readily evaluate the size of invalidity domains in these cases. To do so we require that the neighborhood of the critical angle θ_{cr} be outside of the Fresnel volume of the considered ray (Fig.3.14a). Making use in addition of the expansion $\theta \simeq \theta_{cr}+(\partial\theta/\partial\xi)\Delta\xi$ at $|\Delta\xi| = b_f$, we may evaluate the *limiting angle* of incidence (reflection) bounding the angular sector within which the reflection formulas do not apply (in Fig.3.14a) this sector is shown by stippling):

---

[13] For electromagnetic waves, the Brewster angle exists only for the case of vertical polarization: $\sin\theta_{cr} = n_2(n_1^2 + n_2^2)^{-1/2}$. For the conditions on θ_{cr2} for sound waves, see [3.56].

$$\left|\theta_{\text{lim}} - \theta_{\text{cr}}\right| \simeq a_f \left|1/R_1^i + (1/a_1)\cos\theta\right| , \tag{3.2.31}$$

where on the right we are to set $\theta = \theta_{\text{cr}}$ if $\theta_{\text{cr}} \neq \pi/2$ (the case of $\theta_{\text{cr}} = \pi/2$ will be considered in Sect.3.2.8 below). For a plane incident wave ($R_1^i \rightarrow \infty$) (3.2.31) assumes the form

$$\left|\theta_{\text{lim}} - \theta_{\text{cr}}\right| \simeq \sqrt{\lambda_0/2n_1\left|a_1\right|\cos\theta_{\text{cr}}} . \tag{3.2.32}$$

A more detailed discussion of the results just derived is given in [3.15] where they are compared with the asymptotics of exact solutions. Two of the examples elaborated in [3.15] will be discussed below (see also [3.57]).

3.2.8 The Invalidity Domain in the Vicinity of a Tangent Ray

Looking for the invalidity region of the reflection formulas for substantially distant observation points ($\sigma \gg R_1^i$), we find from (3.1.30, 58), for $\theta \simeq \pi/2$, $n_1 = 1$, that

$$a_f{}^2 = \lambda_0\left|1/R_1^i - 1/\sigma\right| \simeq \lambda_0(\left|a_1\right|/2)\cos\theta ,$$

where instead of a_1 we may take the radius of curvature of the interface at its point of tangency with the ray, rather than at the reflection point (Fig.3.14b). At $\theta \simeq \pi/2$ we may let $\cos\theta \simeq \pi/2-\theta$; then from (3.2.31) it follows

$$\cos\theta_{\text{lim}} = (\lambda_0/2\left|a_1\right|)^{1/3} = (\pi/k_0\left|a_1\right|)^{1/3} . \tag{3.2.33}$$

To the angles of incidence (reflection) $\theta > \theta_{\text{lim}}$ there correspond the rays (Fig.3.14b) traveling within the invalidity region of the reflection formulas[14]

$$\cos\theta > (\pi/k_0\left|a_1\right|)^{1/3} , \tag{3.2.34}$$

which agrees with the condition $\cos\theta > (2/k_0\left|a_1\right|)^{1/3}$ derived by *Fock* with the help of the parabolic equation technique [3.19] and discussed in a number of subsequent papers [3.24, 46]. Note that the penumbra region, where geometrical optics is invalid, represents the region of effective transverse diffusion of the field [3.58, 59].

3.2.9 Wave Diffraction at a Surface of Variable Impedance

Surface impedance Z_s, varying over the surface of a body, brings about a change in the wave reflection coefficient of this body. Hence to make the reflection formulas applicable we should have to impose an additional condition in accord with (3.2.28), namely,

---

[14] Similar geometrical arguments for a specific case of a spherical interface have been given in [3.10,54].

$$\frac{a_f}{\cos\theta}\left|\frac{1}{Z_s}\frac{\partial Z_s}{\partial\xi}\right| \ll 1 \; . \tag{3.2.35}$$

For a plane incident wave, it transforms to

$$\left|\frac{1}{Z_s}\frac{\partial Z_s}{\partial\xi}\right| \ll \sqrt{\frac{\cos\theta}{\lambda_0}\left|\frac{\sigma - a\cos\theta}{\sigma a}\right|} \; , \tag{3.2.36a}$$

whence we have two limiting situations

$$\lambda_0\left|\frac{1}{Z_s}\frac{\partial Z_s}{\partial\xi}\right| \langle \ll \sqrt{\frac{\lambda_0}{|a|}}\cos\theta\; , \quad \sigma \,\rangle\gg |a|\cos\theta \qquad \text{or}$$

$$\lambda_0\left|\frac{1}{Z_s}\frac{\partial Z_s}{\partial\xi}\right| \ll \sqrt{\frac{\lambda_0}{\sigma}}\cos\theta\; , \quad \lambda_0 \ll \sigma \ll |a|\cos\theta \; . \tag{3.2.36b}$$

The last conditions agree, for the normal incidence ($\theta = 0$), with the results which *Felsen* obtained from an exact solution for a sinusoidal variation on the cylinder [3.60, 61].

Note in conclusion that in dealing with interface problems the possibilities of the ray method widen substantially if the treatment involves diffracted rays [3.7, 24, 62] (Sects. 2.2, 2.5 and Chap. 5).

3.3 Rays and Caustics in Plane-Stratified Media

3.3.1 Ray Equations

The model for a plane-stratified medium whose reflective index depends on a single Cartesian coordinate, say z, is often used to study the propagation of electromagnetic waves in the troposphere and ionosphere, of sound waves in the ocean or atmosphere, of elastic waves in the earth's crust, etc. For $n = n(z)$ the differential equations of the rays (2.1.34) take the form

$$\frac{dx}{d\tau} = p_x\; , \quad \frac{dy}{d\tau} = p_y\; , \quad \frac{dz}{d\tau} = p_z\; , \tag{3.3.1a}$$

$$\frac{dp_x}{d\tau} = 0\; , \quad \frac{dp_y}{d\tau} = 0\; , \quad \frac{dp_z}{d\tau} = \frac{1}{2}\frac{dn^2}{dz}\; . \tag{3.3.1b}$$

From these equations and the eikonal equation $p_x^2 + p_y^2 + p_z^2 = n^2$ it follows that

$$p_x = \text{const} = p_x^0\; ,$$

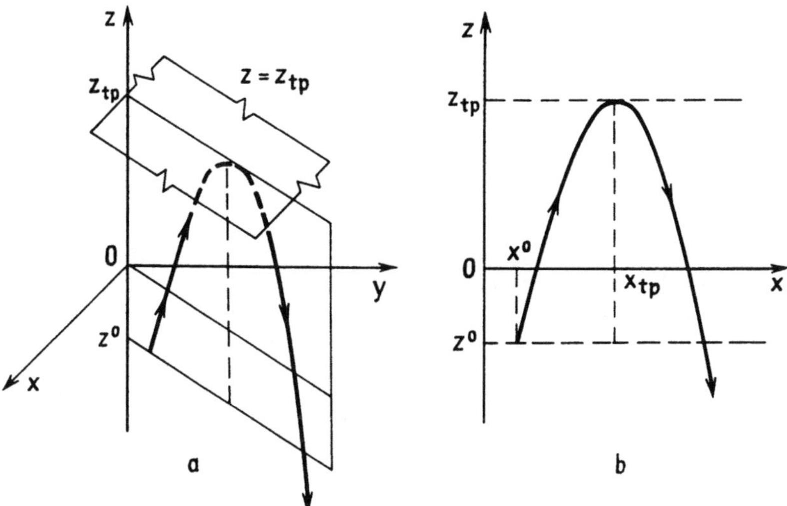

Fig.3.15. (a) In a plane-stratified medium, each ray is a plane curve. (b) If there exists a turning point, the ray has upgoing and downcoming segments

$$p_x = \text{const} = p_y^0 , \qquad (3.3.2)$$

$$p_z = \pm\, n^2(z) - (p_x^0)^2 - (p_y^0)^2 ,$$

where p_x^0 and p_y^0 are the components of the primary moment p^0.

Eliminating τ from (3.3.1) and accounting for (3.3.2), we have the equations

$$\frac{dx}{dz} = \pm\, \frac{p_x^0}{(n^2(z) - (p_x^0)^2 - (p_x^0)^2)^{1/2}} ,$$

$$\frac{dy}{dz} = \pm\, \frac{p_y^0}{(n^2(z) - (p_x^0)^2 - (p_y^0)^2)^{1/2}} , \qquad (3.3.3)$$

which can be integrated by quadratures. The upper sign in these equations refers to the condition $p_z > 0$, which is satisfied on the *upgoing* segment of the ray, i.e., that toward larger positive z, while the lower sign refers to the *downgoing* segment ($p_z < 0$). The condition $p_z = 0$ defines the turning point of the ray, z_{tp}. At this point the ray is tangent to the plane z = const = z_{tp} on which $n^2(z_{tp}) = (p_x^0)^2 + (p_y^0)^2$ (Fig.3.15a). To account for a turning point the ray equations become

$$x = x^0 + \left(\int_{\tilde{z}_>}^{\tilde{z}_>} \mp \int_{z_<}^{z_>} \right) \frac{p_x^0 dz}{(n^2(z) - (p_x^0)^2 - (p_y^0)^2)^{1/2}} \equiv X(z,\xi,\eta) , \qquad (3.3.4a)$$

133

$$y = y^0 + \left(\int_{\tilde{z}_<}^{\tilde{z}_>} \mp \int_{z_<}^{z_<} \right) \frac{p_y^0 dz}{(n^2(z) - (p_y^0)^2 - (p_y^0)^2)^{1/2}} \equiv Y(z,\xi,\eta) , \qquad (3.3.4b)$$

where x^0, y^0, and z^0 are the coordinates of the point of departure of the ray; $\tilde{z}_<$ ($\tilde{z}_>$) is the lowest (largest) value of the two z^0 and z_{tp}, and similarly $z_< \equiv \min(z, z_{tp})$ and $z_> \equiv \max(z, z_{tp})$; and ξ and η are the ray coordinates labeling the rays. The upper sign in (3.3.4) is chosen on the ray segment prior to the turning point, i.e., for $|x^0| < |x| < |x_{tp}|$, $|y^0| < |y| < |y_{tp}|$, and the lower sign behind the turning point for $|x| > |x_{tp}|$, $|y| > |y_{tp}|$, where $x_{tp} \equiv X(z_{tp},\xi,\eta)$ and $y_{tp} \equiv Y(z_{tp},\xi,\eta)$; the coordinate x_{tp} is shown in Fig.3.15b.

Equations (3.3.4) hold true in the absence of a turning point as well – the difference of the integrals in (3.3.4) should then be replaced by one integral between z^0 and z. Specifically, for the upgoing ray path we have ($p_z > 0$)

$$x = x^0 + \int_{z^0}^{z} \frac{p_x^0 dz}{(n^2(z) - (p_x^0)^2 - (p_y^0)^2)^{1/2}} ,$$

$$\tag{3.3.4c}$$

$$y = y^0 + \int_{z^0}^{z} \frac{p_y^0 dz}{(n^2(z) - (p_x^0)^2 - (p_y^0)^2)^{1/2}} ;$$

for the downgoing leg ($p_z < 0$) the limits of integration in (3.3.4) should be reversed. Expressions (3.3.4) thus describe rays of various geometry (Fig. 3.16). One can easily verify that the ray equations (3.3.1) may be derived from the complete integral of the eikonal equation, which was obtained in Sect.2.8 by separation of variables.

We also give the parametric ray equations convenient for calculation of the Jacobian $\mathscr{D}(\tau)$; they follow from (3.3.1,2) and are evidently equivalent to (3.3.4):

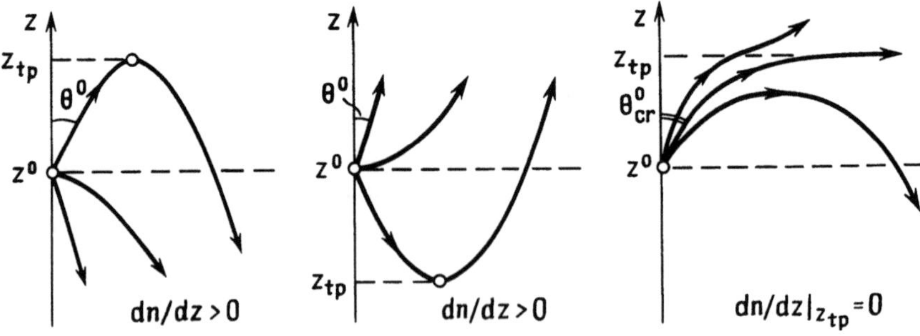

Fig.3.16. Depending on the sign of dn/dz rays bend either (a) downward or (b) upward. (c) If dn/dz falls off to zero at a height z_{tp}, then a critical ray appears having the horizontal asymptote $z = z_{tp}$

$$x = x^0 + p_x^0 \tau \, ,$$

$$y = y^0 + p_y^0 \tau \, , \tag{3.3.5}$$

$$\tau = \left(\int_{\tilde{z}_<}^{\tilde{z}_>} \pm \int_{z_<}^{z_>} \right) \frac{dz}{(n^2(z) - (p_x^0)^2 - (p_y^0)^2)^{1/2}} \, .$$

3.3.2 Ray Tracing in a Plane-Stratified Medium

Introducing the angles θ, ϕ, θ^0, and ϕ^0 by

$$p_x = n\sin\theta\cos\phi \, , \quad p_y = \eta\sin\theta\sin\phi \, , \quad p_z = n\cos\theta \, , \tag{3.3.6a}$$

$$p_x^0 = n^0\sin\theta^0\cos\phi^0 \, , \quad p_y^0 = n^0\sin\theta^0\sin\phi^0 \, , \quad p_z^0 = n^0\cos\theta^0 \, , \tag{3.3.6b}$$

where $n^0 \equiv n(z^0)$, from (3.3.2) we have

$$\tan\phi = p_y/p_x = \text{const} = p_y^0/p_x^0 = \tan\phi^0 \, , \tag{3.3.7a}$$

$$\sqrt{p_x^2 + p_y^2} = n\sin\theta = \text{const} = n^0\sin\theta^0 = \sqrt{(p_x^0)^2 + (p_x^0)^2} \, . \tag{3.3.7b}$$

The first relation implies that in a plane-stratified medium a ray is represented by a *plane* curve, while the second is the *law of refraction* for continuous plane-stratified media with $n(z)$ (θ^0 is the angle of incidence, and θ is the instantaneous angle of refraction). Expressions (3.3.7) are easily deducible from Snell's law (2.5.7) if the continuous medium $n(z)$ is viewed as the limit of a system of infinitesimally thin plane layers; then for $p_y^0 = 0$ the ray equation (3.3.4a) immediately follows from both (3.3.7) and the obvious condition $dx/dz = \tan\theta$.

The turning point $z = z_{tp}$ is the *point of ray reflection* (i.e., the plane of its deepest penetration into the medium); at this point $\theta = \pi/2$ and, as a consequence, $n(z_{tp}) = n^0\sin\theta^0$. We should expect a turning point on an up-going ray only for $dn/dz < 0$, and on a downgoing for $dn/dz > 0$ (Fig.3.16). The ray path is symmetric about the turning point. For $n > n^0\sin\theta^0$ there is no turning point, when the refractive profile is nonmonotonous, the wave propagation assumes *waveguide* character (Sect.3.7) with an infinite number of turning points.

Note that for

$$\frac{d}{dz} n(z_{tp}) = 0 \, ,$$

the horizontal position of the turning point tends to infinity ($x_{tp} \to \infty$, $y_{tp} \to \infty$). This situation corresponds to the appearance of creeping trajectories (Fig.3.16) in antiwaveguide propagation in the media of $n^2(z)$ having an extremized profile, say parabolic. This implication follows from the Taylor expansion

$$n^2(z) - (p_x^0)^2 - (p_y^0)^2 = \frac{d}{dz}n^2(z_{tp})\,(z - z_{tp})$$

$$+ \frac{1}{2}\frac{d^2}{dz^2}n^2(z_{tp})\,(z - z_{tp})^2 + \dots , \qquad (3.3.8)$$

the coefficients of which define the order of singularity for the improper integral in the ray equations (3.3.4).

In accordance with (2.1.4) the ray curvature (3.3.4) equals

$$K = \frac{1}{n}\Big|\frac{dn}{dz}\Big|\sin\theta = \frac{n^0\sin\theta^0}{n^2(z)}\Big|\frac{dn}{dz}\Big| . \qquad (3.3.9)$$

At normal incidence $(\theta^0=0)$ $K = 0$ since the ray is a straight line normal to the layers $n(z)$ = constant.

3.3.3 Equations of Caustics, and the Geometry of the Ray Family

A ray family is described by (3.3.4 or 5) with the coordinates x^0, y^0, and z^0, and the components of momentum p_x^0, p_y^0 (or angles θ^0 and ϕ^0) given on an initial surface Q (Sect.2.2) as functions of the ray coordinates ξ and η. With (3.3.5) the Jacobian $\mathscr{D}(\tau)$ becomes

$$\mathscr{D}(\tau) = p_z\left[\frac{\partial X}{\partial\xi}\frac{\partial Y}{\partial\eta} - \frac{\partial X}{\partial\eta}\frac{\partial Y}{\partial\xi}\right] = p_z\frac{\partial(X,Y)}{\partial(\xi,\eta)} , \qquad (3.3.10)$$

where the functions X and Y are expressed by (3.3.4), and p_z is found from (3.3.2) with the signs are adopted in accord with (3.3.4). For the simplest case when all rays leave the plane $z = z^0$ = constant at equal angles θ and ϕ^0 (a plane wave), we have $\mathscr{D}(\tau) = p_z$. All the rays are parallel and the caustic surface $z = z_c$, on which $\mathscr{D} = 0$, coincides with the locus of the turning points $(p_z = 0)$, i.e., $n(z_c) = n^0\sin\theta^0$ (Fig.3.17). In the general case, at a caustic $p_z \neq 0$,[15] and the position of the caustic is defined both by (3.3.4) and the condition $\mathscr{D}(\tau) = 0$, this yields the implicit function $z = z_c(\xi,\eta)$. In evaluating caustics in a two-dimensional or an axially symmetric problem one should observe that

---

[15] Analysis indicates that in the general case, at turning points where $p_z = 0$, the Jacobian (3.3.10) is different from zero (this can be easily verified taking as an example a source in a linear layer), i.e., the turning point, generally is not a caustic point on the ray. At times the locus of turning points is positionally close to a caustic, though a caustic point on a certain ray may be fairly distant from its turning point.

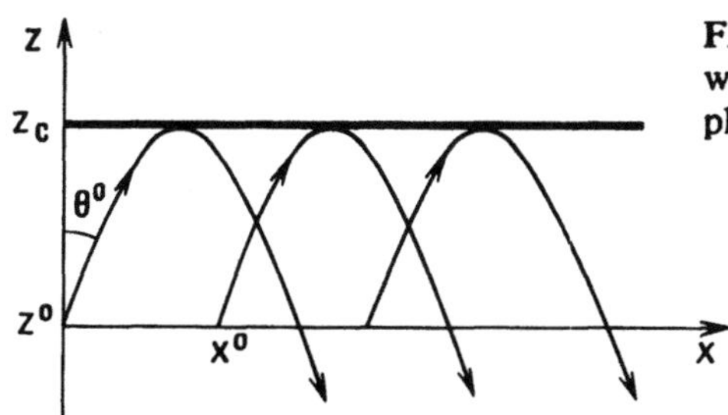

Fig.3.17. Plane caustic $z = z_c$ occurs where a plane wave is incident on a plane-stratified medium

$$\mathscr{D}_{2D}(\tau) = p_z \frac{\partial X}{\partial \xi} \, ,$$

$$D_{axc}(\tau) = p_z r \frac{\partial r}{\partial \xi} \, ,$$

(3.3.11)

where the functions $X(z,\xi)$ and $r(z,\xi)$ describe the ray families, and $r = \sqrt{(x^2+y^2)}$ is a cylindrical coordinate.

Consider, by way of an example, radiation from a point source at $x^0 = y^0 = 0$, $z = z^0$. The ray equations (3.3.4) reduce to the form

$$r = \left(\int_{\tilde{z}_<}^{\tilde{z}_>} \mp \int_{z_<}^{z_>} \right) \frac{n^0 \sin\theta^0 dz}{\sqrt{n^2(z) - (n^0 \sin\theta^0)^2}} \equiv r(z,\theta^0) \, . \tag{3.3.12}$$

In the particular case of $dn/dz < 0$ for the rays with $\theta^0 < \pi/2$ we should substitute $\tilde{z}_< = z^0$, $\tilde{z}_> = z_{tp}$, $z_< = z$, $z_> = z_{tp}$ into (3.3.10) (Fig.3.18a); for $\theta^0 > \pi/2$, (3.3.10) reads with only one integral from z to z^0. The caustic of rays, (3.3.10), is defined by the set of equations ($\xi \equiv \theta^0$)[16]

$$r = r(z,\theta^0) \, , \quad \partial r(z,\theta^0)/\partial\theta^0 = 0 \, . \tag{3.3.13}$$

From (3.3.11) and $\mathscr{D}_{axs} = 0$ it follows that at normal (vertical) incidence ($\theta = 0$) the refractive index becomes zero on the caustic, i.e., $n(z_c) = 0$ (Fig.3.18a). Properties of the caustic (3.3.13) have been studied in [3.16, 57, 63-65]. Following [3.63, 65] we shall next consider the salient features of the caustic (3.3.13) for the case of an inhomogeneous medium ($z < 0$) having a common boundary with a homogeneous half-space ($z < 0$) with $n = n(0) = 1$.

3.3.4 Rays and Caustics due to a Point Source in an Inhomogeneous Medium

For an *interior* source positioned at $z^0 > 0$, the rays and caustics inside an inhomogeneous half-space are the same as in the unbound medium and

---

[16] The condition $\partial r(z,\theta^0)/\partial\theta^0 = 0$ defines the envelope of the ray family $r = r(z,\theta^0)$ [3.2,4,16].

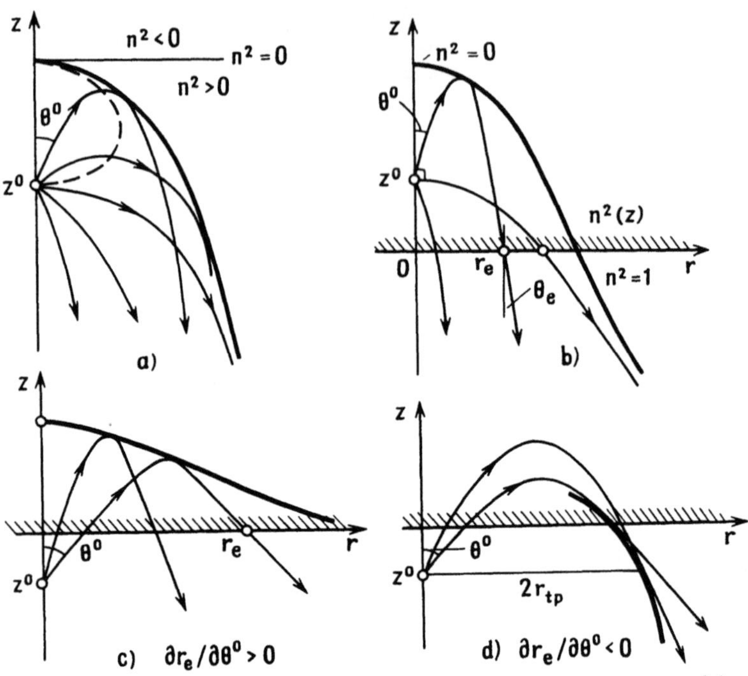

Fig.3.18. Typical caustics occurring in an inhomogeneous space (a) unbounded, and (b,c,d) bounded by a homogeneous halfspace $z < 0$; (b) The source is in the inhomogeneous medium ($z > 0$). (c) The source is within $z < 0$, the caustic appears in the inhomogeneous medium. (d) The source is at $z < 0$; the caustic protrudes into the homogeneous halfspace

defined by (3.12, 13). The ray family $r = r_e - z\tan\theta_e$ emerging into the free half-space forms the caustic

$$r = r_e - \cos\theta_e \sin\theta_e \frac{dr_e}{d\theta_e} ,$$

$$(3.3.14)$$

$$z = \cos^2\theta_e \frac{dr_e}{d\theta_e} ,$$

where θ_e and r_e are, respectively, the angles at which the emanating ray leaves the medium and the coordinate of crossing the boudnary $z = 0$. Here $\sin\theta_e = n(z^0)\sin\theta^0$. According to (3.3.14), this caustic exists only for the rays with $dr_e/d\theta_e < 0$; for $dr_e/d\theta_e = 0$ it crosses the boundary $z = 0$, as shown in Fig.3.18b. The caustic (3.3.14) asymptotically tends to infinity – the asymptote being the ray at $\theta^0 = \pi/2$ (Fig.3.18b) because for this ray $d\theta^0/d\theta_e \rightarrow \infty$ and $dr_e/d\theta_e \rightarrow \infty$. The direction of the caustic's asymptote $\theta_e = \arcsin n(z^0)$ defines the boundary in the radiation pattern of the source (hence this asymptote bears the information on the value $n(z^0)$ [3.66]). For a number of specific $n(z)$ profiles the geometry of the caustic (3.3.14) will be discussed below.

138

When the source is placed in the homogeneous medium at a distance z^0 from the interface $z = 0$ with the inhomogeneous medium (Fig.3.18c, d) the ray equations, in accord with (3.3.12), take the form

$$r = |z^0| \tan\theta^0 + \left(\int_0^{z_{tp}} \mp \int_z^{z_{tp}} \right) \frac{\sin\theta^0 \, dz}{\sqrt{n^2(z) - \sin^2\theta^0}} \equiv r(z,\theta^0) \,, \qquad (3.3.15)$$

where z_{tp} is the root of the equation $n(z_{tp}) = \sin^2\theta^0$. The caustic of the rays (3.3.15) is described by (3.3.14) with the rays, apparently, touching the caustic beyond the turning point, on the downgoing segment of the trajectory (Fig.3.18c). As has been demonstrated [3.63], the caustic (3.3.13) may cross the interface $z = 0$. From (3.3.13) it follows that this crossing point corresponds to the condition $\partial r_e / \partial\theta^0 = 0$, where r_e is the coordinate of the point where the ray emerges (Fig.3.18d):

$$r_e = |z^0| \tan\theta^0 + 2 \int_0^{z_{tp}} \frac{\sin\theta^0 \, dz}{\sqrt{n^2(z) - \sin^2\theta^0}} \,. \qquad (3.3.16)$$

For $z < 0$ the reflected rays can be found from $r = r_e - z\tan\theta^0$. The equation of the caustics for the reflected rays can be obtained from (3.3.16):

$$r = r_e - \sin\theta^0 \cos\theta^0 \frac{dr_e}{d\theta^0} \,,$$

$$\qquad (3.3.17)$$

$$r = \cos^2\theta^0 \frac{dr_e}{d\theta^0} \,.$$

This implies that for $z < 0$ this caustic may form only if $dr_e/d\theta^0 < 0$ (Fig. 3.18d); otherwise $(dr_e/d\theta^0 > 0)$ the caustic will form inside the inhomogeneous medium (Fig.3.19c) and is defined by another set of equations, namely (3.3.13, 15). Evidently, the condition for the caustic (3.3.17) to be at the level of the source at $z = z^0$ is $\partial r_{tp}/\partial r^0 = 0$, where $r_{tp} = r(z_{tp}, \theta^0)$ are the coordinates of the turning points of rays (3.3.15).

3.3.5 Rays and Caustics in a Linear Layer

For a number of characteristic $n^2(z)$ profiles, the integrals stated above are straightforward. The results are especially simple for the linear profile for which

$$n^2(z) = 1 - az = 1 - z/H \,, \qquad (3.3.18)$$

where $H = 1/a > 0$ is the characteristic scale of the layer.

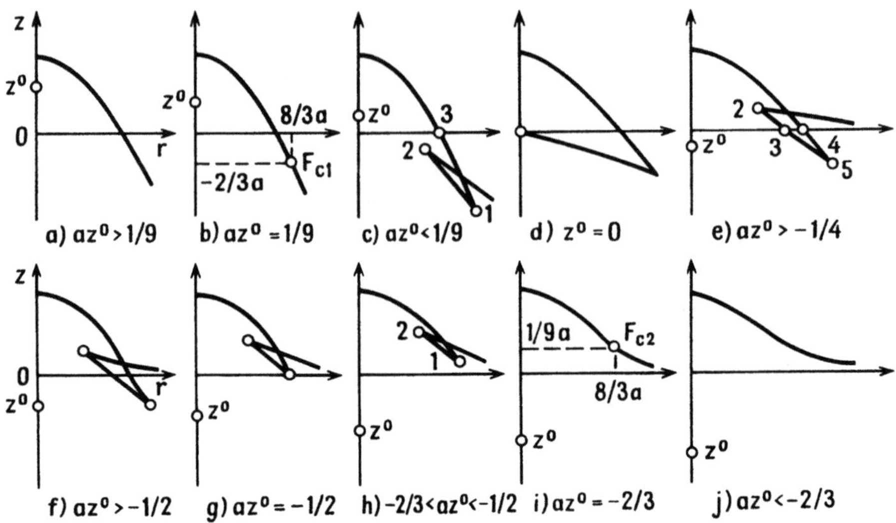

Fig.3.19. Evolution of caustic shapes with the height z^0 of a point source placed inside or outside of the linear layer interfacing a homogeneous halfspace $z < 0$

For an infinite linear layer $-\infty < z < \infty$ the family of rays (3.3.13) becomes the parabolas

$$a(x - x_{tp})^2 = 4(1 - az^0)\sin^2\theta^0(z_{tp} - z) , \qquad (3.3.19)$$

for which the locus of turning points

$$ax_{tp} = (1 - az^0)\sin 2\theta^0 , $$

$$az_{tp} = 1 - (1 - az^0)\sin^2\theta^0 , \qquad (3.3.20a)$$

is the ellipsoid of revolution (the broken line in Fig.3.18a):

$$\left(\frac{ax_{tp}}{1 - az^0}\right)^2 + \left(1 - 2\frac{1 - az_{tp}}{az^0}\right)^2 = 1 . \qquad (3.3.20b)$$

The caustic of rays (3.3.19) is nonsingular and has the form of a paraboloid of revolution (Fig.3.18a)

$$a^2x^2 = 4(1 - az^0)(1 - az) \qquad (3.3.21)$$

with the apex at $z = H = 1/a$.

For a semi-infinite linear layer $0 < z < \infty$ having at $z = 0$ a common boundary with a homogeneous half-space $z < 0$, the form of the caustic complicates unexpectedly [3.63,65]; instead of a smooth caustic we now may find caustics having singularities in the form of a pocket with two cusp points (a section of the swallowtail catastrophe, Sect.2.4). Such singularities have been observed and studied in [3.63,65], respectively, for a source placed outside ($z^0 < 0$) and inside ($z^0 > 0$) the inhomogeneous medium.

For a semi-infinite linear layer of

$$n^2(z) = \begin{cases} 1 - az & \text{for } z \geq 0 \\ 1 & \text{for } z \leq 0 \end{cases}, \tag{3.3.22}$$

the rays are straight lines for $z < 0$, and segments of parabolas for $z > 0$. It turns out that the caustic is smooth when the source is placed either below $z = -2H/3$, or above $z = H/9$; whereas for $-2H/3 < z^0 < H/9$, the caustic acquires two cusp points bounding the *caustic pocket*. Figure 3.19 demonstrates how the caustic changes its form when the source is lowered (or, in the general case, as the parameter za^0 varies). At $z^0 = -2H/3$ (Fig.3.19i) and $z^0 = H/9$ (Fig.3.19b) the pocket at the caustic contracts into the *caustic foci* F_{c1} and F_{c2}. According to the nonmenclature of Sect.2.4 this phenomenon corresponds to the critical section through the origin of the swallowtail catastrophe. Through every point inside the caustic pocket (Fig. 3.19) there travel *four* different rays (of course, the caustic shadow contains no real-valued rays).

Note, in addition, that for $z^0 > -H/2$ the pocket at the caustic closes in the free half-space $z < 0$ (Fig.3.19e,f) and the caustic (3.3.17) may cross the source plane $z = z^0$ at two points if $z^0 > -H/4$ (Fig.3.19e). For $z^0 < -H/4$, the caustic is situated "above" the plane $z = z^0$, and for $z^0 < -H/2$ it is completely within the inhomogeneous medium (Fig.3.19h-j).[17] The formulas needed to determine the coordinates of all singularities of the caustic follow from the general relations of Sect.3.3.4 and can be found in [3.59, 61] (see also [3.70]). Figure 3.20 plots the computational results, thus enabling a search for these points (the subscripts of the coordinates refer to the labeling of the points in Fig. 3.19c,e,h). These results may serve as an

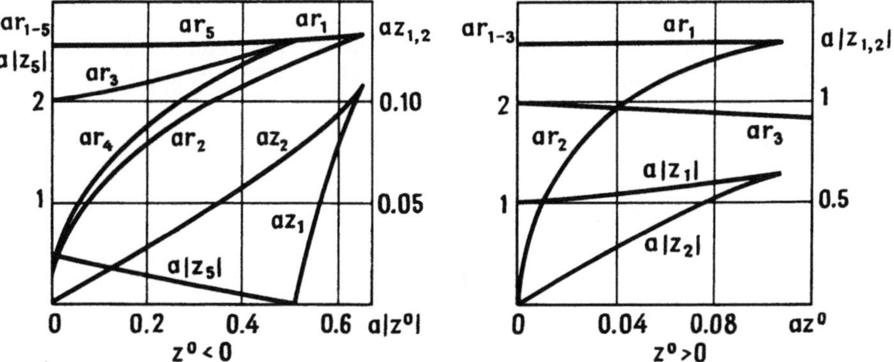

Fig.3.20. Plots of computations to evaluate the location of the characteristic points indicated in Fig.3.19c,e,h.

---

[17] In [3.67] separate branches of the caustic (3.3.17) have been observed experimentally in free half-space, see also [3.68,69]; for an interpretation of these observations, see [3.70].

illustration of the reciprocity theorem for rays and caustics (Sect.2.6, [3.65]); in particular, for the caustic foci F_{c1} and F_{c2} (Fig.3.19b,i) we have

$$z_{c1} = z_{c2}^0 = -2H/3 \, ,$$

$$z_{c1}^0 = z_{c2} = H/9 \, ,$$

$$r_{c1} = r_{c2} = 8H/3 \, .$$

In a linear layer the form of a caustic depends on the shape of the incident phase front. In [3.75] caustics have been studied for a wave passing through a plane sinusoidal phase screen at $z = 0$ (Sect.3.1), and in [3.9,72] those generated by an aperture with the phase distributed according to a square law. The diversity of caustic shapes in these cases is due to the mutual influence of two factors: the focusing effect caused by the inhomogeneity of the medium and the primary phase profile of the field.

3.3.6 Layers of Other Profiles

For a smooth, monotonous function $n^2(z)$, characteristic of a plasma, when there exists an area $z > z_{tp}$ of $n^2(z) < 0$, the caustic shapes for a point source are generally analogous to those presented in Fig.3.19 [3.63]. However, for the quadratic profile $n^2(z) = 1-bz^2$ ($b > 0$), which smoothly matches that of the free half-space, it turns out that only a smooth caustic (for $z^0 < 0$, of the type of Fig.3.18c) is formed. One might think that pocket-type singularities are due to the presence of a weak interface at which the first derivative dn/dz is discontinuous. Analysis shows, however, when dn/dz becomes smooth, the caustic pocket does not disappear, but only deforms slightly. The geometry of caustics for a number of elementary profiles of $n^2(z)$ has been outlined also in [3.16]. So, for example, for $n^2(z) = (1+az)^{-2}$ ($a>0$, $z>0$), which corresponds to a linear variation of sound velocity, the caustic due to a point source will always possess a cusp (Fig.3.21).

Caustics acquire an interesting feature in inhomogeneous media of a piecewise-linear, notably bilinear, profile of $n^2(z)$ [3.73]. In a bilinear layer

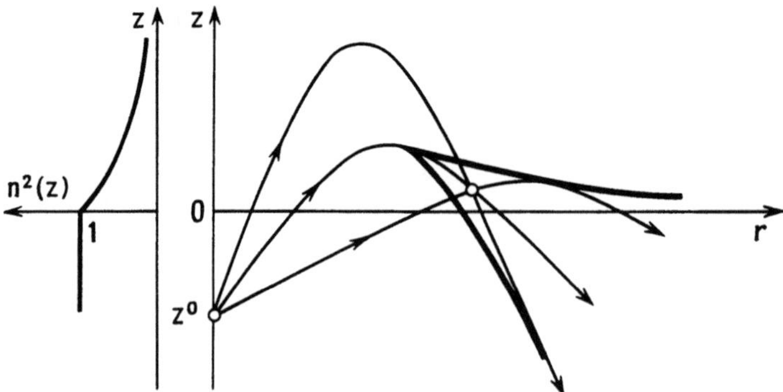

Fig.3.21. Caustic cusp due to a point source placed near a layer of $n^2(z) = (1 + az)^{-2}$

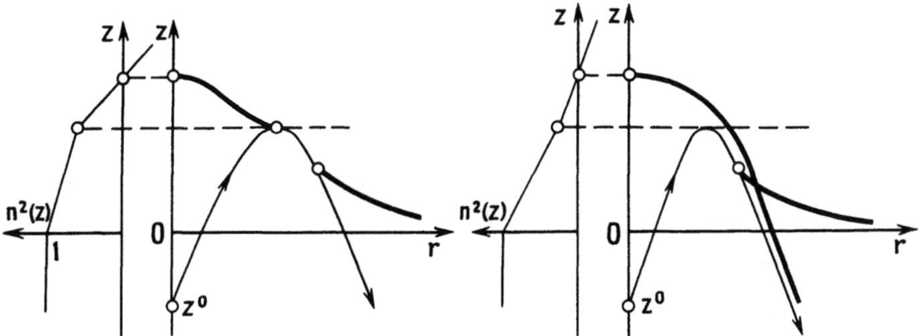

Fig.3.22. Broken caustics occurring in media with piecewise-linear refractive-index profiles

there form *discontinuous* (*broken*) caustics, the discontinuities are located on the critical ray tangent to the interlayer boundary. Figure 3.22 exemplifies the caustics of two characteristic shapes. In geometrical-optics computations one should allow for the possibility of broken caustics to form: for smooth n(z) systems a piecewise-linear approximation is liable to yield imaginary caustics. To avoid them one should employ approximate layers with continuous derivatives dn/dz at the boundaries [3.73] (Sect.3.10).

3.3.7 Plane Waves in a Parabolic Layer

For a parabolic profile

$$n^2(z) = \epsilon_m + \epsilon_2(z - z_m)^2 \tag{3.3.23}$$

it is customary to distinguish (by analogy with quantum mechanics) wave propagation through a potential *barrier*, when $\epsilon_2 > 0$, from waveguided propagation in a potential *well*, when $\epsilon_2 < 0$ (Fig.3.23). In what follows we consider wave propagation through a barrier (guided modes will be discussed in Sect.3.7).

For a plane wave incident from the side of negative z (Fig.3.24), the rays (3.3.4) represent a family of parallel curves:

$$x = x^0 - \ln\left|\frac{1}{z_{tp}}(z^0 - z_m + b\cos\theta^0)\right|$$

$$\pm \ln\left|\frac{1}{z_{tp}}[z - z_m + \sqrt{n^2(z) - (n^0\sin\theta^0)^2}]\right| , \tag{3.3.24}$$

where $b \equiv \epsilon_2^{-1/2} > 0$ and $\epsilon_2 > 0$. The turning points of the rays

$$z_{tp}^{\pm} = z_m \mp b\sqrt{(n^0\sin\theta^0)^2 - \epsilon_m} \tag{3.3.25}$$

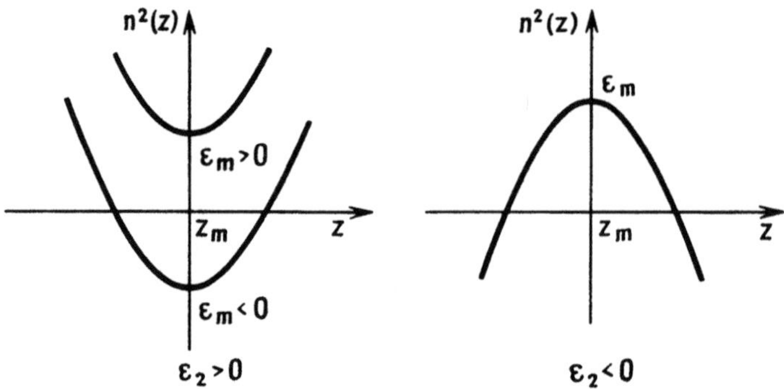

Fig.3.23. Squared refractive index profiles of a parabolic layer (3.3.23) for various ϵ_m and ϵ_2

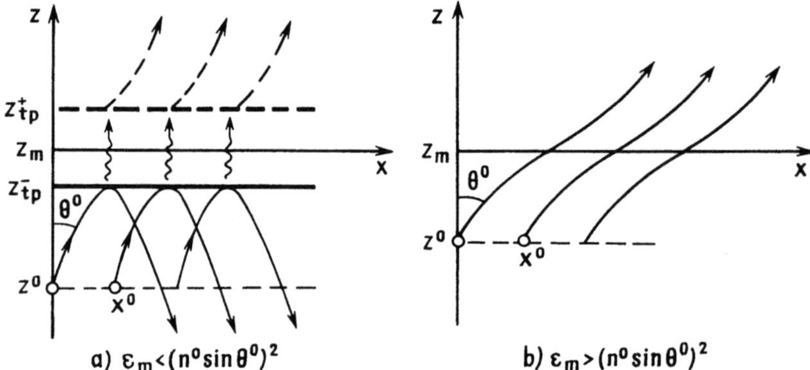

Fig.3.24. Reflection and subbarrier transmission (a) and barrier penetration (b) of rays in the case of a parabolic profile

exist only for $\epsilon_m < (n^0 \sin\theta^0)^2$. In this case the envelope of the rays (3.3.4) is the plane $z = z^-_{tp}$ that separates the two-ray region $z < z^-_{tp}$ from the geometrical shadow $z > z^-_{tp}$. Nonmonotonous profiles of $n^2(z)$ are unique in that the law of refraction, (3.3.7b), again allows for the rays at real-valued angles $\theta = \theta(z)$ to propagate in the part of shadow region with $z > z^+_{tp}$.[18] The caustic for these rays is the plane $z = z^+_{tp}$ (Fig.3.24a). That the region of *secondary* propagation $z < z^+_{tp}$ (Fig.3.24a) has appeared can be explained by the diffraction effect - *sub-barrier penetration* of the wave. The planes $z = z^{\pm}_{tp}$, corresponding to the front and trailing edges of the barrier, are the two branches of a *penetrated caustic* which belongs to a new type of caustic characteristic of the inhomogeneous medium [3.74]. Note that for $\epsilon_m > (n^0 \sin\theta^0)^2$ the wave penetrates *directly* through the layer (the propagation *above the barrier*) with only one ray passing through a point of the plane $\{x, z\}$ (Fig.3.24b); this ray is defined by (3.3.24) with the upper sign.

---

[18] From the viewpoint of the complex geometrical optics [3.75], these rays are complex, since for them one cannot indicate any real coordinates x^0 of the points where they leave the initial plane $z = z^0$ [3.76].

144

3.3.8 A Point Source in a Parabolic Layer

For a point source at $z = y = 0$, $z = z^0 < 0$, the equations of rays in the parabolic layer (3.3.23) have the form

$$r = \tau n^0 \sin\theta^0 \; ,$$

$$z - z_m = (z - z^0)\cosh(\tau\sqrt{\epsilon_2}) + (n^0/\sqrt{\epsilon_2})\cos\theta^0 \sinh(\tau\sqrt{\epsilon_2}) \; . \qquad (3.3.26)$$

For $\epsilon_m < 0$, when the barrier is impenetrable for the vertical ray ($\theta^0 = 0$), the rays (3.3.26) and their caustic are similar to those depicted in Fig.3.18a. In the case of $\epsilon_m > 0$, the ray pattern changes qualitatively: in addition to the rays reflected from the layer, there exist rays penetrated through it. These groups of rays are separated by the *critical ray* with $\sin^2\theta^0_{cr} = \epsilon_m(n^0)^{-2}$, which, as $\tau \to \infty$, approximates asymptotically the barrier's top at $z = z_m$. Figure 3.25 exemplifies the ray pattern for the case of a finite parabolic layer of

$$n^2(z) = \begin{cases} 1 - \delta^{-2}[1 - (1 - z/z_m)^2] & \text{for} \quad 0 \le z \le 2z_m \; , \\ 1 & \text{for} \quad z \le 0 \quad \text{and} \quad z \ge 2z_m \end{cases} \qquad (3.3.27)$$

where $\delta = (1 - \epsilon_m)^{-1/2}$, and z_m is the halfwidth of the layer. The rays at $\theta^0 < \theta^0_{cr}$ form a diverging bundle filling the entire space $z > z_m$. For $\theta^0 > \theta^0_{cr}$, the rays form a caustic which in the simplest case possesses one cusp and signifies the appearance of a *three-ray* region (Fig.3.25).

The rays close to the critical ray form an almost parallel ray pencil. Owing to small refraction they proceed sufficiently long as a dense group near the barrier's top at $z = z_m$, and then leave this unstable region. For large r, the critical ray becomes the axis of an *antiwaveguide*. The described mechanism of antiwaveguide propagation might prove of significance at a large distance from the source, say, in the ionosphere [3.77-80].

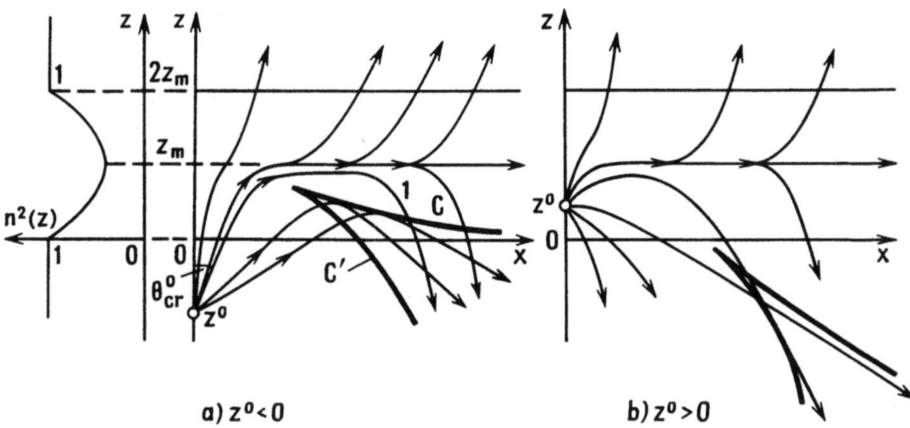

Fig.3.25a,b. Caustic cusp appears as rays are reflected from a parabolic layer: (a) the source is outside the layer, (b) the source is inside the layer

It is worth noting that in the theory of short-wavelength propagation in the ionosphere [3.80] the caustic branch C′ in Fig.3.25a bounds the so-called *dead-zone* for the rays reflected from the ionosphere; for $z^0 < 0$ the "upper" of the two reflected rays reaching the observation point (labeled 1 in the figure), at the smaller angle of incidence, is called Pedersen's ray [3.80-84].

Figure 3.25 illustrates only the simplest caustics which can appear in a bounded parabolic layer (3.3.27) with $\epsilon_m > 0$. A complete caustic pattern would be fairly sophisticated; as has been shown in [3.64,85], some relations between the parameters δ and $|z^0|/z_m$ give rise to caustics possessing multiple pockets - "caustic stars". The maximum number of rays passing through such regions rises from three (Fig.3.25) to *five*. Catastrophe theory (Sect.2.4) would tell us that the marked complication of caustic shapes - as compared with, say, those in a linear layer - is due to the increase of the caustic codimension to four (a parabolic layer would have to be described already by two parameters) so that now various sections of a four-dimensional caustic hypersurface are projected onto the plane $\{x, z\}$.

Note that for $\epsilon_m < 0$ in a parabolic layer the singularities on a caustic are similar to those in a linear layer (Fig.3.19) [3.64,85]. The "rear" branch of the penetrated caustic for $\epsilon_m < 0$, depicted for a plane wave in Fig.3. 24a, shifts toward the complex domain in the case of a plane source [3.76].

A complete classification of caustics for both interior ($z^0 > 0$) and outer ($z^0 < 0$) point sources was given in [3.64]. Accordingly, the simplest-shape caustics - a smooth caustic for $\delta < 1$ and a single-cusp caustic for $\delta > 1$ - are to form with certainty with $\delta > 1.0835$ and either $z^0/z_m < -0.8675$ or $z^0/z_m > 0.156$. Otherwise, additional ray and caustic singularities may appear, e.g., a *five-ray* region or a *butterfly-type* caustic may form (Fig. 3.2c, d); the respective values of δ and z^0/z_m have been listed in [3.64]. For the sake of illustration, Fig.3.26 demonstrates how a caustic changes its shape when the source varies its position (z^0/z_m) for $\delta = 1.05$ - the case leading to a butterfly-type caustic (the Roman numerals in the figure indicate the number of rays). It is obvious how the caustic pocket shifts from one branch to another, and then disappears to leave behind a one-cusp caustic.

The investigation of rays and caustic in media of arbitrary n(z) profile employs numerical techniques (Sect.3.10); however, the perturbation technique may prove useful for some systems.

3.4 Wave Fields in Plane-Stratified Media

3.4.1 The Field of an Arbitrary Wave

Consider a plane-stratified medium with n(z) and assume that on an initial surface $\mathbf{r} = \mathbf{r}^0(\xi,\eta)$ the initial field distribution of an arbitrary wave is given in the form

$$u^0(\xi,\eta) = A^0(\xi,\eta) \exp[ik_0 \psi^0(\xi,\eta)] .$$

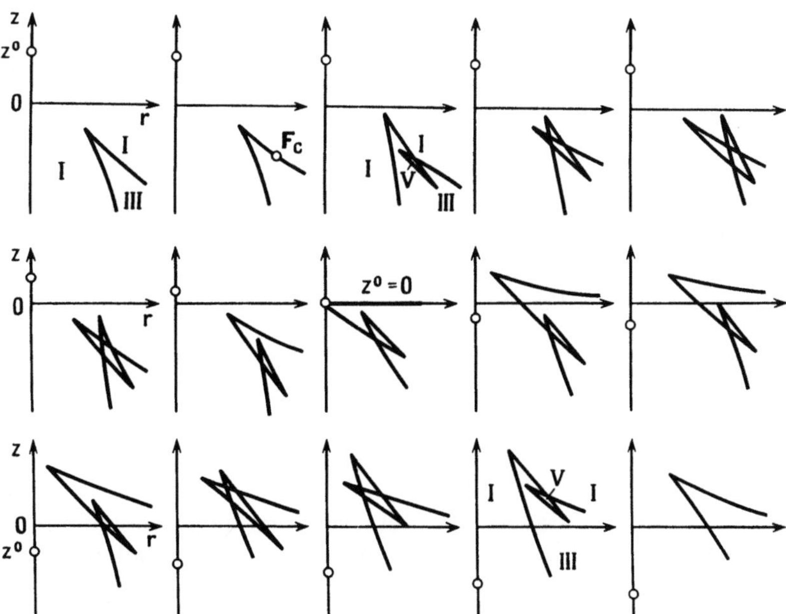

Fig.3.26. Evolution of caustic shape in a parabolic layer with the height of the source z^0 (indicated by an open circle)

In the geometrical-optics approximation, according to Sect. 2.3, the amplitude of the wave is

$$A = A^0 \sqrt{\mathcal{D}(0)/\mathcal{D}(\tau)} \,, \tag{3.4.1}$$

where the Jacobian $\mathcal{D}(\tau)$ is given by (3.3.10) which, as can readily be shown, leads to $\mathcal{D}(0) = p^0(\partial r^0/\partial\xi)(\partial r^0/\partial\eta)$. In taking the square root of the divergence $\mathcal{J} = \mathcal{D}(\tau)/\mathcal{D}(0)$, one should allow for the caustic phase shift (Sect. 2.4).

In a plane-stratified medium with $n(z)$ we may seek the eikonal of a wave either by a separation of variables (Sect. 2.8) or by the method of characteristics (by integrating along the ray, Sect. 2.2). For instance, on the upgoing branch of the ray ($p_z > 0$)

$$\psi = \psi^0 + \int_0^r n^2 d\tau = \psi^0 + \int_{z^0}^z \frac{n^2(z)}{p_z} dz$$

$$= \psi^0 + \int_{z^0}^z \frac{n^2(z)dz}{\sqrt{n^2(z) - (n^0\sin\theta^0)^2}} \,. \tag{3.4.2}$$

on the ray (3.3.4) and similarly

147

$$\psi = \psi^0 + \left(\int_{\tilde{z}_<}^{\tilde{z}_>} \mp \int_{z_<}^{z_>} \right) \frac{n^2(z)\,dz}{\sqrt{n^2(z) - (n^0 \sin\theta^0)^2}} \ . \qquad (3.4.3)$$

The same expression results from the complete integral of the eikonal equation (Sect. 2.8)[19] if we eliminate the coordinates x and y with the help of the ray equations (3.3.4):

$$\psi = \psi^0 + p_x^0(x-x^0) + p_y^0(y-y^0)$$

$$+ \left(\int_{\tilde{z}_<}^{\tilde{z}_>} \mp \int_{z_<}^{z_>} \right) \sqrt{n^2(z) - (n^0 \sin\theta^0)^2}\ dz \ . \qquad (3.4.4)$$

With (3.4.1,3) we can deduce the field variation along an individual ray (3.3.4). The total field is determined by summing over all the rays reaching the observation point (Sect. 2.3.5).

3.4.2 The Field of a Plane Wave

Let the initial field be given on a plane $z = z^0$ as

$$u^0 = A^0 \exp(ik_0 \xi n^0 \sin\theta^0)\ , \qquad (3.4.5)$$

where $n^0 \equiv n(z^0)$, $A^0 = $ constant, $\psi^0 = \xi n^0 \sin\theta^0$ is the initial eikonal, and $\xi = x$ at $z = z^0$; and we look for the field at $z > z^0$ (Fig. 3.17). The initial condition (3.4.5) applies, for example, when a plane wave is incident from a homogeneous half-space $z < z^0$ on a plane-stratified medium $z > z^0$. Therefore, for brevity, we shall refer to any wave of the form (3.4.5) as a plane wave.

As follows from Sect. 3.3, in the situation under consideration

$$\mathscr{D}(\tau) = p_z = \pm \sqrt{n^2(z) - (n^0 \sin\theta^0)^2}\ ,$$

where the plus sign refers to the wave incident on the caustic, $z = z_{tp}$, while the minus sign to the reflected wave (Fig. 3.17). As a result

$$A = \frac{A^0}{\sqrt{J}} = A^0 \sqrt{(p_z^0)/(p_z)} = \frac{A^0 n^0 \cos\theta^0 \exp(-\tfrac{1}{2} i\pi q_{1,2})}{\sqrt[4]{n^2(z) - (n^0 \mathrm{sub}\theta^0)^2}} \equiv A_{1,2}\ , \qquad (3.4.6)$$

where $\theta^0 < \pi/2$, $q_1 = 0$ for the incident wave, and $q_2 = 1$ for the reflected wave. A reflected wave does not arise if everywhere on the ray $n(z) > n^0 \sin\theta^0$. At the caustic $z = z_{tp}$, as usual, the amplitude becomes infinite.

---

[19] Expression (3.4.4) follows from (2.4.14) if one allows for the initial conditions for the eikonal at the point of departure (x^0, y^0, z^0) and the turning point (z_{tp}, y_{tp}, z_{tp}).

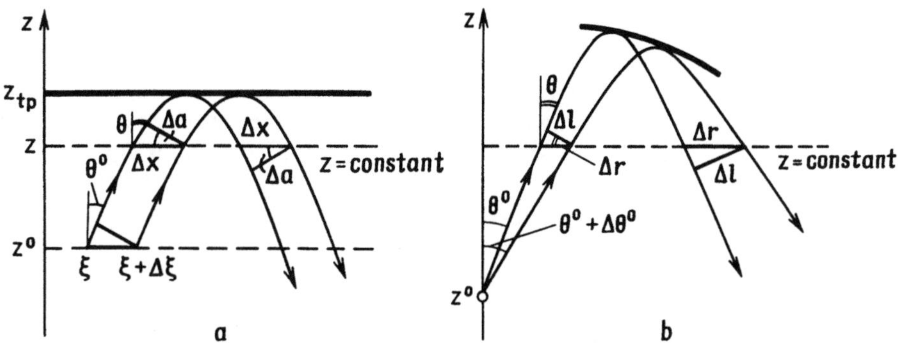

Fig.3.27. Illustration of the field computations for (a) a plane wave, and (b) a point source

The divergence $\mathscr{J}$ the rays in (3.4.6) can easily be obtained geometrically (Fig.3.27) if we observe that by virtue of (2.3.21) $\mathscr{J} = n^0 da^0/nda$ and $\Delta a^0 = \Delta\xi\cos\theta^0$, and

$$\Delta a = \Delta x\cos\theta(z) = \pm\ \Delta\xi\ \frac{1}{n(z)}\sqrt{n^2(z) - (n^0\sin\theta^0)^2}\ .$$

Besides, as the rays (3.3.4) are parallel at $\theta = $ constant, $|\Delta x| = |\Delta\xi|$, and $\theta(z)$ is found from the law of refraction (3.3.7).

The eikonals of two rays reaching the observation point (x, z) are readily derived from (3.4.4) if we recognize that $p_y^0 = 0$, $p_x^0 = n^0\sin\theta^0$, and $x \equiv \xi$, i.e.,

$$\psi_{1,2}(x, z) = xn^0\sin\theta^0 + \left(\int_{z^0}^{z_{tp}} \mp \int_z^{z_{tp}}\right)\sqrt{n^2(z) - (n^0\sin\theta^0)^2}\,dz\ , \qquad (3.4.7)$$

where $\psi_1 \equiv \psi_i$, $\psi_2 \equiv \psi_r$, for the incident and reflected waves, respectively.

Subject to (3.4.6, 7) the resultant wave field is

$$u(x, z) = A_1 e^{ik_0\psi_1} + A_2 e^{ik_0\psi_2}$$

$$= \frac{2A^0 n^0\cos\theta^0}{\sqrt[4]{n^2(z) - (n^0\sin\theta^0)^2}}e^{ik_0\psi_{tp} - i\pi/4}\cos\left[k_0\int_z^{z_{tp}}\gamma(z)dz - \frac{\pi}{4}\right] \quad (3.4.8)$$

where

$$\psi_{tp} = xn^0\sin\theta^0 + \int_{z^0}^{z_{tp}}\gamma(z)dz\ , \quad \gamma(z) \equiv \sqrt{n^2(z) - (n^0\sin\theta^0)^2}\ .$$

At $\theta = 0$ Eq.(3.4.8) coincides with the familiar WKB solution of the one-dimensional problem [3.16, 61, 86, 87]. Note that the WKB method may be employed to corroborate rigorously the geometrical-optics approximation in one-dimensional problems [3.16, 87-90]. Another approach to verify the use of ray optics in a one-dimensional problem, based on a converging expansion, was considered, for example, in [3.17, 90]. Applications of the WKB method to a variety of one-dimensional problems have been detailed in [3.88-90].

3.4.3 Fields of Point and Linear Sources

Let the field of a point source, placed at z = y = 0, z = z^0, at a distance R small compared with the characteristic length L of an n(z) variation (i.e., at $R \equiv \sqrt{x^2 + y^2 + (z-z^0)^2} \ll L$) be of the form (Sect. 2.3)

$$u = (B^0/R)\exp(ik_0 n^0 R) , \quad B^0 = \text{const.} , \quad n^0 \equiv n(z^0) . \qquad (3.4.9)$$

In order to derive the field amplitude at an arbitrary distance from the source, we invoke (2.3.26) and insert into it the elementary solid angle $d\Omega = \sin\theta^0 d\theta^0 d\phi^0$ and the ray-tube cross section $da = n^{-1}\mathcal{D}(\tau)d\theta^0 d\phi^0$, for which, in agreement with (3.3.11 and 6),

$$\mathcal{D}(\tau) = p_z r \partial r/\partial\theta^0 = n\cos\theta r \partial r/\partial\theta^0 . \qquad (3.4.10)$$

Here the role of the ray coordinate ξ is played by the angle of emergence θ^0, and $r = r(z, \theta^0)$ is the equation of the ray family (3.3.12). This leads to

$$A = B^0 \sqrt{\frac{n^0 d\Omega}{n da}} = B^0 \sqrt{\frac{n^0 \sin\theta^0}{\mathcal{D}(\tau)}} = B^0 \sqrt{\frac{n^0 \sin\theta^0}{nr\cos\theta \partial r/\partial\theta^0}} , \qquad (3.4.11)$$

or, observing the law of refraction (3.3.7b),

$$A = B^0 \frac{1}{\sqrt[4]{n^2(z) - (n^0\sin\theta^0)^2}} \sqrt{\frac{n^0\sin\theta^0}{r|\partial r/\partial\theta^0|}} \exp(-\tfrac{1}{2}i\pi q) , \qquad (3.4.12a)$$

where

$$q = \frac{1}{2}(1 - \text{sgn}\mathcal{D}) = \frac{1}{2}\left[1 - \text{sgn}(p_z)\text{sgn}\left(\frac{\partial r}{\partial\theta^0}\right)\right] . \qquad (3.4.12b)$$

We have accounted for the additional phase shift arising upon touching the caustic (3.3.13): for $\mathcal{D} < 0$, q = 1, and the phase shift in (3.4.12) is $-\pi/2$; in front of the tangetial point $\mathcal{D} > 0$ and q = 0 (Sect. 2.4).

In a homogeneous medium, i.e., for constant $n(z) = n^0$, $r = (z-z^0)\tan\theta^0$ follows from (3.3.12). Then $r\partial r/\partial\theta^0 = R^2\tan^2\theta^0$, and (3.4.12) yields the obvious result $A_h = B^0/R$. Restating (3.4.12) in a similar form we have $A = B^0/R_{eff}$, R_{eff} being the *effective distance* from the source; and in free space it will yield the same cross section of ray tube as (3.4.12):

$$R_{eff} = \sqrt{\frac{\sqrt{n^2(z) - (n^0\sin\theta^0)^2}}{n^0\sin\theta^0}\, r \left|\frac{\partial r}{\partial\theta^0}\right|} \; . \tag{3.4.13a}$$

The ratio of A to the amplitude A_h for a homogeneous medium, i.e.,

$$f = \left|\frac{A}{A_h}\right|^2 = \frac{n^0\sin\theta^0 R^2}{r\sqrt{n^2(z) - (n^0\sin\theta^0)^2}\,|\partial r/\partial\theta^0|} = \frac{R^2}{R_{eff}^2} \tag{3.4.13b}$$

shows at a certain point of the inhomogeneous medium how many times the intensity exceeds that which would exist if the medium were homogeneous. It is known as the *focusing factor* [3.16,57]. For ionospheric conditions f was calculated in [3.91,92]; for similar studies in seismology, see [3.93].

The eikonal (3.4.4) for a point source is

$$\psi = rn^0\sin\theta^0 + \left(\int_{\tilde{z}_<}^{\tilde{z}_>} \mp \int_{z_<}^{z_>}\right)\sqrt{n^2(z) - (n^0\sin\theta^0)^2}\, dz \; . \tag{3.4.14}$$

The total field of the source is given by the sum over all contributions from the rays reaching the observation point (r,z), i.e.,

$$u(r,z) = \sum_\nu A_\nu\exp(ik_0\psi_\nu) \; , \tag{3.4.15}$$

where A_ν and ψ_ν are given by (3.4.12–14), and the angles of emergence $\theta_\nu^0 = \theta_\nu^0(r,z)$ ($\nu = 1,2,...,m$) are to be defined from the ray-family equation (3.3.12).

Expressions similar to (3.4.12–14) also result for a linear source (in a two-dimensional problem on the x,z plane). As follows from (3.3.11), the wave amplitude is then determined by (3.4.12) mulitplied by $\sqrt{r/\sin\theta^0}$ with r replaced by x. Upon similar substitution (x for r), (3.4.14) yields the eikonal.

Similarly to the case of a plane wave (Sect. 3.4.2) we note that (3.3.27) and (3.4.1,2) have a simple geometrical interpretation: due to the axial symmetry the cross section of the ray tube is $\Delta a = r\Delta\phi^0\Delta l$, and with reference to Fig. 3.27b $\Delta l = \cos\theta\Delta r$.

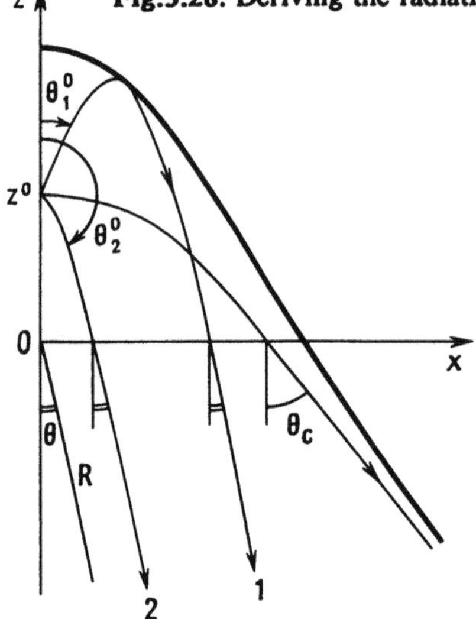

Fig.3.28. Deriving the radiation pattern of a source in a confined linear layer

The formulas derived above are valid for the sources both placed in an infinite medium and in the presence of a weak interface with a homogeneous half-space (Sect. 3.3.4).

3.4.4 A Point Source in a Linear Layer

Let an omnidirectional source be placed inside a linear layer defined by (3.3.22), at a distance z^0 from the interface $z = 0$ ($z^0 > 0$). We wish to find the radiation pattern in the free half-space $z < 0$. The ray pattern has been analyzed in Sect. 3.3.5: for $z^0 < H/9$, a four-ray region arises inside the caustic pocket (Fig.3.19). However, in the far field, an observation point will have only two rays passing through it. These rays emanate from the source at different angles $\theta_{1,2}^0$ (Fig.3.28) and are obtained subject to the condition

$$\sin\theta = \sqrt{1 - az^0}\sin\theta_{1,2}^0 , \qquad (3.4.16)$$

where θ is the angle of observation (we will measure it from the normal toward the interface, and assume $\theta < \pi/2$). The angles θ_1^0 and θ_2^0 coincide ($\theta_1^0 = \theta_2^0 = \pi/2$) at $\theta = \arccos\sqrt{(az^0)} \equiv \theta_c$, which corresponds to the direction of the caustic's asmymptote (3.3.15), as shown in Fig.3.28. The condition $\theta < \theta_c$ defines the bright region where (3.4.16) has two solutions, and $\theta > \theta_c$ the shadow region where (3.4.16) has no solution.

The explicit expression for the ray coordinates $\theta_{1,2}^0$ leads, by means of (3.4.14,16), to the analytic eikonals $\psi_{1,2}$ and amplitudes $A_{1,2}$ of two rays reaching the observation point (R,θ) in the far field:

$$\psi_{1,2} = r\sin\theta - z\cos\theta + \frac{2}{3a}[\cos^3\theta \pm (\cos^2\theta - az^0)^{3/2}]$$

152

$$= R + \frac{2}{3a}[\cos^3\theta \pm (\cos^2\theta - az^0)^{3/2}] \,, \tag{3.4.17}$$

$$A_{1,2} = \frac{B^0}{R} \frac{\sqrt{\cos\theta}}{\sqrt[4]{\cos^2\theta - az^0}} \exp[-\tfrac{1}{4}i\pi(1\pm 1)] \,, \tag{3.4.18}$$

where $R = \sqrt{r^2+z^2}$ is the distance to the observer (Fig.3.28).

In deriving the amplitude with (3.4.12) we have allowed for the possibility of simplifying the equation of rays, emanated into the free space, in the far field:

$$r = r_e - z\tan\theta \simeq -z\tan\theta = R\sin\theta = Rn^0\sin\theta^0_{1,2} \,,$$

whence

$$\frac{\partial r}{\partial\theta^0} = Rn^0\cos\theta^0_{1,2} = \pm R\sqrt{\cos^2\theta - az^0} \,,$$

while the term (1 ± 1) shows up in (3.4.18) due to $q = 1+\mathrm{sgn}(\partial r/\partial\theta^0) = 1\pm 1$.
The complete field of the source equals

$$u(R,\theta) = A_1 e^{ik_0\psi_1} + A_2 e^{ik_0\psi_2}$$

$$= B^0 \frac{e^{ik_0 R}}{R} \frac{2\sqrt{\cos\theta}}{\sqrt[4]{\cos^2\theta - az^0}} \cos\left[\frac{2k_0}{3a}(\cos^2\theta - az^0)^{3/2} - \frac{\pi}{4}\right]$$

$$\times \exp\left[i\frac{2k_0}{3a}\cos^3\theta - \frac{i\pi}{4}\right] \equiv B^0 \frac{e^{ik_0 R}}{R}F(\theta) \,, \tag{3.4.19}$$

where $F(\theta)$ is the required radiation pattern of the source deduced in the geometrical-optics approximation.

Figure 3.29 plots the normalized characteristic radiation patterns $|F_n(\theta)| = |F(\theta)|/|F_{\max}(\theta)|$ calculated with (3.4.19) (dashed lines) and with the exact expression (solid lines). The exact solution has been obtained by the method of separation of variables [3.8, 17]. The plots indicate values of k_0/a equal to 5 (curves 2 in Fig.3.29a), 10 (curves 1 in Fig.3.29a), and 50 (curves in Fig.3.29b). A comparison with the exact solution (see also Fig. 3.30) proves that computations with the geometrical-optics formula (3.4.19) in the region where it is valid ($\theta < \theta_c$, especially for $k_0/a \gg 1$) are fairly accurate. As is apparent from Fig.3.29, for $k_0/a \gg 1$, the radiation pattern in the bright area ($\theta < \theta_c$) has an oscillating character due to the interference of the two ray fields. Beyond the caustic, i.e., for $\theta > \theta_c$, is the area of the geometrical shadow where the field diminishes substantially (it is zero in the geometrical optics approximation). As usual, in the caustic direction $\theta = \theta_c$, geometrical optics yields an infinite value for $F_n(\theta)$ suggesting that

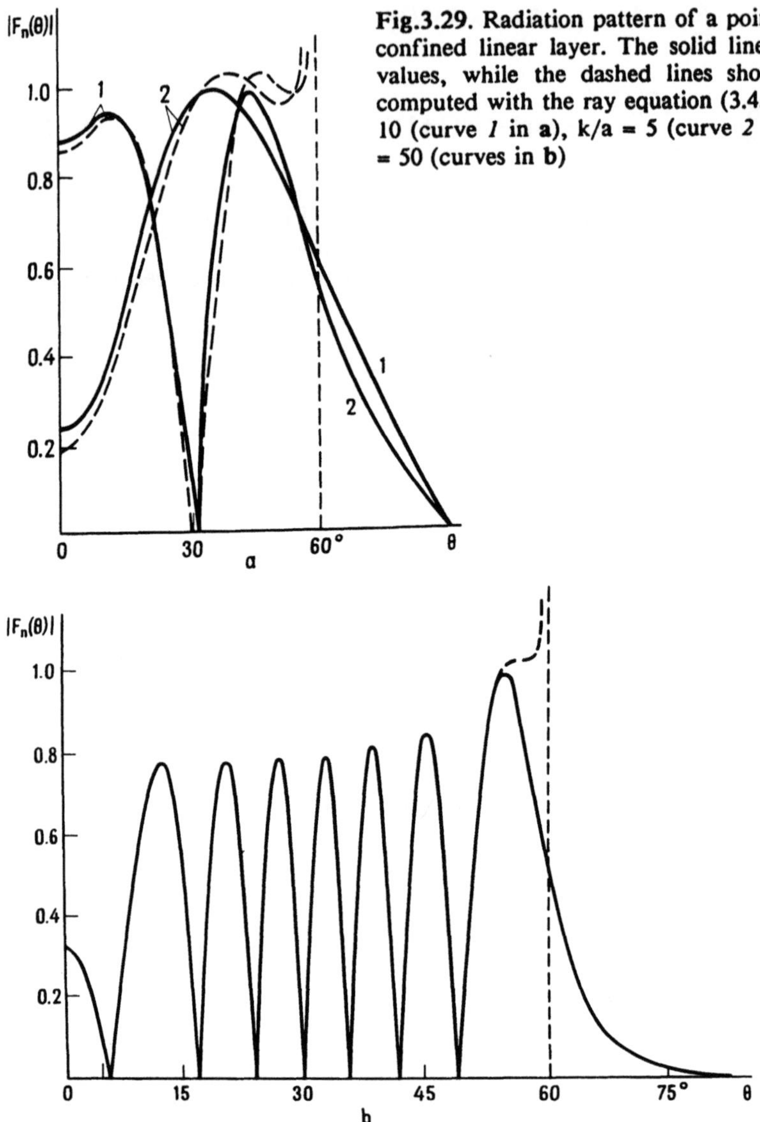

Fig.3.29. Radiation pattern of a point source in a confined linear layer. The solid lines show exact values, while the dashed lines show the values computed with the ray equation (3.4.19) for $k/a = 10$ (curve *1* in **a**), $k/a = 5$ (curve *2* in **a**) and $k/a = 50$ (curves in **b**)

(3.4.19) should be refined. This can be easily done with the formulas of the uniform caustic asymptotics [3.94–96].

Figure 3.30 illustrates how the radiation pattern changes when the source varies its position. The results are plotted for $k_0/a = 20$ and three values of az^0, namely, 0.8 (curves 1 in Fig.3.30a), 0.5 (curves 2 in Fig. 3.30a), and 0.1 (curves in Fig.3.30b). It is seen that as the source moves deeper into the medium, i.e., with increasing parameter az^0, the illuminated sector diminishes (the shadow angle θ_c decreases) and the wrinkling of the pattern also decreases. A fewer number of lobes is consistent with the smaller difference between the optical path length ψ_1 and ψ_2 (along the rays 1 and 2, Fig.3.28).

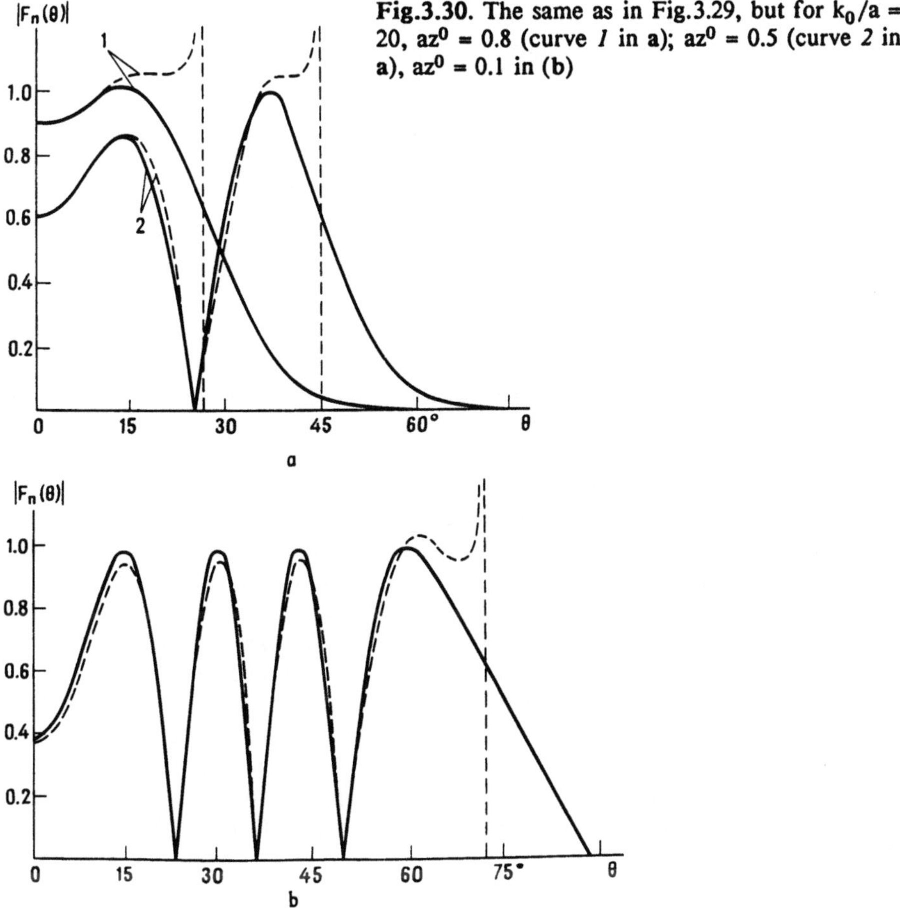

Fig.3.30. The same as in Fig.3.29, but for $k_0/a = 20$, $az^0 = 0.8$ (curve *1* in a); $az^0 = 0.5$ (curve *2* in a), $az^0 = 0.1$ in (b)

The demonstrated results may also serve as a good illustration for the observability criteria which we have already discussed in Sect.2.10.9. Specifically, as can be seen from Figs.3.29b and 30a, the presence of a caustic can readily be proved by the pronounced shadow region where the wave field falls off sharply.

3.4.5 A Point Source in a Parabolic Layer

The field is difficult to calculate because, in contrast to Sect.3.4.4, we cannot obtain the ray coordinates in analytic form. Therefore we confine ourselves to finding the amplitude on the critical ray in the parabolic layer (3.3.23) far form the source.

According to (3.3.29)

$$\mathscr{D}(\tau) = \frac{r\partial(r,z)}{\partial(\theta,\tau)}$$

$$= rn^0[\tau p_z \cos\theta^0 + (n^0/\sqrt{\epsilon_2})\sin^2\theta^0 \sinh(\tau\sqrt{\epsilon_2})] . \tag{3.4.20}$$

155

On the critical ray ($z \to z_m$ as $\tau \to \infty$) we have $p_z \to 0$ and $\tau = r/n^0 \sin\theta_{cr} = r/\sqrt{\epsilon_m}$. As a result, as $\tau \to \infty$ and $\theta^0 = \theta^0_{cr}$

$$\mathscr{D}(\tau) \simeq (r/\sqrt{\epsilon_2})(n^0 \sin\theta^0_{cr})^2 \sinh(r\sqrt{\epsilon_2/\epsilon_m})$$

$$\simeq (r\epsilon_m/2\sqrt{\epsilon_2})\exp(r\sqrt{\epsilon_2/\epsilon_m}) \; ,$$

and in accordance with (3.4.11) we get

$$A = B^0 \sqrt{n^0 \sin\theta^0/D(\tau)} \simeq B^0 \sqrt{2/rh}\exp(-r/2h) \; , \tag{3.4.21}$$

where $h = \sqrt{\epsilon_m/\epsilon_2}$. Expression (3.4.21) holds for $r/h \gg 1$ and indicates that the amplitudes decay *exponentially* with the distance in antiwaveguide propagation (Sect. 3.3.8). This pattern is obviously due to the "instability" of the rays on the barrier's top (Figs. 3.25, 31): the more the amplitude decreases, the flatter the barrier's top (the lower E_2) and the greater the square of the refractive index ϵ_m at the top. The exponent in (3.4.21) is defined by the dimension of the layer $h = \sqrt{(\epsilon_m/\epsilon_2)}$, equal to the distance $|z_m - z|$ at which $n^2(z)$ increases twofold compared with ϵ_m, i.e., $n^2(z = z_m \pm h) = 2\epsilon_m$. The region where expressions of the type (3.4.21) are applicable can be readily determined with the criteria of Sect. 2.10 (see also Sect. 3.4.7).

An investigation of the field in the vicinity of the critical ray has specifically been carried out in [3.78]; the antiwaveguide mechanism of wave propagation has been the focus of several studies, particularly [3.16, 97-99]. A large number of papers have been devoted to various aspects of wave-field propagation in plane stratified media [3.16, 44, 57, 59, 61, 68, 69, 80, 81, 86, 100-103].

3.4.6 The Fresnel Volumes in Plane-Stratified Media

The shape of the Fresnel volumes in plane-stratified media was analyzed in [3.15]. We present some results from [3.15] for the "plane" phase front (a linear eikonal distribution $\psi^0 = \xi n^0 \sin\theta^0$ at $z = z^0$, see also Sect. 3.4.2); the ray pattern for this case was discussed in Sect. 3.3.3 (Fig. 3.17). For a caustic with $z = z_{tp} =$ constant we have two *Fresnel volumes*, FI and FII, (Fig. 3.32) corresponding to the two rays reaching the observation point (x, z) (Fig. 3.17).

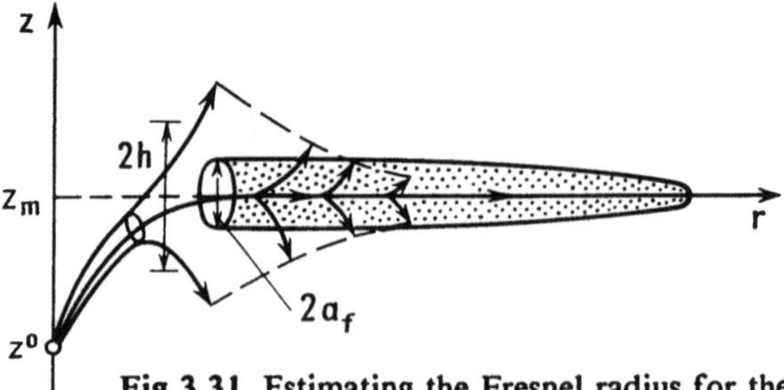

Fig. 3.31. Estimating the Fresnel radius for the critical ray in a parabolic layer

156

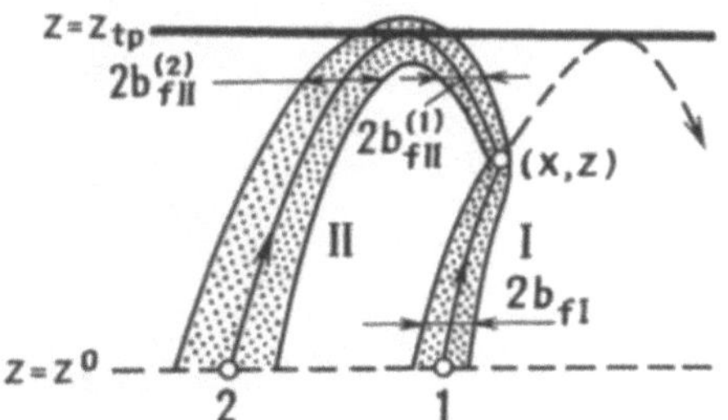

Fig.3.32. Fresnel volumes of rays in a plane-stratified medium

The horizontal dimension of FI is

$$b_{fI}(z') = |x' - x'_0| = \sqrt{\lambda_0 \frac{d}{ds^0} \int_{z'}^{z} \frac{s^0 dz}{\sqrt{n^2(z) - (s^0)^2}}} \equiv \sqrt{\lambda_0 L_I}, \qquad (3.4.22)$$

where $s^0 \equiv n^0 \sin\theta^0$, (x, z) is the observation point, (x', z') is a certain point of the Fresnel-volume boundary, and x'_0 is the coordinate of the point where "reference" ray 1 crosses the plane $z' = $ constant (Fig.3.33a). The general view of FI is depicted in Fig.3.33b. The salient feature of FII corresponding to ray 2 reflected from the caustic, is its double valuedness: for $z' > z$ it crosses the plane $z' = $ const. twice, giving rise thereby to the two sizes

$$b_{fII}^{1,2}(z') = \sqrt{\lambda_0 \frac{d}{ds^0} \frac{\left(\int_{z'}^{z_{tp}} \mp \int_{z}^{z_{tp}}\right) s^0 dz}{\sqrt{n^2(z) - (s^0)^2}}} \equiv \sqrt{\lambda_0 L_{II}^{1,2}}. \qquad (3.4.23)$$

Formulas (3.4.22, 23) have been derived for a square approximation in $\Delta x = |x' - x'_0|$ (Sect.2.10). Evidently, in the absence of the caustic only one Fresnel volume appears (Fig.3.33a) with the size defined by (3.4.22).

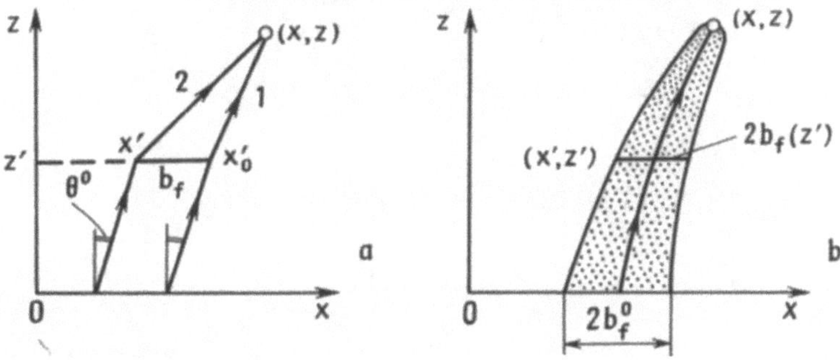

Fig.3.33. Illustration of the Fresnel volume estimation for a ray incident on a stratified medium

At $\theta^0 = 0$ we get from (3.4.23)

$$b_{fI}(z') = \sqrt{\lambda_0 \int_{z'}^{z} \frac{dz}{n(z)}} \quad ,$$

(3.4.24)

$$b_{fII}^{1,2}(z') = \sqrt{\lambda_0 \left(\int_{z'}^{z_{tp}} \mp \int_{z}^{z_{tp}} \right) dz/n(z)} \quad .$$

At constant $n(z) = n^0$ Eq.(3.4.24) yields $b_{fI} = [\lambda_0(z-z')/n^0]^{1/2}$, which certainly coincides with (2.10.15). For the linear layer (3.3.18) we have by virtue of (3.4.24)

$$b_{fI}(z') = \{(2\lambda_0 H[n(z') - n(z)]\}^{1/2} \ , \quad z' < z \ ,$$

$$b_{fII}^{1,2}(z') = \{(2\lambda_0 H[n(z) \mp n(z')]\}^{1/2} \ , \quad z' > z \ .$$

(3.4.25)

The Fresnel volumes in the linear layer (at $\theta^0 = 0$) are illustrated in Fig.3.34a; FII has the form of an upside-down bag with the bottom fallen inside - illustrated separately in Fig.3.34b.

By analogy with (3.4.24) we have for a point source (for the ray at $\theta^0 = 0$)

$$b_{fI} = \sqrt{\lambda_0 \left(\frac{1}{L_1} + \frac{1}{L_2} \right)^{-1}} = \sqrt{\lambda_0 \frac{L_1 L_2}{L_1 + L_2}} \ , \qquad \text{where}$$

(3.4.26)

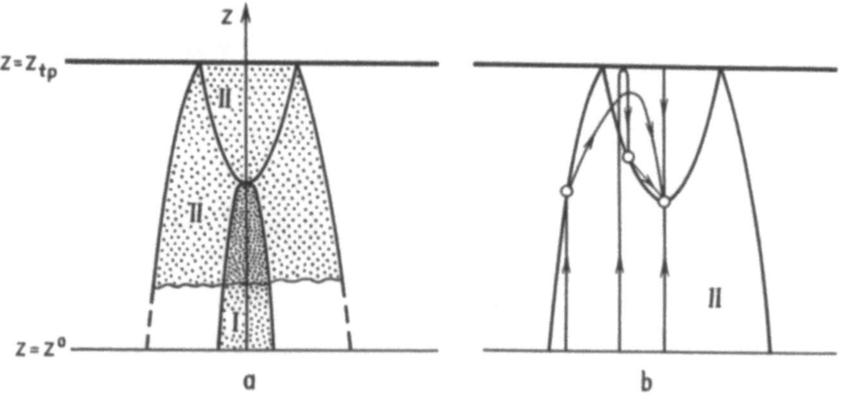

Fig.3.34. (a) Fresnel volume profiles for the incident (I) and reflected (II) rays (*stippled*). (b) Reflected virtual rays that form the Fresnel volume

$$L_1 = \int_{z^0}^{z'} \frac{dz}{n(z)} , \quad L_2 = \int_{z'}^{z} \frac{dz}{n(z)} .$$

A similar formula is found for $b_{fII}^{1,2}$ as well. Needless to say, to determine the Fresnel volume in the general case, one would have to resort to numerical techniques.

3.4.7 Validity Conditions
of the Geometrical-Optics Approximation

In the one-dimensional problem of plane-wave propagation (Sect. 3.4.4) these conditions coincide with the familiar conditions for the applicability of the WKB approximation [3.16, 86, 89, 90] whereas in the general case we should use the criteria of Sect. 2.10.4. Consider, for instance, the case of a plane phase front when the size of Fresnel volume b_f is known analytically (Sect. 3.4.6). For a nonuniform initial amplitude distribuiton $A^0 = A^0(x)$ at $z = z^0$ and $-b < x < b$, which is characteristic of the ray bundle of width $2b$, the validity criterion of the ray-optical solution (Sect. 2.10) takes the form

$$b_f^0 \left| \frac{1}{A^0} \frac{dA^0}{dx} \right| \ll 1 , \quad b_f^0 \ll b , \tag{3.4.27}$$

where b_f^0 is the size of the Fresnel zone on the initial surface, i.e., at $z = z^0$. Using the generalized notation L_{eq} for the equivalent distances L_I, $L_{II}^{1,2}$ at $z' = z^0$ in (3.142, 143), we rewrite (3.4.27) as

$$L_{eq} \ll \frac{b^2}{\lambda_0} , \quad L_{eq} \ll \frac{1}{\lambda_0} \left| \frac{1}{A^0} \frac{dA^0}{dx} \right|^{-2} , \tag{3.4.28}$$

imposing the restrictions on the equivalent distance L_{eq}. Specifically, for $\theta^0 = 0$, in the absence of caustics these conditiosn imply that

$$\left| \int_{z^0}^{z} \frac{dz}{n(z)} \right| \ll \frac{b^2}{\lambda_0} , \quad \left| \int_{z^0}^{z} \frac{dz}{n(z)} \right| \ll \frac{1}{\lambda_0} \left| \frac{1}{A^0} \frac{dA^0}{dx} \right|^{-2} , \quad \text{or} \tag{3.4.29a}$$

$$\left| \int_{z^0}^{z} \lambda(z)dz \right| \ll b^2 , \quad \left| \int_{z^0}^{z} \lambda(z)dz \right| \ll \left| \frac{1}{A^0} \frac{dA^0}{dx} \right|^{-2} , \tag{3.4.29b}$$

where $\lambda(z) = \lambda_0/n(z)$ is the local wavelength in the medium. In addition to these conditions one needs to follow the variation of n(z) within the Fresnel volume (Sect. 2.10).

With the help of these criteria we determine, by way of example, the validity region for the ray approximation (3.4.21) for the wave field on the critical ray in a parabolic layer. Near the barrier's top where $|z-z_m| < h$, the medium can be deemed homogeneous, i.e., $n(z) \simeq \sqrt{\epsilon_m}$ (Sect. 3.4.5), so that the radius of the Fresnel zone can be taken equal to $a_f \simeq \sqrt{\lambda_0 r/n} \simeq (\lambda_0 r)^{1/2}/\epsilon_m^{1/4}$. Subject to the condition $a_f \leq h$, which assumes homogeneity within a_f (Fig. 3.28), we find the estimate

$$r \ll \frac{\sqrt{\epsilon_m}}{\lambda_0} h^2 = \frac{\epsilon_m^{3/2}}{\lambda_0 \epsilon_2} = \frac{k_0 \epsilon_m^{3/2}}{2\pi\epsilon_2} \, ,$$

consistent with the result of [3.77] obtained by the method of related functions.

Other examples which employ the above-derived applicability conditions were discussed in [3.15, 16]. An interesting aspect of the near and far fields of an antenna (wave pencil) in an inhomogeneous medium has been evaluated in [3.104–106].

3.5 Waves in Radially Inhomogeneous Media

3.5.1 Ray Equations for Spherically Stratified Media

Radially inhomogeneous media of $n = n(r)$ are encountered as frequently as plane-stratified media. The equations of rays for a spherically stratified medium with $n = n(r)$, according to (2.2.44), take the form

$$\frac{dr}{d\tau} = \tilde{p}_r \, , \qquad \frac{d\tilde{p}_r}{d\tau} = n \frac{dn}{dr} + \frac{1}{r}(n^2 - \tilde{p}_r^2) \, ,$$

$$\frac{d\theta}{d\tau} = \frac{1}{r}\tilde{p}_\theta \, , \qquad \frac{d\tilde{p}_\theta}{d\tau} = \frac{1}{r}(\tilde{p}_\phi^2 \cot\theta - \tilde{p}_r\tilde{p}_\theta) \, , \qquad (3.5.1)$$

$$\frac{d\phi}{d\tau} = \frac{1}{r\sin\theta}\tilde{p}_\phi \, , \qquad \frac{dp_\phi}{d\tau} = -\frac{\tilde{p}_\phi}{r\sin\theta}(\tilde{p}_r\sin\theta + \tilde{p}_\theta\cos\theta) \, ,$$

where the components of momentum $\tilde{p}_r$, $\tilde{p}_\theta$, and $\tilde{p}_\phi$ are connected by the eikonal equation $p_r^2 + p_\theta^2 + p_\phi^2 = n^2(r)$. Simple manipulations transform (3.5.1) into

$$\frac{d}{d\tau}[r^2(n^2 - \tilde{p}_r^2)] = 0 \, , \qquad \frac{d}{d\tau}(r\tilde{p}_\phi\sin\theta) = 0, \qquad \text{whence}$$

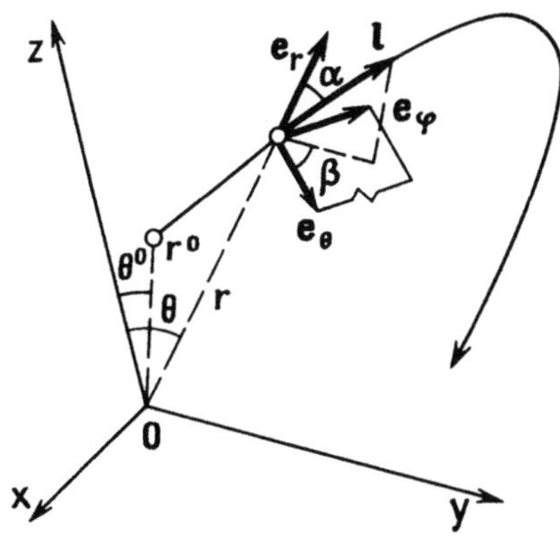

Fig.3.35. Illustrating the definition of the refraction angles α and β in a radially inhomogeneous medium

$$r^2(n^2 - \tilde{p}_r^2) = \text{const.} = (r^0)^2\,[(n^0)^2 - (\tilde{p}_r^0)^2] \equiv \rho^2 \;,$$

$$r\tilde{p}_\phi \sin\theta = \text{const.} = r^0 \tilde{p}_\phi^0 \sin\theta^0 \equiv \rho_1 \;. \tag{3.5.2}$$

If by analogy with (3.3.6) we introduce further the angles of refraction α, β, and α^0, β^0 (Fig.3.35) by

$$\tilde{p}_r = n\cos\alpha \;,\quad \tilde{p}_\theta = n\sin\alpha\sin\beta \;,\quad \tilde{p}_\theta = n\sin\alpha\cos\beta \;,$$

$$\tilde{p}_r^0 = n^0\cos\alpha^0 \;,\quad \tilde{p}_\phi^0 = n^0\sin\alpha^0\sin\beta^0 \;,\quad \tilde{p}_\theta^0 = n^0\sin\alpha^0\cos\beta^0 \;, \tag{3.5.3}$$

then from (3.5.2) we determine the laws of refraction of a wave in a spherically stratified medium[20]

$$rn(r)\sin\alpha = \text{const.} = r^0 n(r^0)\sin\alpha^0 \equiv \rho \;, \tag{3.5.4a}$$

$$\sin\theta\sin\beta = \text{const.} = \sin\theta^0\sin\beta^0 \;. \tag{3.5.4b}$$

If $\beta^0 = 0$ ($\tilde{p}_\phi^0 = 0$) and the rays lie in the plane of constant $\phi = \phi^0$, then the condition (3.5.4b) is satisfied identically, and therefore would drop out from further consideration.

Taking into account (3.5.2-4) the ray equations (3.5.1) can be integrated by quadrature. To demonstrate this, we have from (3.5.2)

$$\tilde{p}_r = \pm\,\sqrt{n^2(r) - \rho^2/r^2} \;,$$

$$\tilde{p}_\phi = \rho_1/r\sin\theta \;, \tag{3.5.5}$$

$$\tilde{p}_\theta = \pm\,\frac{1}{r}\sqrt{\rho^2 - \rho_1^2/\sin^2\theta} \;, \qquad \text{and therefore}$$

---

[20] The parameter ρ defined by (3.5.4) is proportional to the shortest distance $r^0 \sin\alpha^0$ from the medium center $r = 0$ to the ray. Therefore it shall be called the generalized impact parameter of the ray.

$$\frac{dr}{d\theta} = \frac{r\tilde{p}_r}{\tilde{p}_\theta} = \pm \frac{r\sin\theta\sqrt{r^2 n^2(r) - \rho^2}}{\sqrt{\rho^2\sin^2\theta - \rho_1^2}} \,,$$

$$\frac{d\phi}{d\theta} = \frac{\tilde{p}_\phi}{\tilde{p}_\theta \sin\theta} = \pm \frac{\rho_1}{\sin\theta\sqrt{\rho^2\sin^2\theta - \rho_1^2}} \,.$$

$$(3.5.6)$$

To simplify the discussion, in what follows we consider the *axially symmetric* case where $\rho_1 = 0$ (i.e., $\tilde{p}_\phi^0 = 0$) and the initial conditions are independent of ϕ. Now from (3.5.6), similar to how we have done it for (3.3.4), we have ($\rho > 0$)

$$\theta = \theta^0 + \left(\int_{\tilde{r}_<}^{\tilde{r}_>} \mp \int_{r_<}^{r_>}\right) \frac{\rho\, dr}{r\sqrt{r^2 n^2(r) - \rho^2}} \,,$$

$$(3.5.7)$$

where $\tilde{r}_<$ and $\tilde{r}_>$ are, respectively, the lowest and largest values of r^0 and r_{tp}, respectively, whereas $r_< \equiv \min(r, r_{tp})$, $r_> \equiv \max(r, r_{tp})$. The coordiante r_{tp} satisfies the equation $r_{tp} n(r_{tp}) = \rho$ and defines the position of the turning point on the ray in which $\tilde{p}_r = 0$ and $\alpha = \pi/2$, the quantity $\tilde{p}_r$ being positive on the upgoing and negative on the downgoing leg of the ray (Fig. 3.36). Equation (3.5.7) is written for a ray having a single turning point (Fig.3.36), that is, when the quantity $rn(r)$ is a monotonous function of r. If $rn(r) > \rho$, there is no turning point on the ray and (3.5.7) should be taken with the upper sign, and the difference of the integrals, as in Sect.3.3, replaced by one integral from r^0 to r.

Introduction of the function $n_M(r) = (r/r^0)n(r)$, referred to as the *modified* refractive index, reduces formally the refraction law (3.5.4) to that in a plane-stratified medium (3.3.7). Let $r = r^0 + z$ be distance from the center of the sphere, and $z \ll r^0$ be the altitude measured from the level $r = r^0$. Then the use of the modified refractive index $n_M(z) \simeq n(z)(1+z/r^0)$

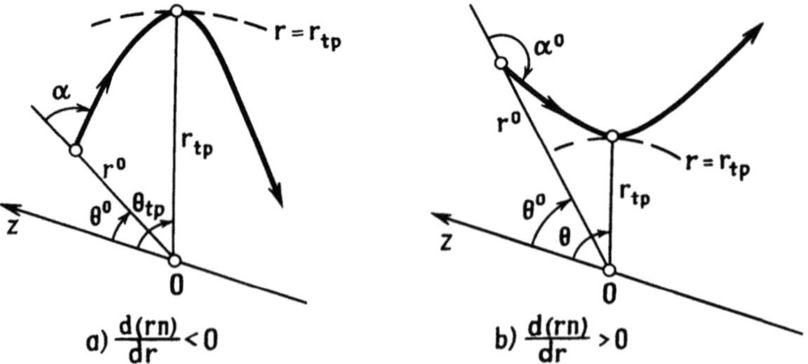

Fig.3.36. Bending of the ray path (a) toward the center and (b) away from the center of symmetry in a radially inhomogeneous medium

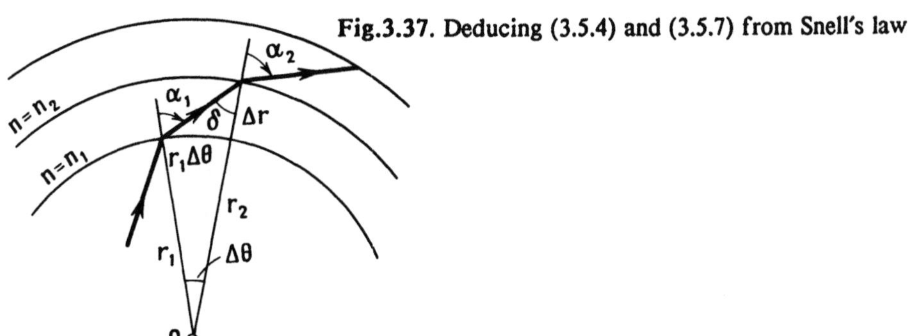

Fig.3.37. Deducing (3.5.4) and (3.5.7) from Snell's law

or, which is equivalent, the modified permittivity $\epsilon_M(z) \simeq \epsilon(z)(1+2z/r^0)$, approximately reduces the problem of the propagation of waves in spherically stratified media to the problem of a plane-stratified medium. This approach is widely applied in the theory of radio-wave propagation in the earth's troposphere and ionosphere [3.17, 82, 107]. We shall employ this approach in Sect.3.7 by discussing the possible types of ionospheric waveguides.

Note that (3.5.4, 7) derived above could be easily deduced on geometrical grounds (Fig.3.37). From Snell's law (2.4.8) it follows that $n_1 \sin\delta = n_2 \sin\alpha_2$ for spherical, homogeneous layers, but from the laws of sines $r_2 \sin\delta = r_1 \sin\alpha_1$ (Fig.3.37) so that $n_1 r_1 \sin\alpha_1 = n_2 r_2 \sin\alpha_2 = $ constant, which corresponds to (3.5.4a). From Fig.3.37 it is obvious that $\tan\alpha = rd\theta/dr$. Taken together with (3.5.4a) this equality leads to the ray equation (3.5.7).

3.5.2 The Eikonal Function for Spherically Stratified Media

The complete integral of the eikonal equation in a spherically stratified medium with $n = n(r)$ is, according to Sect.2.8,

$$\psi = \rho_1 \phi \pm \int^r \sqrt{n^2 - (\rho^2)/(r^2)} \, dr \pm \int^\theta \sqrt{\rho^2 - \rho_1^2/\sin^2\theta} \, d\theta + \psi^0 \; . \qquad (3.5.8)$$

Noting that $\mathbf{p} = \nabla\psi$, from (3.5.8) we immediately obtain (3.5.5) for $\tilde{p}_r$, $\tilde{p}_\phi$, and $\tilde{p}_\theta$, the conservation law (3.5.2) following from the fact that the coordinate ϕ is cyclic. Under the same conditions ($\rho_1 = 0$) that have been used in deriving the ray equation (3.5.7) we have

$$\psi = \psi^0 + \rho(\theta-\theta^0) + \left(\int_{\tilde{r}_<}^{\tilde{r}_>} \mp \int_{r_<}^{r_>} \right) \sqrt{n^2(r) - \rho^2/r^2} \, dr \; , \qquad (3.5.9)$$

from which we find that on the ray (3.5.7)

$$\psi = \psi^0 + \left(\int_{\widetilde{r}_<}^{\widetilde{r}_>} \mp \int_{r_<}^{r_>} \right) \frac{rn^2(r)dr}{\sqrt{r^2 n^2(r) - \rho^2}} \ . \tag{3.5.10a}$$

The same equation results if we integrate $n^2(r)$ along the ray (3.5.7) since, subject to (3.5.1), $d\tau = dr/\widetilde{p}_r$. To illustrate, on the upgoing segment of the ray $(\widetilde{p}_r > 0)$

$$\psi = \psi^0 + \int_0^\tau n^2 d\tau$$

$$= \psi^0 + \int_{r^0}^r \frac{n^2}{p_r} \, dr = \psi^0 + \int_{r^0}^r \frac{rn^2(r)dr}{\sqrt{r^2 n^2(r) - \rho^2}} \ . \tag{3.5.10b}$$

3.5.3 Cylindrically Stratified Media

For $n = n(r)$ the ray equation (3.2.6) becomes in the cylindrical coordinates r, ϕ, z:

$$\frac{dr}{d\tau} = \widetilde{p}r \ , \qquad \frac{d\phi}{d\tau} = \frac{1}{r}\widetilde{p}_\phi \ , \qquad \frac{dz}{d\tau} = \widetilde{p}_z \ ,$$

$$\frac{d\widetilde{p}_r}{d\tau} = n\frac{dn}{dr} + \frac{1}{r}\widetilde{p}_\phi^2 \ , \qquad \frac{d\widetilde{p}_\phi}{d\tau} = -\frac{1}{r}\widetilde{p}_r\widetilde{p}_\phi \ , \qquad \frac{d\widetilde{p}_z}{d\tau} = 0 \ , \tag{3.5.11}$$

where $\widetilde{p}_r^2 + \widetilde{p}^2 + \widetilde{p}_z^2 = n^2(r)$. From these expressions we find immediately

$$\widetilde{p}_z = \text{const.} = \widetilde{p}_z^0 \ , \qquad r\widetilde{p}_\phi = \text{const.} = r^0\widetilde{p}_\phi^0 \equiv \rho \ , \tag{3.5.12a}$$

and, consequently,

$$\widetilde{p}_r = \pm \sqrt{n^2(r) - \rho^2/r^2 - (\widetilde{p}_z^0)^2} \ . \tag{3.5.12b}$$

Then the ray equations (3.5.11) may be represented as

$$\frac{d\phi}{dr} = \pm \frac{\rho}{r\sqrt{r^2(n^2 - \widetilde{p}_z^{0\,2}) - \rho^2}} \ , \qquad \frac{dz}{dr} = \pm \frac{r\widetilde{p}_z^0}{\sqrt{r^2(n^2 - \widetilde{p}_z^{0\,2}) - \rho^2}} \ , \tag{3.5.13}$$

to be integrated by quadrature. At $\widetilde{p}_\phi = 0$ $(\rho=0)$, (3.5.13) coincide with the ray equations for a plane-stratified medium, whereas at $\widetilde{p}_z = 0$ the ray equation (3.5.7) for a spherically stratified medium. An explanation for this

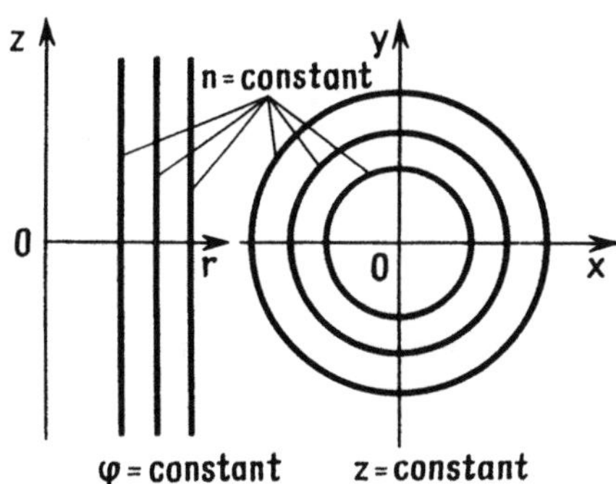

Fig.3.38. Lines of equal n of a radially inhomogeneous medium in planes ϕ = constant and z = constant

can be furnished by the fact that in the two principal planes ϕ = const. and z = const. the loci of constant n of the cylindrically stratified medium are, respectively, plane and circular layers (Fig.3.38). Indeed, for $\tilde{p}_z = 0$ the rays are in the plane of constant $z = z^0$ and are defined by

$$\phi = \phi^0 + \left(\int_{\tilde{r}_<}^{r_>} \mp \int_{r_<}^{r_>} \right) \frac{\rho\,dr}{r\sqrt{r^2 n^2(r) - \rho^2}}, \tag{3.5.14}$$

where the notation used is the same as in (3.5.7). Equation (3.5.14) and the refraction law $r\tilde{p}_\phi$ = constant coincide, respectively, with (3.5.7 and 4) if for $p_z^0 = 0$ we define the angle of refraction, α, by the relations $\tilde{p}_\phi = n\sin\alpha$ and $p_r = n\cos\alpha$ similar to (3.5.3). In the general case, when $\tilde{p}_\phi^0 \neq 0$ ($\rho \neq 0$) and $p_z^0 \neq 0$, the rays (3.5.13) are no longer represented by plane curves and possess torsion [3.108]. As follows from Sect.2.8 the eikonal of a wave in a cylindrically stratified medium is as follows:

$$\psi = \tilde{p}_z^0(z - z^0) + \rho(\phi - \phi^0)$$

$$+ \left(\int_{\tilde{r}_<}^{\tilde{z}_>} \mp \int_{r_<}^{r_>} \right) \sqrt{n^2(r) - p^2/r^2 - (\tilde{p}_z^0)^2}\; dr + \psi^0 , \tag{3.5.15}$$

whence for $p_z^0 = 0$ we find the expressions which differ from (3.5.9, 10) only replacing $\theta - \theta^0$ by $\phi - \phi^0$.

3.5.4 Ray Geometry

Here we confine ourselves to considering the pattern of rays in a spherically stratified medium at ϕ = const. or in a cylindrically stratified medium at z = constant. Many properties of such rays are analogous to those of rays in a plane-stratified medium (Sect.3.3). In particular, the rays are plane curves possessing symmetric branches with respect to the turning point (r_{tp}, θ_{tp}) (Fig.3.36). The radially inhomogeneous medium is outstanding in

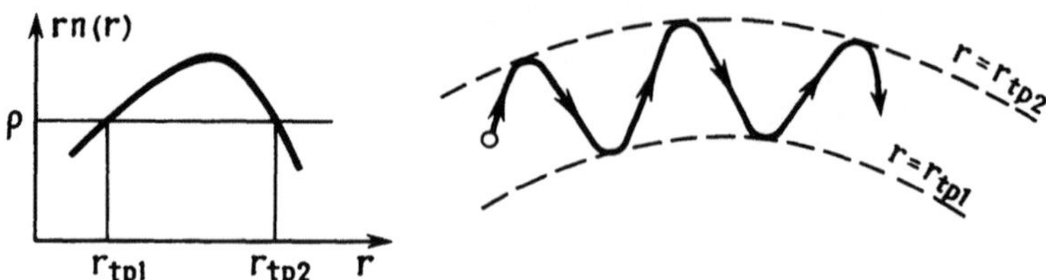

Fig.3.39. Waveguiding profile of rn(r) (*left*) confining rays in a radially inhomogeneous channel (*right*)

that it has a central point $r = 0$ around which ray paths can revolve various times.

It is customary to distinguish between the *infinite* and the *finite* trajectories[21] of which the first approach infinity in r (as $r \rightarrow \infty$), and the second concentrate in the annular region $r_{tp1} \leq r \leq r_{tp2}$, where $r_{tp1,2}$ satisfy the equation $r_{tp1,2} n(r_{tp1,2}) = \rho$. The finite ray paths correspond to waveguide propagation in a radially inhomogeneous medium (Sect.3.7). They arise only when the function rn(r) possesses a maximum; besides on the trajectory there is always $rn(r) \geq \rho$ (Fig.3.39). In their simplest form the infinite trajectories are illustrated in Fig.3.36; in the general case they can form several loops around the center $r = 0$, as shown in Fig.3.40a [3.115].

Looking at finite trajectories we may single out among them the ray for which at $r = r_{tp} = r_{cr}$, the function rn(r) possesses a *minimum* (Fig.3.40b). For this ray the angular coordinate θ_{tp} of the turning point becomes infinite (Fig.3.40c)[22] and the ray starts spiraling around the point

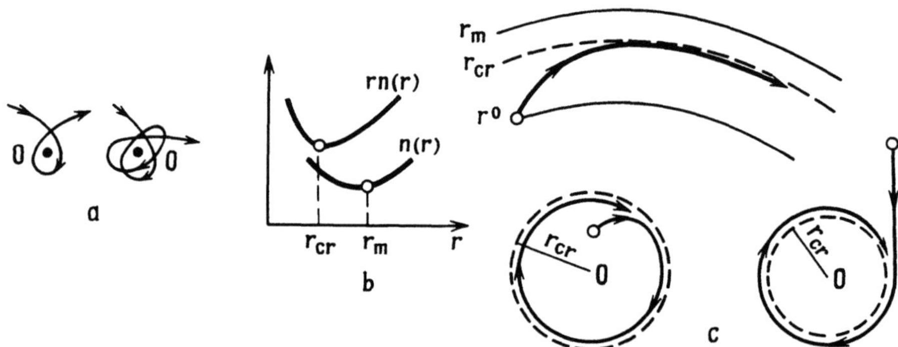

Fig.3.40. (a) Single and multiple-loop ray paths in a radially inhomogeneous medium. (b) rn(r) profile which ensures the existence of a critical ray. (c) Specific situations of rays approaching the circle of critical radius

---

[21] The names are borrowed from mechanics [3.109,110]. In many works on mechanics [3.109-114] the trajectories of particles moving in a central force field U(r) have been given extensive study. By virtue of the analogy existing between optics and mechanics (Sect.2.2) these results can be easily transferred onto wave problems (geometrical optics).

[22] By a Taylor expansion, similar to (3.38) for $r^2 n^2(r)$, one may readily prove that if $d(rn)/dr = 0$ at $r = r_{tp}$, then the improper integral in (3.5.7) diverges logarithmically as r $\rightarrow r_{tp}$.

166

$r = 0$ nearing the level $r = r_{cr}$ asymptotically (in quantum mechanics the spiraling gives rise to spiral scattering phenomena [3.116, 117]. This ray is analogous to the critical ray near the top of a plane layer (Fig.3.25, Sects.3.3, 4), but in contrast to the latter it arises at $dn/dr = -n/r < 0$ (rather than at $dn/dr = 0$).

As follows from Sect.2.2, the ray curvature in a radially inhomogeneous medium is

$$K = \frac{1}{n}\left|\frac{dn}{dr}\right|\sin\alpha = \frac{\rho}{rn^2}\left|\frac{dn}{dr}\right| , \tag{3.5.16}$$

where $\alpha = \alpha(r)$ is angle of refraction, and $\rho = r^0 n^0 \sin\alpha^0$ is the impact parameter of the ray. Note that properties of ray trajectories in radially inhomogeneous media have been treated in a large number of publications [3.63, 80-84, 86, 91, 107, 118-121]; *Krasnushkin* [3.122] has used the method of the phase plane for his study.

3.5.5 The Field due to a Point Source

The ray family generated by a point source with the coordinates r^0, $\theta^0 = 0$ in a spherically stratified medium is defined by (3.5.7), whence

$$\theta = \left(\int_{\tilde{r}_<}^{\tilde{r}_>} \mp \int_{r_<}^{r_>}\right)\frac{\rho\,dr}{r\sqrt{r^2 n^2(r) - \rho^2}} \equiv \theta(r,\alpha^0) , \tag{3.5.17}$$

where $\rho = r^0 n^0 \sin\alpha^0$, and α^0 is the angle at which the ray leaves the source. Observing that rays (3.5.17) do not depend on ϕ and lie in the plane of constant $\phi = \phi^0$, we get

$$\mathscr{D}(\tau) = \frac{\partial(x,y,z)}{\partial(\alpha^0,\phi^0,\tau)}$$

$$= r^2\sin\theta\,\frac{\partial(r,\theta)}{\partial(\alpha^0,\tau)} = r^2\tilde{p}_r\sin\theta\,\frac{\partial}{\partial\alpha^0}\theta(r,\alpha^0) , \tag{3.5.18}$$

where $\tilde{p}_r$ is given by (3.5.5).

Taking into account (3.5.18) and (3.3.3) we may find the field amplitude due to a point source

$$A = B^0\sqrt{\frac{n^0\sin\alpha^0}{\mathscr{D}(r)}} = B^0\sqrt{\frac{\rho}{r^2 r^0\sin\theta\tilde{p}_r\,\dfrac{\partial}{\partial\alpha^0}\theta(r,\alpha^0)}} .$$

From this, similar to (3.4.12), we have

$$A = \frac{B^0}{\sqrt[4]{r^2 n^2(r) - \rho^2}} \sqrt{\frac{\rho}{rr^0 \sin\theta \left| \partial\theta/\partial\alpha^0 \right|}} \, \exp(-\tfrac{1}{2} i\pi q) \, , \qquad (3.5.19a)$$

where

$$q = \tfrac{1}{2}(1 - \mathrm{sgn}\mathscr{D}) = \frac{1}{2}\left[1 - \mathrm{sgn}(\tilde{p}_r)\mathrm{sgn}\frac{\partial\theta}{\partial\alpha^0} \right] . \qquad (3.5.19b)$$

Representing (3.5.19a) in the form $|A| = B^0/R_{\mathrm{eff}}$, we define the effective distance R_{eff} and the focusing factor f (Sects.2.4 and 3.4.3) in a spherically stratified medium:

$$R_{\mathrm{eff}} = \sqrt{rr^0 \sin\theta \frac{1}{\rho} \sqrt{r^2 n^2(r) - \rho^2} \left| \frac{\partial\theta}{\partial\alpha^0} \right|} \, , \qquad (3.5.20a)$$

$$f = \left| \frac{A}{A_h} \right|^2 = \frac{\rho R^2}{rr^0 \sqrt{[r^2 n^2(r) - \rho^2] \left| \dfrac{\partial\theta}{\partial\alpha^0} \right|}} \equiv (R/R_{\mathrm{eff}})^2 \, . \qquad (3.5.20b)$$

The eikonal (3.5.9) for a point source is given by

$$\psi = \rho\theta + \left(\int_{\tilde{r}_<}^{\tilde{r}_>} \mp \int_{r_<}^{r_>} \right) \sqrt{n^2(r) - \rho^2/r^2} \, dr. \qquad (3.5.21)$$

For a linear source in a cylindrically stratified medium (a two-dimensional problem) we have

$$A = B^0 \sqrt{\frac{n^0 d\alpha^0}{n \, da}} = B^0 \sqrt{\frac{n^0}{r\tilde{p}_r \, \partial\phi/\partial\alpha^0}}$$

$$= \frac{B^0 \sqrt{n^0}}{\sqrt[4]{r^2 n^2(r) - \rho^2}} \left| \frac{\partial\phi}{\partial\alpha^0} \right|^{-1/2} \exp(-\tfrac{1}{2} i\pi q) \, , \qquad (3.5.22)$$

where $\phi = \phi(r, \alpha^0)$ is the equaiton of the ray family (3.5.14), and the quantity q is analogous to (3.5.19b) with ϕ substituted for θ. The same substitution is to be done in (3.5.21) for the eikonal.

168

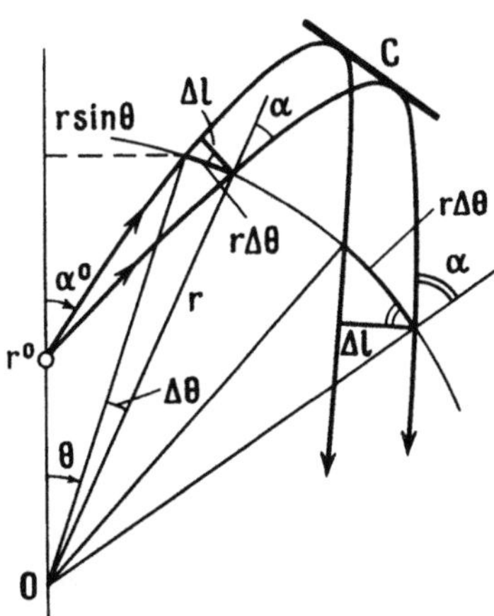

Fig.3.41. Geometrical approach to the equations for amplitude in a radially inhomogeneous medium

Formula (3.5.18) for $\mathscr{D}(r)$, the respective formula for the two-dimensional problem $\mathscr{D}_{2D}(r) = r\tilde{p}_r \partial\phi/\partial\alpha^0$, and (3.5.19, 22) for the amplitudes can readily be deduced from a geometrical argument (Fig.3.41) if we observe that the cross-sections of the ray tube are $\Delta a = r\sin\theta\,\Delta\phi^0\,\Delta l$, $\Delta l = r\cos\alpha\,\Delta\theta$, and that $n\cos\alpha = \tilde{p}_r$.

3.5.6 The Field of a Plane Wave

When $n^2(r) \to 1$ as $r \to \infty$, we may define, at infinity, a plane wave by $u^0 = A^0\exp(-ik_0 z)$, incident on the center of a spherically stratified medium (Fig.3.42). The ray equations becomes

$$\theta = \left(\int_{r_{tp}}^{\infty} \mp \int_{r_{tp}}^{r}\right) \frac{\rho\, dr}{r\sqrt{r^2 n^2(r) - \rho^2}} \equiv \theta(r,\rho) , \tag{3.5.23}$$

where ρ is the impact parameter of the ray (Fig.3.42).[23] The computations like those for a point source yield the following amplitude and eikonal:

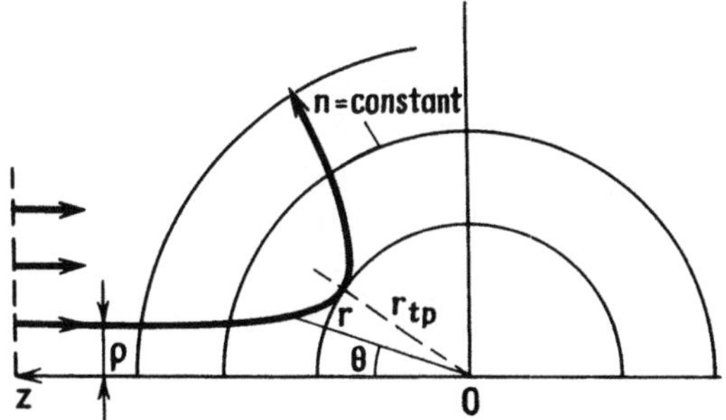

Fig.3.42. Path of a ray experiencing reflection in a radially inhomogeneous medium (ρ is the impact parameter)

---

[23] The equations of rays can be also readily written for the case when several turning points, rather than one, are present (the trajectory forms several loops around the medium's center $r = 0$, Fig.3.40).

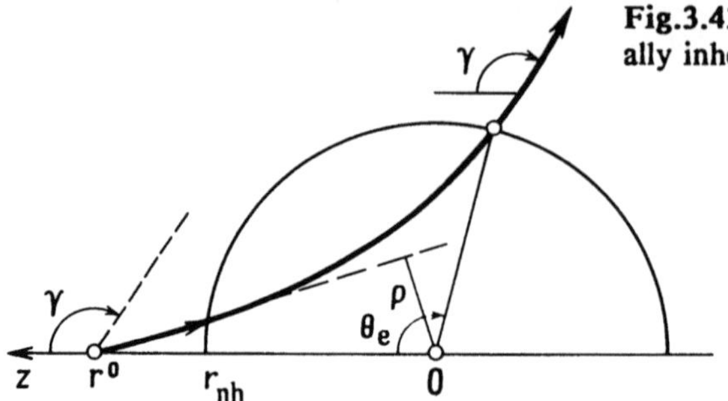

Fig.3.43. Ray path in a bounded radially inhomogeneous medium

$$A = \frac{A^0}{\sqrt[4]{r^2 n^2(r) - \rho^2}} \sqrt{\frac{\rho}{r\sin\theta |\partial\theta(r,\rho)/\partial\rho|}} \exp(-\tfrac{1}{2}i\pi q) , \qquad (3.5.24a)$$

$$\psi = \rho\theta + \left(\int_{r_{tp}}^{\infty} \mp \int_{r_{tp}}^{r}\right) \sqrt{n^2(r) - \rho^2/r^2} \, dr , \qquad (3.5.24b)$$

where $q = [1 + \mathrm{sgn}(\tilde{p}_r)\mathrm{sgn}(\partial\theta/\partial\rho)]/2$. The expression for the wave amplitude in a cylindrically stratified medium differs from (3.5.24) only by the factor $(\rho/r\sin\theta)^{1/2}$ which we have ignored. Obviously, the formulas for the amplitudes may be deduced on geometrical grounds (Fig.3.42).

3.5.7 Caustics

In a spherically stratified medium a point source generally produces two caustics according to (3.5.18): an axial caustic at $\theta = 0$ or π, and a caustic of revolution defined by the condition $\partial\theta(r,\alpha^0)/\partial\alpha^0 = 0$. If we characterize the ray family (3.5.17) by the impact parameter ρ rather than the angle α^0, i.e., $\theta = \theta(r,\rho)$, then the equations for the caustic of revolution have the form

$$\theta = \theta(r,\rho) , \quad \partial\theta(r,\rho)/\partial\rho = 0 . \qquad (3.5.25)$$

These same equations apply to the caustic of a plane wave if we express the function $\theta(r,\rho)$ by (3.5.23). If we devise for a plane wave, a law n(r) such that the caustic (3.5.25) contracts to a point - a focal point - then by the reciprocity theorem for rays (Sect.2.6) we get an ideal radially inhomogeneous lens antenna [3.14, 120]. For a given law of refraction n(r) the shape of the caustic (3.5.25) can be fairly complicated both for plane and spherical waves.

Let a point source be situated outside a bounded, spherically stratified inhomogeneity of radius r_{nh} (Fig.3.43). Consider, following [3.62], the caustics formed by this inhomogeneity in free space. Since for $r > r_{nh}$, $n^2(r) = 1$, the equations of rays (3.5.17) for this case have the form

$$\theta = \theta_e - \arccos(\rho/r_{nh}) + \arccos(\rho/r)$$

$$= \gamma - \arcsin(\rho/r) \equiv \theta(r,\rho) . \qquad (3.5.26a)$$

Here θ_e is the coordinate where the ray emanates from the medium, i.e.,

$$\theta_e = \arccos\frac{\rho}{r^0} - \arccos\frac{\rho}{r_{nh}} + 2\int_{r_{tp}}^{r_{nh}} \frac{\rho\,dr}{r\sqrt{r^2 n^2(r) - \rho^2}}, \qquad (3.5.26b)$$

and $\gamma = \theta_e + \arcsin(\rho/r_{nh})$ is the angle which the ray makes with the z axis (Fig.3.43). The caustic (3.5.25) formed by the rays (3.5.26) is defined by

$$\theta = \gamma - \arcsin\frac{\rho}{r} \equiv \theta_c(\rho) , \quad r = \sqrt{\rho^2 + (dy/d\rho)^{-2}} \equiv r_c(\rho) , \qquad (3.5.27a)$$

or in the cylindrical coordinates $z = r\cos\theta$, $R = r\sin\theta$:

$$z = \rho\sin\gamma + (dy/d\rho)^{-1}\cos\gamma \equiv z_c(\rho) ,$$

$$R = -\rho\cos\gamma + (dy/d\rho)^{-1}\sin\gamma \equiv R_c(\rho) . \qquad (3.5.27b)$$

The caustic (3.5.27) exists only subject to the conditions $d\gamma/d\rho > 0$ and $d\theta_e/d\rho < 0$ (for $d\theta_e/d\rho > 0$, the caustic is situated inside the inhomogeneity and governed by other equations). At $d\theta_e/d\rho = 0$ it crosses the boundary $r = r_{nh}$, and at $d\gamma/d\rho = 0$ it asymptotically approaches infinity ($r \to \infty$). The condition for cusp points to appear on caustic (3.5.27) has the form $\gamma'' = \rho(\gamma')^3$, where $\gamma' \equiv d\gamma/d\rho$, and so on. Examples for a variety of $n(r)$ profiles have been given in [3.62]. The caustics can exhibit pockets (cross sections of a swallowtail catastrophe) and other caustic singularities which deform as the position of the source and the $n(r)$ parameters vary; see other examples in [3.39, 122-125].[24] To conclude, we remark that a substantial number of publications have been devoted to various wave problems in radially inhomogeneous media; see, e.g., [3.44, 91, 103, 108, 124-138]. In particular, [3.135-139] compared the results derived by geometrical optics and other asymptotic methods with those obtained by numerical computation. Section 3.8 will devote some space to the problems of wave scattering at localized, radially inhomogeneous formations.

---

[24] The evaluation of caustics is of interest, for example, in estimating the effects of ionospheric focusing, the focusing in radio occultation of planetary atmospheres, and solar corona; various scattering problems, etc. (Sect.3.8 and [3.39, 83, 84, 119, 123-126]).

3.6 Tapered and Other Inhomogeneous Media

3.6.1 The Eikonal and Rays in a Tapered Medium

The geometrical optics of a tapered medium in which the refractive index n is a function of the polar angle ϕ was considered in [3.140]. In a medium with $n = n(\phi)$ the lines of constant n are inclined to each other. This makes the model advantageous for approximating media with complex structure, specifically, in accounting for the horizontal gradients in stratified media. Evidently, the vicinity of the apex of the taper $r \leq (1/k_0 n^2)|dn/d\phi|$, where the gradient $|\nabla n|$ turns out large, should be excluded from the consideration because here geometrical-optics approximation fails.

The tapered-medium model is also useful in that the eikonal equation for $n = n(\phi)$ can be solved by a special (*multiplicative* rather than additive) technique of separation of variables, whereas in the initial wave equation (2.1) the variables are nonseparable. For the sake of convenience, consider the two-dimensional eikonal equation

$$\frac{1}{r^2}\left(\frac{\partial\psi}{\partial\phi}\right)^2 + \left(\frac{\partial\psi}{\partial r}\right)^2 = n^2(\phi) \; . \tag{3.6.1}$$

Substituting $\psi(r,\phi) = rn(\phi)\sin\theta(\phi)$ into (3.6.1) leads to the following variation of θ (law of refraction):

$$\frac{d\theta}{d\phi} = 1 - \frac{1}{n}\frac{dn}{d\phi}\tan\theta \; . \tag{3.6.2}$$

The geometrical meaning of the angle θ can be deduced from the expression for the momentum $\mathbf{p} = \nabla\psi$ in polar coordinates: $\tilde{p}_r = n\sin\theta$, $\tilde{p}_\phi = n\cos\theta$; i.e., the angle of refraction θ is determined by the vector $\mathbf{p}$, tangent to the ray, and the normal $\mathbf{e}_\phi$ to the layer of constant n, namely, $\cos\theta = (1/n)\mathbf{p}\cdot\mathbf{e}_\phi = \mathbf{l}\cdot\mathbf{e}_\phi$. Here $\phi < \pi/2$ on the upgoing ($\tilde{p}_\phi$) and $\theta > \pi/2$ on the downgoing ($\tilde{p} < 0$) branches of the ray (Fig.3.44a). Both branches merge at the turning point where $\tilde{p}_\phi = 0$ and $\theta = \pi/2$. The law of refraction (3.6.2) could easily be deduced by applying Snell's law to a set of infinitesimally thin homogeneous prisms (Fig.3.44b): $n\sin\theta = (n+\Delta n)\sin\theta'$, but $\theta' = \theta+\Delta\theta-\Delta\phi$, whence $\Delta n\sin\theta = n\cos\theta(\Delta\phi-\Delta\theta)$, which, upon passing to the limit, as $\Delta\phi \to 0$, coincides with (3.6.2).

The differential equations of rays, (2.2.47), for a tapered medium with $n = n(\phi)$ have the form ($\theta = \frac{1}{2}\pi-\alpha$)

$$\frac{dr}{d\tau} = n\sin\theta \; ,$$

$$\frac{d\phi}{d\tau} = \frac{1}{r}n\cos\theta \; , \tag{3.6.3}$$

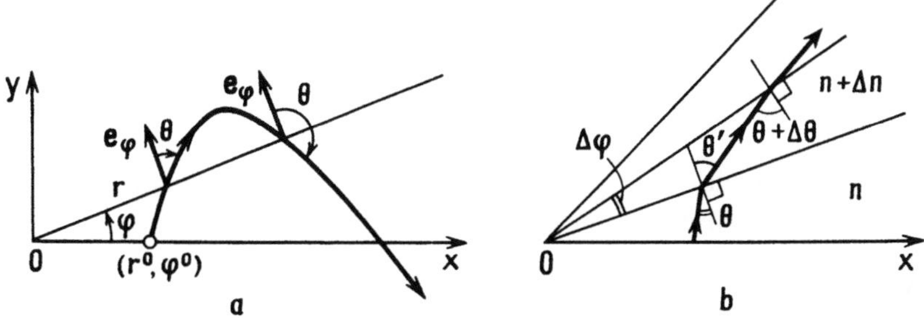

Fig.3.44. (a) A ray path in a tapered medium. (b) Illustrating the derivation of (3.6.2) from Snell's law

$$\frac{d\theta}{dr} = \frac{1}{r}\left[n\cos\theta - \frac{dn}{d\phi}\sin\theta\right] .$$

Dividing the first and third equations by $d\phi/dr$, we again arrive at the law of refraction (3.6.2), augmented by the equation $dr/d\phi = r\tan\theta$, solved by quadratures[25]

$$r = r^0\exp\left[\int_{\phi^0}^{\phi}\tan\theta(\phi)d\phi\right] . \tag{3.6.4}$$

Here $\theta(\phi)$ is the solution of (3.6.2) satisfying the boundary condition $\theta = \theta^0$ at $\phi = \phi^0$, and r^0 and ϕ^0 are the initial coordinates of the ray. From (3.6.2) it follows that the ray paths in a tapered medium are *unsymmetric* curves due to a horizontal gradient of n.

For the eikonal in a tapered medium, we have

$$\psi = \psi^0 + \int_0^r n^2 dr$$

$$= \psi^0 + \int_{\phi^0}^{\phi} n^2 \frac{dr}{d\phi}d\phi = \psi^0 + \int_{\phi^0}^{\phi}\frac{rn}{\cos\theta}d\phi . \tag{3.6.5a}$$

Considering

$$rn = \cos\theta\frac{d}{d\phi}(rn\sin\theta) ,$$

---

[25] This equation can be derived independently of (3.6.3), as well as an equation defining the angle of tangent to the ray in a polar system of coordinates.

which can be easily verified with the help of (3.6.2,4), we find from (3.6.5a)

$$\psi = \psi^0 + rn\sin\theta - r^0 n^0 \sin\theta^0 \,, \tag{3.6.5b}$$

which obviously coincides with the solution of the eikonal equation (3.6.1) by multiplicative separation of variables.

3.6.2 The Field of a Plane Wave

Deriving the Jacobian $\mathscr{D}(\tau) = r\partial(r,\phi)/\partial(\xi,\tau)$ by means of (3.6.3,4) it is a simple matter to find that

$$\mathscr{D}(\tau) = \tilde{p}_\phi \frac{\partial}{\partial\xi} r(\phi,\xi) = n\cos\theta \frac{\partial}{\partial\xi} r(\phi,\xi) \,, \tag{3.6.6}$$

where $r = r(\phi,\xi)$ is the equation of the ray family (3.6.4), and ξ is a ray coordinate labeling the rays. Now in accordance with the general formula (2.3.9) we may define the amplitude of an arbitrary wave in a tapered medium

$$A = A^0 \sqrt{\frac{\mathscr{D}(0)}{\mathscr{D}(\tau)}} = A^0 \left[\frac{n^0 \cos\theta^0}{n\cos\theta} \frac{\partial r}{\partial\xi}\Big|_{\phi^0} \left(\frac{\partial r}{\partial\xi}\right)^{-1} \right]^{1/2} \,, \tag{3.6.7}$$

where the vertical bar with the subscript ϕ^0 means the derivative $\partial r/\partial\xi$ must be taken at $\phi = \phi^0$.

Let the initial field $u^0 = A^0 \exp(ik_0\psi^0)$ be given on the straight line $\phi = \phi^0 = \text{const.}$ so that $\psi^0 = r^0 n^0 \sin\theta^0$, $\theta^0 = \text{const.}$, and $A^0 = \text{const.}$ (as in Sect.3.1.2, we will refer to this wave as a plane wave). Since θ^0 is constant, the rays (3.6.4) are parallel curves (Fig.3.45a) such that $r^0 \equiv \xi$ and $\partial r/\partial\xi = r/r^0$, and $\partial r/\partial\xi|_{\phi^0} = 1$; therefore, by virtue of (3.6.7 and 5b),

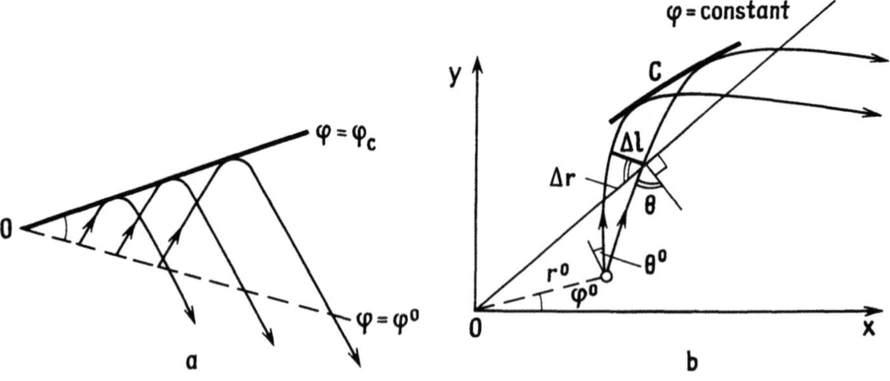

Fig.3.45. (a) A straight-line caustic occurred as a result a "plane" wave being incident on a tapered medium. (b) Illustrating the derivation of equation (3.6.9) for the point source field in a tapered medium

$$u(r,\phi) = A^0 \sqrt{\frac{r^0 n^0 |\cos\theta^0|}{rn|\cos\theta|}}\, \exp(ik_0 rn\sin\theta - \tfrac{1}{2}i\pi q) \,, \tag{3.6.8}$$

where $q = [1-\mathrm{sgn}(\cos\theta\cos\theta^0)]/2$, and $\theta = \theta(\phi)$, is the solution of (3.6.2) subject to the boundary condition $\theta = \theta^0$ at $\phi = \phi^0$. The caustic of a plane wave is the plane of constant $\phi = \phi_c$ on which $\theta = \pi/2$, i.e., the caustic is the locus of turning points for the rays (3.6.4). The coordinate ϕ_c of the caustic is determined from (3.6.2) by $\theta(\phi_c) = \pi/2$.

3.6.3 The Field due to a Linear Source

In the case of a source placed at a point r^0, ϕ^0 (two-dimensional problem) we take for the ray coordinate the angle θ^0 at which the ray has emerged (Fig.3.45b). Then, according to (3.6.6.), $\mathscr{D}(\tau) = n\cos\theta\partial r(\phi,\theta^0)/(\partial\theta^0)$. In order to determine field amplitudes by (2.3.29) we should observe that in the two-dimensional problem the differential solid angle $d\Omega$ reduces to $d\theta^0$:

$$A = B^0 \sqrt{\frac{n^0 d\Omega}{n da}}$$

$$= B^0 \sqrt{\frac{n^0}{\mathscr{D}(\tau)}} = B^0 \left(\frac{n}{n^0}\cos\theta\, \frac{\partial}{\partial\phi^0} r(\phi,\theta^0)\right)^{-1/2} , \tag{3.6.9a}$$

where $n^0 \equiv n(\phi^0)$. As a result,

$$u = B^0 \left(\frac{n}{n^0}\left|\cos\theta\, \frac{\partial r}{\partial\phi^0}\right|\right)^{-1/2} \exp[ik_0(rn\sin\theta - r^0 n^0\sin\theta^0) - \tfrac{1}{2}i\pi q] \,, \tag{3.6.9b}$$

where the quantity $q = [1-\mathrm{sgn}(\cos\theta\,\partial r/\partial\theta^0)]/2$ defines the phase shift on the caustic given by

$$r = r(\phi,\theta^0) \,, \frac{\partial}{\partial\theta^0} r(\phi,\theta^0) = 0 \,. \tag{3.6.10}$$

The relation $r = r(\phi,\theta^0)$ describes the family of rays (3.6.4) radiated by the source. Note also that the expression for the ray-tube cross section

$$\cos\theta\, \frac{\partial r}{\partial\theta^0} d\theta^0$$

used in (3.6.9a) has a simple geometrical meaning: with reference to Fig.3.45b one may easily verify that $\Delta l = \cos\theta\Delta r$.

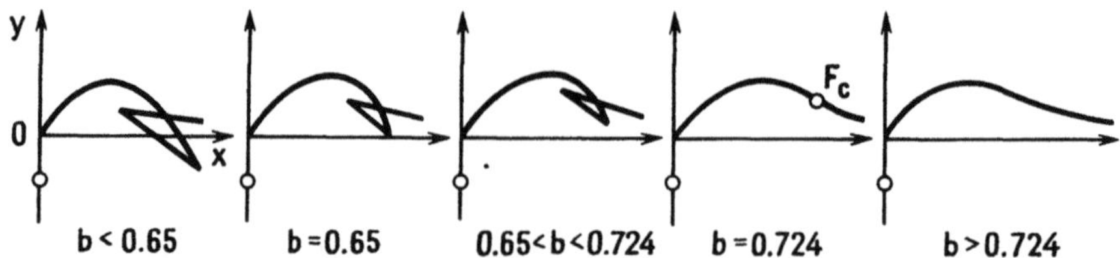

Fig.3.46. Evolution of the caustic in a tapered medium of exponential profile $n^2(\phi) =$ exp($-b\phi$)

The features of the caustic (3.6.10) have been studied in [3.140] to indicate that this caustic is close in shape to that appearing, e.g., in a linear plane layer (Fig.3.19). For the sake of illustration, Fig.3.46 plots the caustics computed in [3.140] as they appear in illuminating, by a linear source, a bounded tapered layer with an exponential profile of $n^2(\theta) = $ exp($-b\phi$), $0 < \phi < \phi_m$, $b > 0$; $n^2(\phi) = 1$ for $\phi < 0$, and the source is situated at $\phi^0 = 3\pi/2$ (Fig.3.47). As is seen from Fig.3.46, the structure of the caustic is determined by the parameter b and is independent of the source coordinate r^0 (r^0 governs only the position of the singular points on the caustic [3.140]). This feature of the caustic evolution and, consequently, of the ray pattern is due to the horizontal gradients of n in the tapered medium [3.140].

3.6.4 Ray Equations in a Two-Dimensional Medium with a Special Profile

Various models of two- and three-dimensional inhomogeneous media allowing for a separation of variables in the eikonal equation have been considered in Sect.2.8. An example of a medium separable in spherical coordinates is that of the refracting index varying as

$$n^2(r,\theta) = \epsilon_1(r) + \epsilon_2(\theta)/r^2 \; . \tag{3.6.11}$$

This model is of value in connection with the problems of wave propagation in plasma formation [3.91,141-143], in the spherical ionosphere with horizontal gradients [3.144-146], in wide-angle inhomogeneous lens antennas [3.120], etc.

By virtue of Sect.2.8 the components of the momentum $\mathbf{p} = \nabla\psi$ in a medium of the refractive index (3.6.11) are ($\alpha_2 \equiv \beta^2$)

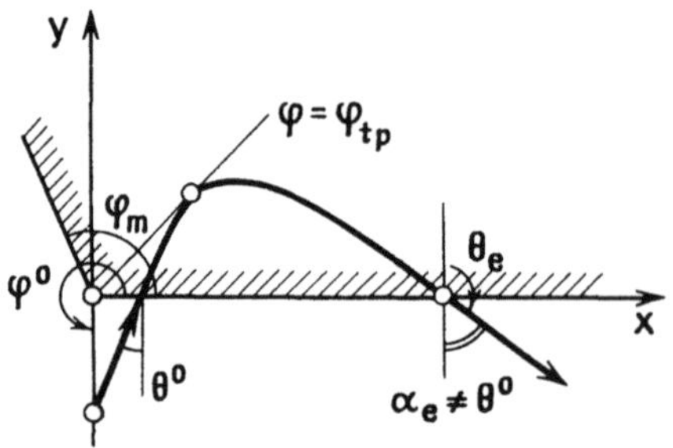

Fig.3.47. The position of the source and the shape of a ray in a tapered medium

$$\tilde{p} = \mp \frac{1}{r} \sqrt{r^2 \epsilon_1(r) - \alpha_1} \ ,$$

$$\tilde{p} = \pm \frac{1}{r\sin\theta} q(\theta) \ , \qquad\qquad\qquad (3.6.12)$$

$$\tilde{p} = \frac{\beta}{r\sin\theta} \ , \qquad \text{where}$$

$$q(\theta) = \sqrt{[\epsilon_2(\theta) + \alpha_1]\sin^2\theta - \beta^2} \ .$$

Together with (2.2.44) this leads us to the ray equations

$$\frac{d\theta}{d\phi} = \sin\theta \, \frac{\tilde{p}_\theta}{\tilde{p}_\phi} = \frac{1}{\beta} q(\theta)\sin\theta\,\mathrm{sgn}(\tilde{p}_\theta \tilde{p}_\phi) \ ,$$

$$\qquad\qquad\qquad\qquad\qquad\qquad (3.6.13)$$

$$\frac{d\theta}{dr} = \frac{1}{r} \, \frac{\tilde{p}_\phi}{\tilde{p}_r} = \frac{q(\theta)\mathrm{sgn}(\tilde{p}_\theta \tilde{p}_\phi)}{r\sin\theta \sqrt{r^2\epsilon_1(r) - \alpha_1}} \ ,$$

which are integrated by quadratures. The salient feature of the ray paths (3.6.13) is that they generally possess turning points of two types consistent with the condition $\tilde{p} = 0$ and $\tilde{p}_r = 0$ (Fig.3.48). At the turning points the ray changes its direction, and $\tilde{p}_\theta$ or $\tilde{p}_r$ in (3.6.12,13) change their signs, so that we may restate the ray equation (3.6.13) in integral form for various ray branches. For instance, if for the ray depicted in Fig.3.48 $\tilde{p}_\phi^0 < 0$, then the equations of its first branch where $\tilde{p}_r < 0$ and $\tilde{p}_\theta > 0$ have the form

$$\phi = \phi^0 + \int_{\theta^0}^{\theta} \frac{|\beta|\,d\theta}{q(\theta)\sin\theta} \ , \quad \int_r^{r^0} \frac{dr}{r\sqrt{r^2\epsilon_1(r) - \alpha_1}} = \int_{\theta^0}^{\theta} \frac{\sin\theta}{q(\theta)}d\theta \ , \qquad (3.6.14)$$

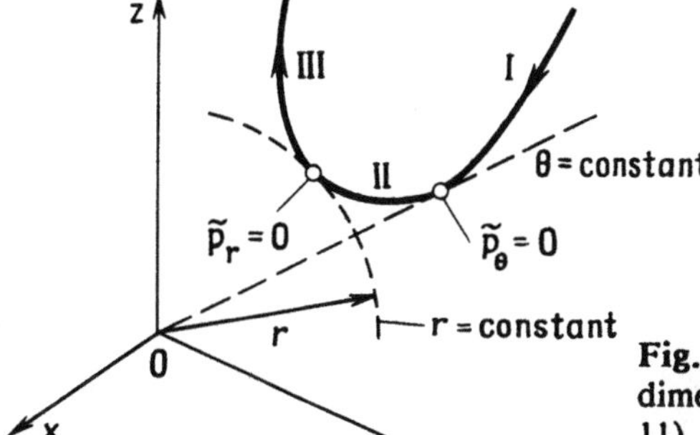

Fig.3.48. Turning points of a ray in a two-dimensionally inhomogeneous medium (3.6.11) allowing for separation of variables in the spherical system of coordinates

where r^0, θ^0, and ϕ^0 are the coordinates of the ray origin. Whether one or another ray geometry is realized depends on the actual definition of the problem.

3.6.5 The Field of a Point Source (Axially Symmetric Problem)

Let a source placed on the z-axis in free space (at $r = r^0$ and $\theta^0 = \pi$) irradiate a *two-dimensionally inhomogeneous cone* the interface surface of which is defined by the angle $\theta_b < \pi/2$ so that for $\theta \leq \theta_b$ the refractive index squared is described by

$$n^2(r,\theta) = 1 - b^2 f(\theta)/r^2 \ , \quad b = \text{const.} > 0 \ , \tag{3.6.15}$$

where $f(\theta) \equiv 0$ for $\theta \geq \theta_b$. In addition, we assume $f(0) = 1$ and $f'_\theta(0) = 0$. Various $n^2 = \text{const.}$ profiles are shown schematically in Fig.3.49. This type of inhomogeneity is used as a model of plasmas [3.141-143], the parameter b measuring the longitudinal extension of the region with $n^2 = 0$ (Fig.3.49).

From (3.6.12) for the model under consideration (3.6.15), we have at $\alpha_1 \equiv \rho^2$

$$\tilde{p}_r = \frac{1}{r}\sqrt{r^2 - \rho^2} \ , \quad \tilde{p}_\theta = \pm \frac{1}{r}\sqrt{\rho^2 \, b^2 F(\theta)} \ , \quad \tilde{p}_\phi = 0 \tag{3.6.16}$$

and the ray equations (3.6.13) become[26]

$$r = \rho \sec\left[\arccos\frac{\rho}{r^0} - \theta_b + \left(\int_{\theta_{tp}}^{\theta_b} \mp \int_{\theta_{tp}}^{\theta} \right) \frac{\rho \, d\theta}{\sqrt{\rho^2 - b^2 f(\theta)}} \right] \equiv r(\theta,\gamma^0) \ . \tag{3.6.17}$$

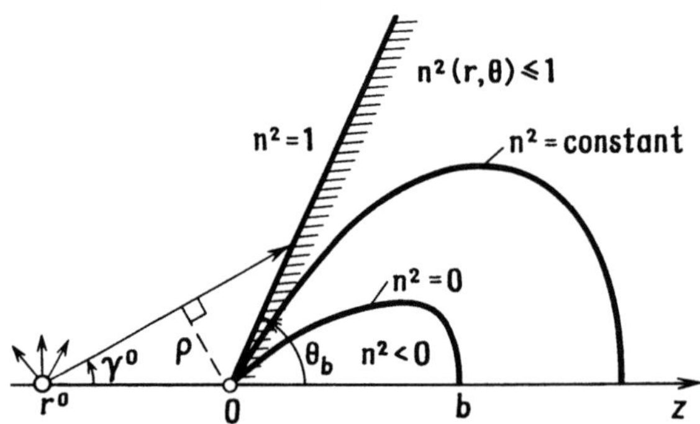

Fig.3.49. Lines of constant n^2 for the model of two-dimensionally inhomogeneous media with the refractive index (3.6.15)

---

[26] The signs and cosntants in ray equation (3.6.17) are chosen subject to the continuity of components (3.6.16) and the components of the momentum of incident ray $p = \{\tilde{p}_r^0, \tilde{p}_\theta^0, \tilde{p}_\phi^0\}$ at the medium interface $\theta = \theta_b$.

Here $\rho = r^0\sin\gamma^0$ is the impact parameter of the ray (i.e., the shortest distance from the center 0 to the ray (Fig.3.49); γ^0 is the angle at which the ray has been emitted by the source, defined as $\cos\gamma^0 = \mathbf{p}^0\mathbf{e_z}$; and θ_{tp} is the coordinate of the turning point, where $\tilde{p}_\theta = 0$ and $b^2 f(\theta_{tp}) = \rho^2$. By analogy with Sects.2.3-5 we adopt the convention that if a turning point is absent on the ray, we take (3.6.17) with the upper sign and replace the integral difference by one integral from θ to θ_b. Evidently, the rays (3.6.17) are nonsymmetric curves in a plane $\phi = $ constant.

In order to compute the amplitude of the point-source field, we are to determine the Jacobian $\mathscr{D}(\tau)$, so from (3.6.17), similar to the procedure of (3.5.16), we obtain

$$\mathscr{D}(\tau) = r^2\sin\theta\,\frac{\partial(r,\theta)}{\partial(\gamma^0,\tau)} = r\sin\theta\tilde{p}_\theta\,\frac{\partial}{\partial\gamma^0}r(\theta,\gamma^0)\ , \tag{3.6.18}$$

where $\tilde{p}_\theta$ is given by (3.6.16). Then in accordance with (2.3.26a) and (3.4.9), like (3.4.12) and (3.5.19), we find

$$A = B^0\sqrt{\frac{\sin\gamma^0}{\mathscr{D}(\tau)}}$$

$$= \frac{B^0}{\sqrt[4]{p^2 - b^2 f(\theta)}}\sqrt{\frac{\sin\gamma^0}{\sin\theta|\partial r(\theta,\gamma^0)/\partial\gamma^0|}}\exp(-\tfrac{1}{2}i\pi q)\ , \tag{3.6.19}$$

where $q = [1-\mathrm{sgn}(\tilde{p}_\theta\partial r/\partial\gamma^0)]/2$. The geometrical meaning of (3.6.18,19) becomes evident from Fig.3.50 as soon as we observe that the ray-tube cross section is $\Delta a = r\Delta\phi^0\Delta l\sin\theta$, where $\Delta l = (\Delta r/n)(\nabla\psi\mathbf{e}_\theta) = \tilde{p}_\theta\Delta r/n$, and $\mathscr{D}(\tau) = nda/d\phi^0 d\gamma^0$.

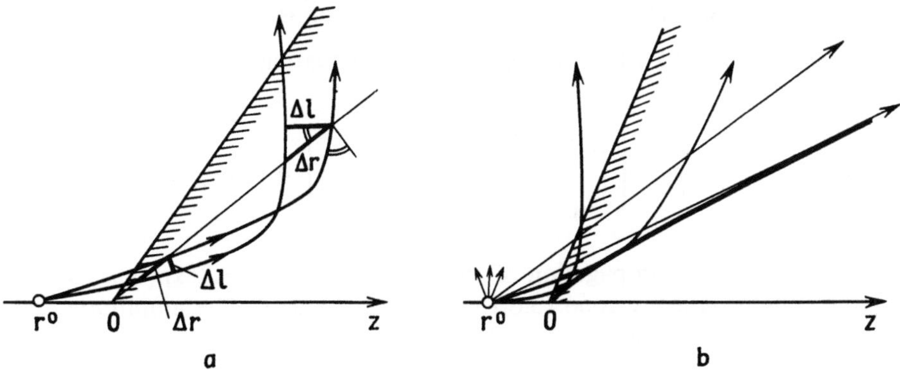

Fig.3.50. (a) Diagram to shed light on the geometrical meaning of (3.6.18,19). (b) The formation of a caustic in the two-dimensionally inhomogeneous medium of refractive index (3.6.15)

It is a simple matter to demonstrate that the eikonal for the point source considered is

$$\psi = \left(\int_\rho^{r^0} + \int_\rho^r \right) \sqrt{1 - \rho^2/r^2}\, dr + \rho(\pi - \theta_b)$$

$$+ \left(\int_{\theta_{tp}}^{\theta_b} \mp \int_{\theta_{tp}}^{\theta} \right) \sqrt{\rho^2 - b^2 f(\theta)}\, d\theta \qquad (3.6.20)$$

(the integrals in r can, of course, be calculated explicitly).

The ray family (3.6.17) for $\partial n/\partial\theta > 0$ is schematized in Fig.3.50b: upon refraction some rays cross the cone interface $\theta = \theta_b$; the others extend to infinity inside the cone asymptotically (as $n^2 \to 1$ as $r \to \infty$). The caustic for this ray family, according to (3.6.18), is defined by

$$r = r(\theta, \gamma^0)\,, \qquad \frac{\partial}{\partial\gamma^0} r(\theta, \gamma^0) = 0 \qquad (3.6.21)$$

and has a branch tending asymptotically to infinity (Fig.3.50b). The rays may touch on the caustic both infront and behind a turning point. An actual analysis of the shape of the caustic (3.6.21) is facilitated by the fact that the integrals in (3.6.17) are calculated analytically for a wide class of functions $f(\theta)$. Note that media of the type (3.6.11) have been used for analysis of ray patterns in the spherical ionosphere with horizontal gradients [3.144]; this time, in contrast to (3.6.15), the radial dependence $\epsilon_1(r)$ plays an important role.

3.6.6 A Plane Wave Incident
on the Two-Dimensionally Inhomogeneous Medium

If we let $r^0 \to \infty$ in (3.6.17), we obtain the ray equation

$$r = \rho \sec\left[\frac{\pi}{2} - \theta_b + \left(\int_{\theta_{tp}}^{\theta_b} \mp \int_{\theta_{tp}}^{\theta} \right) \frac{d\theta}{\sqrt{\rho^2 - b^2 f(\theta)}} \right] \equiv r(\theta, \rho)\,, \qquad (3.6.22)$$

which describes the propagation of the plane wave $u^0 = A^0 \exp(ik_0 z)$ striking the inhomogeneous cone (3.6.15) along the z axis. The amplitude and eikonal for this case are

$$A = A^0 \sqrt{\frac{\rho}{\sin\theta \left| \frac{\partial}{\partial\rho} r(\theta, \rho) \right|}} \; \frac{1}{\sqrt[4]{\rho^2 - b^2 f(\theta)}} \exp(-\tfrac{1}{2} i\pi q) \quad \text{and} \qquad (3.6.23a)$$

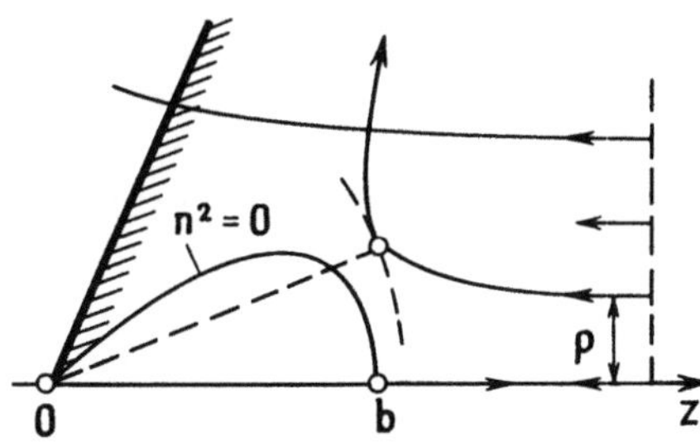

Fig.3.51. Ray paths in the two-dimensionally inhomogeneous medium of refractive index (3.6.15) (ρ is the impact parameter)

$$\psi = \rho(\tfrac{1}{2}\pi - \theta_b) + \int_\rho^r \sqrt{1 - \rho^2/r^2}\; dr + \left(\int_{\theta_{tp}}^{\theta_b} \pm \int_{\theta_{tp}}^{\theta}\right) \sqrt{\rho^2 - b^2 f(\theta)}\; d\theta \;, \quad (3.6.23b)$$

where $q = [1 - \mathrm{sgn}(\tilde{p}_\theta\, \partial r/\partial\rho)]/2$.

If the plane wave is incident on an inhomogeneous cone (3.6.15) from the side of positive z, i.e., $p^0 = -e_z$ (Fig.3.51), then unlike (3.6.22) the rays acquire only the turning points of the second kind, at which $\tilde{p} = 0$, under the conditions

$$\tilde{p}_r = \mp\, \frac{1}{r}\, \sqrt{r^2 - \alpha_1}\;, \quad \tilde{p}_0 = \frac{1}{r}\, \sqrt{\alpha_1 - b^2 f(\theta)} \;. \quad (3.6.24)$$

The separation constant α_1 is determined by the matching condition on $\tilde{p}_\theta$ and $\tilde{p}_\theta^0 = \sin\theta^0 = \rho/r^0$ (ρ is the impact parameter of the ray, Fig.3.51) at a sufficiently great distance $r = r^0$ where the medium (3.6.15) may be deemed almost homogeneous. As a result, as $r^0 \to \infty$ ($\theta^0 \to 0$), recalling that we have assumed $f(0) = 1$, we obtain $\alpha_1 = b^2 + \rho^2 \equiv \kappa_0^2$.

The ray family (3.6.3) is governed by

$$r = \kappa_0 \operatorname{cosec}\left[\int_0^\theta \frac{\kappa_0\, d\theta}{[\kappa_0^2 - b^2 f(\theta)]^{1/2}}\right] \equiv r(\theta,\kappa_0) \;. \quad (3.6.25)$$

We will not elaborate on the field due to these rays. We only note that a specularly reflected beam obviously corresponds to the case of $\rho = 0$; Sect.3.8 discusses the topics related to the far-field evaluation (see also [3.143]). Examples for the field of a plane wave, being obliquely incident on an inhomogeneous medium, can be found, for instance, in [3.141, 142].

3.6.7 Weakly Inhomogeneous, Quasi-Stratified, and Random Media

We shall consider a weakly inhomogeneous medium whose permittivity $\epsilon = n^2$ varies insignificantly within a considered region of space V. If $\tilde{\epsilon}$ is the mean permittivity over V, then the above definition implies that in the weakly inhomogeneous medium the deviation $\nu = \epsilon - \tilde{\epsilon}$ is small compared

181

with ϵ, or $|\nu| \ll \bar{\epsilon}$. A geometrical-optical description media would be natural to conduct by means of perturbation theory (Sect.2.9).[27] Such theory has been utilized, specifically, to compute the refration of light and radio waves in the earth's troposphere where $|\nu| \leq 10^{-2}$ [3.107, 147], and the refraction of microwaves in the ionosphere [3.147-149].

When the mean permittivity $\bar{\epsilon}$ corresponds to a stratified medium, then at $\epsilon = \bar{\epsilon} + \nu(\mathbf{r})$ with $|\nu| \ll \bar{\epsilon}$, we deal with a *quasi-stratified* medium. A perturbation approach for rays in such a medium has been constructed in [3.150].

Finally, when $\nu(\mathbf{r})$ is a random function of the coordiantes, then we speak of a *random inhomogeneous* medium. A perturbation theory for this case has been proposed in [3.151, 152], to be followed by a vast amount of literature [3, 153-156].

It is important to note that numerous results of geometrical optics remain valid even when diffractional effects are no longer negligible. First of all, this refers to calculations of the phase fluctuation and to the calculation of the second-order coherence function whose behavior is primarily governed by phase fluctuations. We point out that a number of fluctuation phenomena, related to the appearance and distortion of caustics, readily allow for a ray description [3.71, 156-159].

3.7 Geometrical Optics of Waveguides and Resonators

3.7.1 Geometrical Optics of Waveguides

In any type of waveguide the rays undergo multiple reflections and therefore form a complicated ray pattern with numerous caustics. Figure 3.52 demonstrates a family of rays launched by a point source in a two-dimensional waveguide produced by two parallel metal planes. Inside a waveguide, caustics do not arise, but extensions of rays meet at virtual focal points coinciding with the mirror images of the source in the waveguide walls. The period of a ray (hop or ray cycle length), X, in a waveguide depends on the angle at which the ray is launched to the waveguide axis

$$X = 4a\cot\beta , \tag{3.7.1}$$

where 2a is the width of the guide. Rays emanating from a source at various angles propagate in the waveguide with different speeds and result in a sophisticated ray pattern with many rays (indefinitely many for a point source) impinging onto a given point. A discussion of the ray method for both uniform and nonuniform guides was presented in [3.160].

In refraction (grade-index) waveguides, a point source produces an even more complicated pattern with numerous ray crossings and caustics. The behavior of a ray bundle in a graded-index guide is exemplified in Fig.3.53b for the case of a point source placed at the waveguide axis where

---

[27] These media must, of course, be also smoothly inhomogeneous; i.e., the inequality $\lambda|\nabla\epsilon| = \lambda|\nabla\nu| \ll \epsilon$ must be valid.

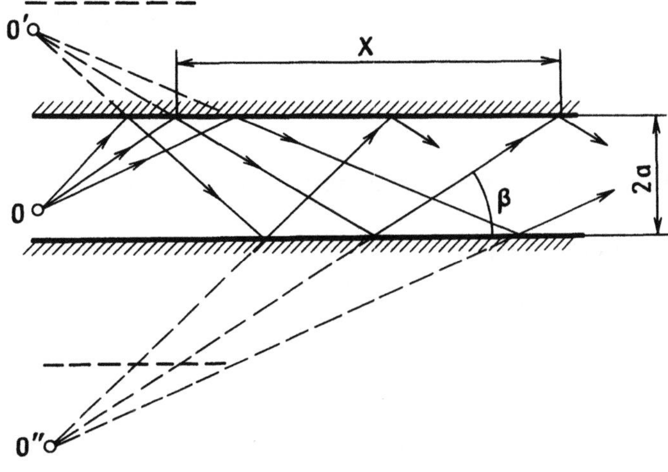

Fig.3.52. To the rays undergoing reflections from the guide walls there correspond the mirror images O', O", and so on

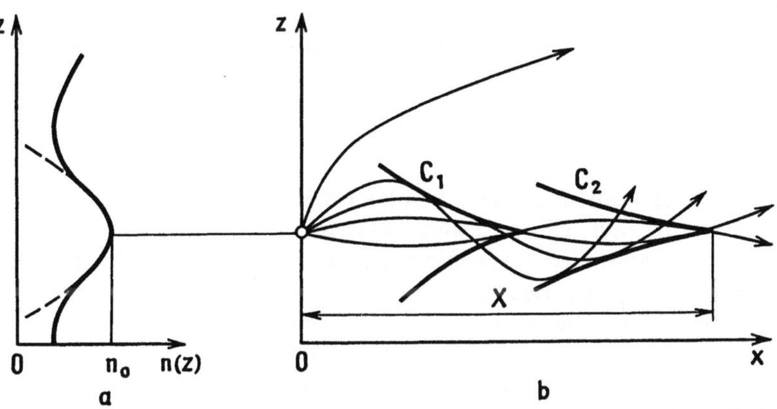

Fig.3.53. Point source placed near a maximum n value in a refraction waveguide gives rise to a sequence of caustic cusps C_1, C_2, etc.

the refractive index is largest; the respective $n(z)$ profile is plotted in Fig. 3.53a. A guided ray in a sandwiched waveguide possesses an infinite number of turning points corresponding to the lower z levels and the upper $\bar{z}$ ones at which $n(z) = n(\bar{z}) = p_x^0$ (for a 2-D problem $p_y = 0$). The ray period X in such a uniform (in x) waveguide can readily be found from the ray equation (3.3.4):

$$X = 2\underline{n} \int_{\underline{z}}^{\bar{z}} [n^2(z) - \underline{n}^2]^{-1/2} dz , \qquad (3.7.2)$$

where $\underline{n} \equiv n(\underline{z}) = n(\bar{z}) \equiv \bar{n}$.

Near the axis of the guide the n(z) profile may be approximated by the parabola

$$n^2(z) \simeq n_0^2 - z^2/b^2 \qquad (3.7.3)$$

(the dashed line in Fig.3.53a); for the paraxial rays we have from (3.7.2)

$$X \simeq 2\pi \underline{n} b , \quad \underline{n} = n_0 \sin\theta_0 \simeq n_0 . \qquad (3.7.4)$$

Here θ_0 is the incidence angle close to $\pi/2$ for paraxial rays. From (3.7.4) it follows that the periods of the paraxial rays are close to each other. As a result such rays are periodically focused on the axis giving rise to a system of caustic cusps [3.161] (solid lines in Fig.3.53b). Profiles n(z) are possible that cannot be such as not to confine all the rays entering the guide. One leaving ray is shown in Fig.3.53b.

The equations of rays in a sandwiched waveguide were given, for example, in [3.17,57]; the caustics formed in the guide are governed by the equations of Sect.3.3. Numerous examples of hop and caustic patterns were considered in [3.17,57,69]. The geometrical-optics formulas for determining amplitude and phase of the fields are similar to those of Sect.3.4. Graphic examples of phase-front computations for a sandwiched graded-index waveguide can be found in [3.162]. The results exemplify features that a geometrical-optical field shows due to the multipath propagation and the formation of caustics.

The general relationships derived in Sect.3.5 pave the way for analyzing the properties of rays, caustics, and fields in a uniform (with respect to θ and ϕ) spherically stratified waveguide with n = n(r). In a similar manner one may consider various two-dimensionally inhomogeneous guides for which the variables in the eikonal equation may be separated, say, in Cartesion or spherical coordinates (Sect.2.8). For waveguides with an arbitrary n(r) profile, the ray and field computations call for the techniques of numerical analysis (Sect.3.10).

In dealing with field computations for either uniform or nonuniform waveguides, ray optics encounters at least two complications. First, the complexity of ray and caustic configurations makes it cumbersome to compute the total wave field at the point of observation, which is the sum of partial ray fields. Second, as the observation point moves farther away from the source the region of near-caustic zones (caustic volumes, Sects.2.4,10) sharply increases so that the validity domain for geometrical optics shrinks. Therefore, at large distances from the source, it is preferable to utilize the method of normal modes rather than ray analysis ([3.17,57,69,103] and Sect.3.7.2).

3.7.2 Ray Description of Modes in Uniform Waveguides

As will be recalled, each wave in a uniform (in x) waveguide may be represented as a sum of *normal modes*, or *modes* for short. The ray field, in particular, may also be decomposed into modes of the waveguide. The general aspects of the two ways of describing the wave field - by rays and by modes - have been studied by *Krasnushkin* [3.103, 163], *Bremmer* [3.164], *Budden* [3.165], and others [3.166]. We will here be concerned only with the ray treatment of some normal modes [3.17, 57, 160, 167].

Consider a two-dimensional waveguide formed by the walls at $z = \pm a$ and loaded by a homogeneous medium with $n = 1$. If the wave field $u(x, z)$ vanishes at the walls, $u(x, \pm a) = 0$, the modes are given by the functions

$$u_m(x, z) = \exp(ih_m x)\sin[\kappa_m(z + a)] \, , \qquad\qquad (3.7.5)$$

where $\kappa_m = m\pi/2a$ is the transverse wave number, and $h_m = (k_0^2 - \kappa_m^2)^{1/2}$ the longitudinal wave number (propagation constant).

The wave field (3.7.5) may be represented as the sum of two plane waves, called the *Brillouin waves*,

$$u_m(x, z) = \frac{1}{2i}\Big(\exp[ih_m x + i\kappa_m(z+a)] - \exp[ih_m x - i\kappa_m(z+a)]\Big). \quad (3.7.6)$$

The angle β between the x axis and the wave vectors of the Brillouin waves $\mathbf{k}_\pm^m = \{\pm\kappa_m, h_m\}$ takes on the discrete set of values

$$\sin\beta_m = \frac{\kappa_m}{k_0} = \frac{\pi m}{2k_0 a} \, . \qquad\qquad (3.7.7)$$

Figure 3.54a depicts the phase front and the rays corresponding to one of the Brillouin waves. On reflecting from the walls one Brillouin wave gives rise to the other.

Similar interpretation may be given to the waves of the form $u(x, z) = U(z)\exp(ihx)$ traveling in graded-index waveguides with $n = n(z)$. The preexponent factor $U(z)$ satisfies the equation

$$U''(z) + k_0^2 [n^2(z) - h^2/k_0^2]U(z) = 0 \qquad\qquad (3.7.8)$$

and can be determined by geometrical optics (the WKB approach). Consider $U(z)$ as the sum of two oppositly traveling waves

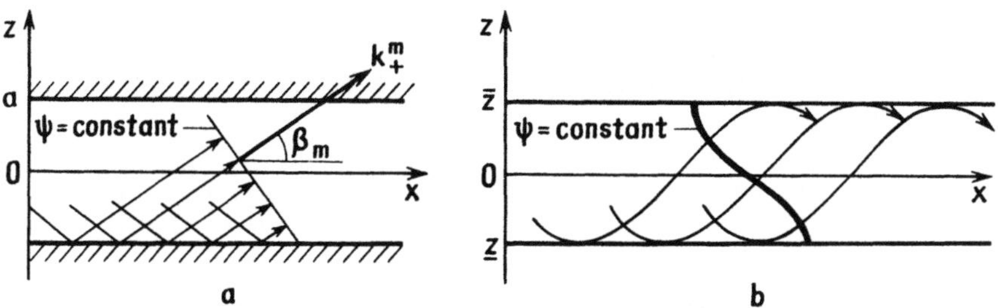

Fig.3.54. Ray interpretation of a Brillouin wave in the waveguide (a) filled with a homogeneous medium, and (b) in a graded index waveguide

$$U(z) = \frac{1}{\sqrt{\gamma}}\left[A_1 \exp\left(ik_0 \int_{\underline{z}}^{z} \gamma\, dz\right) + A_2 \exp\left(ik_0 \int_{z}^{\bar{z}} \gamma\, dz\right)\right] , \qquad (3.7.9)$$

where $\gamma(z) = |p_z| = [n^2(z)-(p_x^0)^2]^{1/2}$, $p_x^0 = h/k_0$, and the turning points $\underline{z}$ and $\bar{z}$ are to be determined from the condition $\gamma(z) = |p_z(z)| = 0$.

Considering the downgoing wave [the second term in (3.7.9)] as the result of reflection of the upgoing wave, we should let, in accordance with Sect. 3.4,

$$A_2 = A_1 \exp\left(ik_0 \int_{\underline{z}}^{\bar{z}} \gamma\, dz - i\frac{\pi}{2}\right), \qquad (3.7.10a)$$

where $-\pi/2$ is the caustic phase shift. A similar argument for the lower turning point leads to

$$A_1 = A_2 \exp\left(ik_0 \int_{\underline{z}}^{\bar{z}} \gamma\, dz - i\frac{\pi}{2}\right). \qquad (3.7.10b)$$

A nontrivial solution to the set (3.7.10) exists provided that

$$2k_0 \int_{\underline{z}}^{\bar{z}} \gamma(z)\, dz - \pi = 2\pi m , \quad m = 0,1,2,\dots , \qquad (3.7.11)$$

which implies that the phase variations for a complete up-and-down cycle must be multiples of 2π, as in the Bohr-Sommerfeld quantization rules.[28]

Condition (3.7.11) is used to determine the propation constant h_m so that the mth-mode field can be represented as

$$u_m(x,z) = U_m(z)e^{ih_m x} = B\gamma^{-1/2}\cos\left(k_0 \int_{\underline{z}}^{z} \gamma\, dz - \frac{\pi}{4}\right)e^{ih_m x} , \qquad (3.7.12)$$

where $B = 2A_1 \exp(i\pi/4)$. Like (3.7.5), it can also be treated as the sum of Brillouin waves

$$u_m(x,z) = \frac{1}{2}B\gamma^{-1/2}\left[\exp\left(ih_m x + ik_0 \int_{\underline{z}}^{z} \gamma\, dz\right) + \exp\left(ih_m x - ik_0 \int_{\underline{z}}^{z} \gamma\, dz\right)\right] ,$$

$$\qquad (3.7.13)$$

---

[28] For a metal-clad waveguide, the analog to (3.7.11) is the relation $4\kappa_m a = 2\pi m$. Both times m is the number of field oscillations in the transverse section of the waveguide.

though now these waves are not planar. The phase fronts of the first wave of (3.7.13) and the corresponding ray family are illustrated in Fig.3.54b. The field (3.7.13) becomes infinite on the caustics $z = \underline{z}$ and $z = \bar{z}$.

3.7.3 Adiabatic Modes of Smoothly Nonuniform Waveguides

In the case of a symmetric, two-dimensional, metal-clad waveguide [the equations of the walls $z = \pm a(x)$], it is intuitively clear that as the width $a(x)$ varies smoothly so that

$$\frac{da}{dx} \simeq \frac{a}{L} \ll 1 \; , \tag{3.7.14}$$

where L is the scale of the $a(x)$ variation, we should expect that in the cross section the field structure will be conserved; specifically, the number of oscillations m remains unchanged. By virtue of (3.7.7), this gives rise to the *adiabatic invariant*:[29]

$$I_m = 2a(x)\sin\beta_m(x) = \text{const.} = \pi m/k_0 \; . \tag{3.7.15}$$

This condition can be deduced directly from the ray equations for the waveguide, assuming that the cycle length along the ray is small compared with L [3.169–172].

From the condition $I_m = $ constant we are able to find the longitudinal wave number $h_m(x) = k_0 \cos\beta_{m(x)}$. Then we arrive at the adiabatic modes of the smoothly nonuniform waveguide

$$u_m(x) = A_m(x)\sin[\kappa_m(z + a)] \exp\left[i \int_{x^0}^{x} h_m(x)dx \right] \; . \tag{3.7.16}$$

The variable amplitude factor $A(x)$ can be determined from the conserving of power flow through the waveguide cross section

$$\frac{1}{ik_0} \int_{-a(x)}^{+a(x)} \text{Im}\left\{ u_m \frac{\partial u_m}{\partial x} \right\} dz = \frac{h_m(x)}{k_0} A_m^2(x) \frac{a(x)}{2} = \text{const.}$$

$$= h_m^0 (A_m^0)^2 a^0 /2k_0 \; ,$$

where $h_m^0 = h_m(x^0)$, $a^0 = a(x^0)$, $A_m^0 = A_m(x^0)$ are the initial values of the parameters. Hence, $A_m = A_m^0 (h_m^0 a^0 /h_m a)^{1/2}$.

---

[29] We recall that the adiabatic invariants are those physical characteristics of oscillatory or wave processes which remain constant on adiabatic (i.e., slow compared to the characteristic periods of the processes) variations of the parameters [3.109, 168].

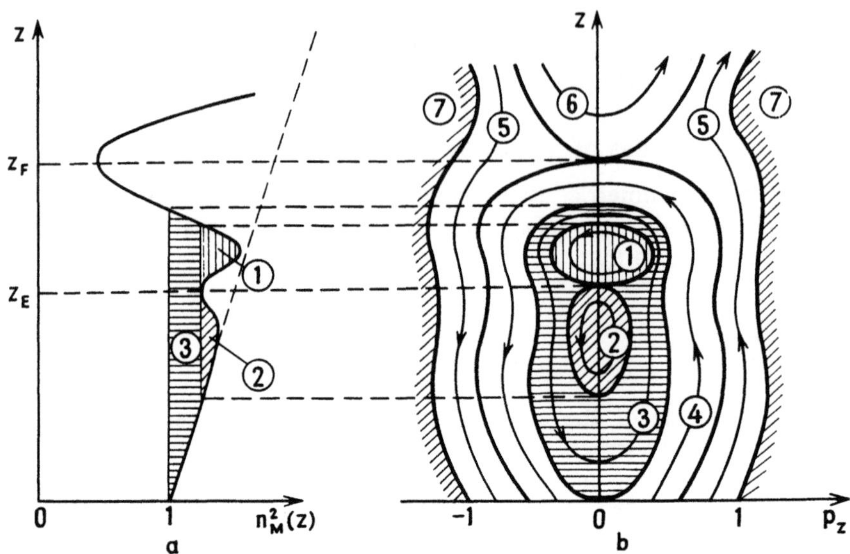

Fig.3.55. Vertical profile of the modified permittivity $n^2_M(z)$ of the earth's ionosphere (a), and the respective ray pattern in the phase plane z, p_z. Areas *1-3* in (b) correspond to channels *1-3* in (a). Area *4* contains rays returning to earth upon one hop. The rays from area *5* leave the earth forever. In area *6* the rays never drop below the z_F level. Area *7* is inaccessible for rays

By virtue of (3.7.15) the field (3.7.16) satisfies the boundary condition $u_m[x, \pm a(x)] = 0$. Eq.(3.7.16) can be refined on calculating the corrections to the phase due to the bending of the phase front at nonuniform segments of the waveguide [3.169, 173, 174].

A waveguide being smoothly nonuniform implies that the inequality $|\partial n/\partial x| \ll |\partial n/\partial z|$ holds true. An adiabatic invariant, by analogy with (3.7.15) and (3.7.11), is the quantity

$$I = \int_{\underline{z}}^{\bar{z}} \sqrt{n^2(x,z) - \underline{n}^2(x)} \, dz = \text{const.} = (m + \tfrac{1}{2}) \frac{\pi}{k_0} . \qquad (3.7.17)$$

The associated adiabatic modes have been analyzed in [3.74, 79, 169].

3.7.4 Ionospheric Wave Channels. The Adiabatic Invariant Method

Consider some salient features of waveguides produced by the earth's ionosphere. Figure 3.55a presents the characteristic altitudinal profile of the modified permittivity $\epsilon_M = n^2_M(z) \simeq \epsilon(z)(1+2z/R)$ for a frequency somewhat below the critical one. The permittivity $\epsilon_M(z)$ maxima z_E and z_F correspond to the largest electron densities of the ionospheric E and F layers, respectively. The valley between the E and F layers is associated with the *interlayer EF channel* (1). The *sublayered E waveguide* (2) is situated below

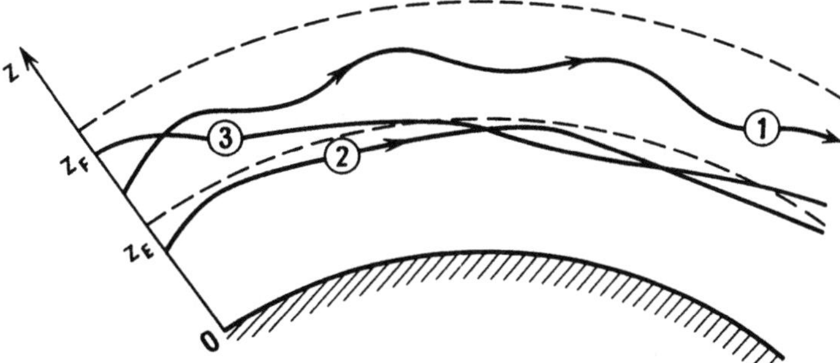

Fig.3.56. Three types of guided ionospheric rays consistent with channels *1-3* in Fig.3.55 (*1*: interlayer EF channel; *2*: sublayered E channel; *3*: sublayered F waveguide)

the E layer; it owes its existence to the curvature of the ionosphere, and vanishes as $R \to \infty$. In addition, the *sublayered F waveguide* (3) can also appear under certain conditions. Figure 3.56 depicts all three types of guided rays. The rays accepted into the E guide (2) are characteristic in their appearance - they reflect successively from the E layer, so they are similar to whispering-gallery rays in nature ([3.103] and Sect.3.7.8).

It is convenient to represent the whole multitude of rays in the ionospheric duct system on the phase plane $\{z,\ p_z=dz/d\tau\}$ [3.119]. Integral curves, satisfying $p_z^2 - \epsilon_M(z) = \text{const.} = (p_x^0)^2$, are plotted in Fig.3.55b. Its area 1,2, and 3 correspond to the ducts 1,2, and 3 in Fig.3.55a. Region 4 confines sky hops which, on reflection from the F, layer return to the earth. The rays from region 5 leave the atmosphere, whereas those of region 6 never descend below the maximum of the F layer. Region 7 is inaccessible for rays [there $(p_x^0)^2 < 0$]. To normal modes there correspond the ray families satisfying the quantization condition (3.7.11).[30]

In a horizontally nonhomogeneous ionosphere of $\epsilon_m = \epsilon_m(x,z)$ the rays may pass from one duct to another, the sky hops may be accepted into the wave channels, or rays may leave the channels. An efficient method of wave-ducting analysis in the horizontally inhomogeneous ionosphere - and the adiabatic invariant approach - has been suggested in [3.175]. It relies on the validity of (3.7.17); assesses the boundaries of the waveguide, $z(x)$ and $\bar{z}(x)$; and indicates whether or not a given mode (or a group of modes) will be accepted or confined in a nonuniform guide, without having recourse to ray calculations. This makes the technique highly adaptable and powerful. The technique proved valuable in investigating the global characteristics of long-range radio-wave propagation and designing long-distance radio channels. For a more detailed explanation of the method, see [3.79].

---

[30] Geometrically this condition implies that the area enclosed by a phase trajectory, $A = \oint |p_z|\ dz = \oint p_z\ dz$, equals $(m + \tfrac{1}{2})\lambda_0$.

3.7.5 Underwater-Sound Ducts.
Summing-Up Incoherent Wave Fields

Acoustic ducts form in the ocean owing to the variability of sound velocity with depth. They can be depth and subsurface sound ducts. Their sound profiles and the behavior of rays in various types of them was outlined in [3.17, 57].

The multipath propagation nature of waveguides is here especially pronounced. As farther from a source the observer receives a proportionally larger number of rays, the wave field computations complicate dramatically. Therefore, in some situations, the interference of ray fields is neglected and the total field is found as an incoherent sum, that is, by adding up the intensities [3.57, 176]. This way of finding the total intensity works better the larger the number of rays, N, arriving at the observation point; the relative error of evaluating the field by this technique diminishes approximately as $N^{-1/2}$.

3.7.6 Optical Fibers

In graded-index (or step-index) optical fibers, rays generally travel in spiral trajectories, either with sharp bends or with continuous bending, keeping the rays within the waveguide. Various nonregularities - guide bends, optical-fiber inhomogeneities - may cause the rays to leave the fiber thus incurring loss of power. The ray theory of these phenomena was treated, for example, in [3.177].

3.7.7 Mode Conversion in Smoothly Nonuniform Waveguides

The adiabatic modes mentioned in Sect.3.7.3 do not describe the conversion of one mode into another one, as they coincide with the exact one-mode solution both before and after passing a nonuniform segment. Recently several approaches have been suggested for describing mode conversion of waves; they were reviewed in [3.7, 169, 174]. An essentially ray-optical approach to the problem developed in [3.170-172, 178] aims at evaluating the distortion of the phase front of the Brillouin wave in a nonregular segment (Fig.3.57) and then developing the ray field $A\exp(ik_0\Psi)$ in the modes of the nonregular guide. This expansion yields the coefficients of conversion from an initial, say mth, to the (m+p)th mode. The results of a geometrical-optics computation well agree with those derived by other methods, specifically, by the method of cross sections [3.7, 169, 174, 178].

We may not elaborate here on either computational details or actual results of the ray approach, so we touch only on some qualitative implications of the ray theory, which demonstrate the heuristic validity of ray optics for the mode conversion problem.

First, ray theory predicted and described the *ray swing effect* [3.7, 171]. Figure 3.58 demonstrates the effect as a wavelike deviation of the ray tilt from the adiabatic curve $\beta_{ad}(x) = \arcsin[\pi m/2k_0 a(x)]$; the plot corresponds to an expanding waveguide. The ray swing arises because in a nonuniform segment rays reflect form places on the wall surface having vari-

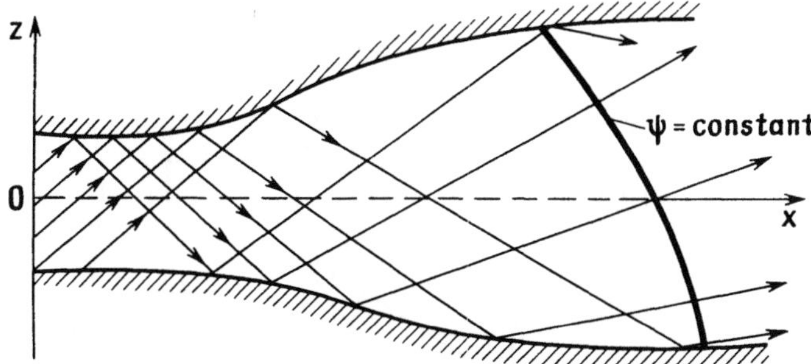

Fig.3.57. Distortion of the plane phase front of a plane Brillouin wave passing through a nonregular segment

ous inclination to the guide axis, and the wavelike trace is due to the different cycle length for various rays. Indeed, for an equal number of reflections, the rays of gradual slope propagate much farther than those steeply incident on the walls. Multivaluedness of the propagation-angle function $\beta(x)$ reflects the multipath propagation in the guide. In a graded-index waveguide, the ray swing causes the region occupied by the rays to increase [3.179].

Secondly, the ray approach enables one to estimate the number of modes, M, excited on a nonuniform segment by simply dividing the swing amplitude $2\Delta\beta$ by the angle distance between the adjacent modes $\delta\beta = \beta_{m+1} - \beta_m \simeq \pi\cos\beta_m/2k_0 a$ [3.171], namely

$$M \simeq 2\Delta\beta/\delta\beta .$$

Thirdly, optics provides guidelines for the design of *nonconverting couplers*: from the ray-optics point of view, a coupler is deemed nonconverting if it launches a Brillouin wave with a plane front [3.170, 180].

Geometrical optics also describes the effect of "interference of the contributions from inhomogeneities": every susequent inhomogeneity - it can be a bend of the duct, an expansion, etc. - can either double the swing amplitude of ray oscillations or cancel it [3.170, 172]. It can be used to check the possibility of exciting subcritical modes in a waveguide which confines only higher modes, and the reverse effect: reflection from a coupler which is not a critical one fo the given mode [3.170]. The ray-optics concept may also interpret the effect of radiation loss inflicted by the rays leaving a graded-index waveguide limited by a low barrier [3.179].

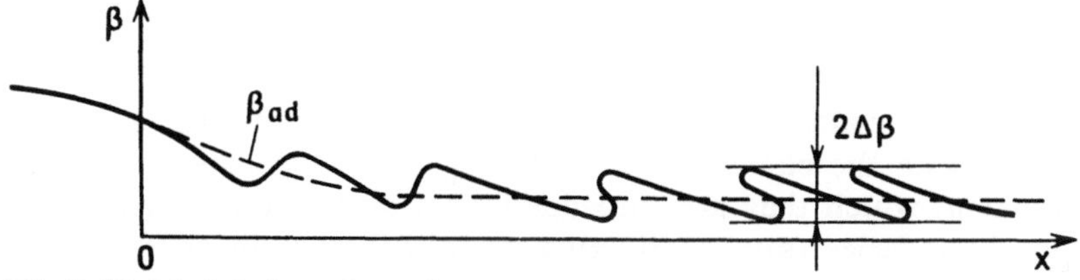

Fig.3.58. Modulation of ray tilt β relative to the adiabatic value β_{ad}

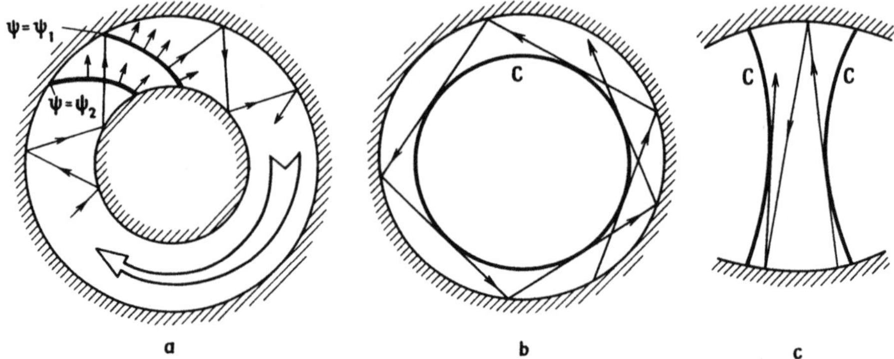

Fig.3.59. Certain types of rays in resonators: (a) rays and phase fronts in a ring resonator; (b) whispering-gallery modes; (c) jumping ball type of rays

3.7.8 Normal Modes in Cavity Resonators

The simple ideas underlying the ray description of normal modes in cavity resonators can be illustrated takign as an example a spherical-mirror resonator (a traveling-wave resonator) sketched in Fig.3.59a (the metallic walls of the resonator are indicated by hatching). Let ψ_1 be the initial phase front of a wave. After the wave completes a roudn trip its phase front ψ_2 does not generally coincide with ψ_1. There exists, though, a (natural) phase fornt ψ_1 such that making a complete round trip the wave repeats the shape of ψ_1 accurate to within an integer number of wavelengths; i.e., the phase front $\psi = \psi_2 = \psi_1 + m\lambda_0$ coincides with ψ_1. It is to be emphasized that hte rays do not trace clsoe paths in general.

This general principle has been stated in [3.181] along with the most important types of natural waves allowing for the ray interpretation: the *whispering-gallery* modes and the waves of the *jumping-ball* type. In the former case the wave propagates in the ring bound by a wall and the casutic, as shown in Fig.3.59b; and in the latter case, the wave being bound by the caustics, in turn, reflects from the upper and lower concave walls (Fig.3.59c).

The natural oscillations in 3-D systems have been studied in [3.46, 48, 182].[31] The studies have been stimulated primarily by the wide use of open resonators in laser and microwave systems.

3.8 Wave Scattering at Localized Inhomogeneities

3.8.1 Effective Scattering Surface

Inhomogeneities are called localized if their spatial dimensions, l_{nh}, are small compared with the characteristic scale L_m over which the properties of the medium, in which they are embedded, vary significantly. They may possess sharp boundaries (say, conducting or solid bodies in an inhomo-

---

[31] The quasi-modal behavior of ray fields is discussed in [3.183].

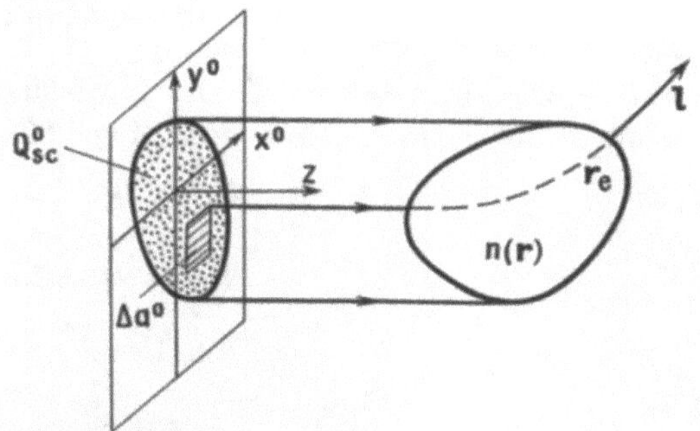

geneous medium), or they may exhibit diffused shapes, as, for example, do smooth inhomogeneities in a terrestrial plasma or plasma formations in intersteller space [3.91]. The localized inhomogeneities with diffuse boundaries are described by a refractive index n(r) that tends to a constant level as r → ∞, so rays at a far distance from the inhomogeneity become straight lines. The problem of scattering of a plane wave at localized inhomogeneities has been treated in [3.91,130,131,141,184-186]. The scattered field is normally characterized by an effective scattering surface, for which according to (3.2.15) we obtain

$$\Sigma = \lim_{r \to \infty} 4\pi r^2 \left| \frac{A}{A^0} \right|^2 = \lim_{r \to \infty} 4\pi r^2 \left| \frac{da^0}{da} \right|$$

$$= 4\pi \left| \frac{da^0}{d\Omega} \right| = 4\pi \left| \frac{\mathscr{D}(0)}{G} \right| , \tag{3.8.1}$$

where $da^0 = \mathscr{D}(0)d\xi d\eta$ is the cross section of the incident ray tube, $d\Omega = r^{-2}da = Gd\xi d\eta$ is the elementary solid angle of scattered rays, and ξ and η are arbitrary ray coordinates labeling the rays.

For scattered rays in the far field of a localized inhomogeneity bound by some interface surface Q, we have

$$r = r_e + \sigma l = r_e + (\psi - \psi_e)l \simeq \psi l ,$$

where $l = l(\xi,\eta)$ is the unit vector defining the direction of scattering, ψ is the eikonal, and the subscript "e" refers to the point where the ray emanates from the inhomogeneity (Fig.3.60). From this we get

$$G = \frac{d\Omega}{d\xi d\eta} = \frac{1}{r^2}\mathscr{D}(r) = \left| \frac{\partial l}{\partial \xi} \times \frac{\partial l}{\partial \eta} \right|$$

$$= \sqrt{\left(\frac{\partial l}{\partial \xi}\right)^2 \left(\frac{\partial l}{\partial \eta}\right)^2 - \left(\frac{\partial l}{\partial \xi} \cdot \frac{\partial l}{\partial \eta}\right)^2} . \tag{3.8.2}$$

For an inhomogeneity with diffused boundaries, $\mathbf{l}$ is taken as the asymptote of the ray, $\mathbf{l} = \lim \nabla \psi$ as $|\mathbf{r}| \to \infty$.

As $da^0 = dx^0 dy^0$, $\mathscr{D}(0) = \partial(x^0, y^0)/\partial(\xi, \eta)$, where x^0 and y^0 are the Cartesian coordinates in the plane of the incident wave front (Fig.3.60). Hence,

$$\Sigma = 4\pi \left| \frac{\partial(x^0, y^0)}{\partial(\xi, \eta)} \right| \left| \frac{\partial \mathbf{l}}{\partial \xi} \times \frac{\partial \mathbf{l}}{\partial y^0} \right|^{-1} , \tag{3.8.3a}$$

or, choosing $\xi \equiv x^0$ and $\eta \equiv y^0$,

$$\Sigma = 4\pi \left| \frac{\partial \mathbf{l}}{\partial x^0} \times \frac{\partial \mathbf{l}}{\partial y^0} \right|^{-1} . \tag{3.8.3b}$$

We introduce the spherical angles of scattering, θ and ϕ, such that

$$l_x = \sin\theta\cos\phi , \quad l_y = \sin\theta\sin\phi , \quad l_z = \cos\theta . \tag{3.8.4}$$

Then (3.8.3) can be represented as

$$\Sigma(\theta, \phi) = \frac{4\pi}{\sin\theta} \left| \frac{\partial(\theta, \phi)}{\partial(x^0, y^0)} \right|^{-1} . \tag{3.8.5}$$

Similarly, in a two-dimensional scattering problem (Fig.3.61a)

$$\Sigma = \lim_{r \to \infty} 2\pi r \left| \frac{A}{A^0} \right|^2 = 2\pi \left| \frac{d\theta}{dx^0} \right|^{-1} , \tag{3.8.6}$$

where $\theta = \theta(x^0)$ is the scattering angle between the incident and scattered rays.

In axially symmetric scattering

$$\frac{\partial(\theta, \phi)}{\partial(x^0, y^0)} = \frac{1}{\rho} \frac{d\theta}{d\rho} ,$$

so that by virtue of (3.8.5)

$$\Sigma = \frac{4\pi\rho}{\sin\theta} \left| \frac{d\theta}{d\rho} \right|^{-1} , \tag{3.8.7a}$$

where ρ is the impact parameter of the ray, and $\theta = \theta(\rho)$ is the scattering angle (Fig.3.61b).

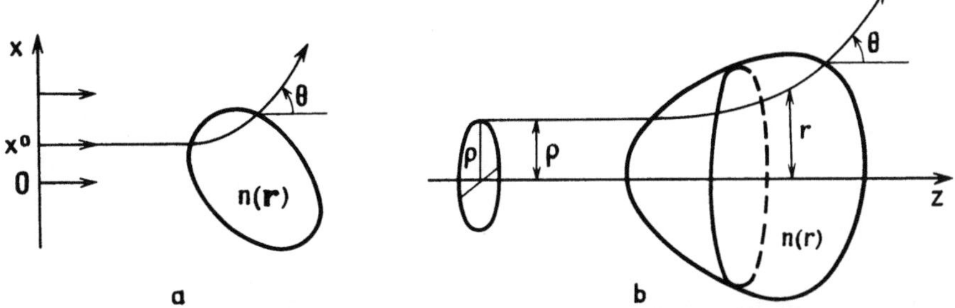

Fig.3.61. Illustrating the ray analysis of the scattering cross section for a localized in-homogeneity in (a) two-dimensional case and (b) 3-D axially symmetrical case

If there exists a specularly reflected ray ($1 = 1^0$), then the derived formulas may be used to determine the *backscattering* ($\theta = \pi$) cross section. To do so one should insert into them the ray coordinates, x^0 and y^0 (or ξ and η) associated with the ray at $\theta = \pi$. Specifically, if in the problem on axially symmetric scattering the specularly reflected ray corresponds to $\rho = 0$, then

$$\Sigma^{\text{bsc}} = 4\pi \lim_{\rho \to 0} \frac{\rho}{\sin\theta} \left| \frac{d\theta}{d\rho} \right|^{-1} = 4\pi \left| \frac{d\theta}{d\rho} \right|^{-2} \quad \text{at } \rho=0 \ . \tag{3.8.7b}$$

The *total* (*integral*) scattering cross section in the geometrical-optics approximation is

$$\Sigma^{\text{tot}} = \frac{1}{4\pi} O \int \Sigma d\Omega = \frac{1}{4\pi} \int 4\pi \frac{da^0}{d\Omega} d\Omega = \int da^0 = Q^0_{\text{sc}} \ . \tag{3.8.8}$$

where Q^0_{sc} is the area of projection of the scattering inhomogeneity onto the plane of the incident wave front (Fig.3.60). It follows then that for inhomogeneities of transversely diffused boundaries with $Q^0_{\text{sc}} = \infty$ the integral geometrical-optical scattering cross section tends to infinity [3.117].

It is a simple matter to show that in multiple ray scattering when, say, two rays pass through the observation point, i.e.,

$$\Sigma = (\Sigma_1 + \Sigma_2) + 2 \sqrt{\Sigma_1 \Sigma_2} \cos[k_0(\psi_1 - \psi_2) + \delta\psi_1 - \delta\psi_2] \ , \tag{3.8.9}$$

where Σ_j are the partial, effective scattering surfaces determined by (3.8.3,5-7), ψ_j is the eikonal of the jth ray as $r \to \infty$, and $\delta\psi_j$ is the caustic phase shift found from (2.4.12,13). The first term in (3.8.9) corresponds to the incoherent (energy) summing of rays, while the second describes the interference effects in scattering. The mean values of the effective scattering surface is $\Sigma_{\text{mean}} = \Sigma_1 + \Sigma_2$.

The caustics of the scattered field are governed by

$$\theta = \theta(\xi,\eta) \ , \quad \phi = \phi(\xi,\eta) \ , \quad \frac{\partial(\theta,\phi)}{\partial(\xi,\eta)} = 0 \ . \tag{3.8.10}$$

In axially symmetric scattering, one caustic degenerates into the axis of symmetry, and the other corresponds to the extremum of the function $\theta = \theta(\rho)$, i.e., where $d\theta/d\rho = 0$. In the vicinity of the caustics the scattered field may markedly increase in intensity (focusing). Well-known examples of this phenomenon are rainbows and halos in light scattering [3.117]. Similar effects are known to appear in quantum-mechanical scattering, too [3.14, 116-123].

3.8.2 Scattering by a Body in an Inhomogeneous Medium

The geometrical-optics formulas derived in the previous section are also valid for reflecting interfaces in an inhomogeneous medium; specifically, they may be applied to evaluating the effective scattering surface for bodies of large electrical size surrounded by a localized inhomogeneity. This situation differs from that of Sect.3.8.1 only in the actual form of the dependancies $l(\xi,\eta)$ or $\theta = \theta(\xi,\eta)$ and $\phi = \phi(\xi,\eta)$, which are now governed not only by the ray equation in the inhomogeneous medium, but also by the laws of reflection at the interface (Sects.2.4, 3.2)

If the size of the scatterer l_{sc} is only a small fraction of the characteristic dimension L of the incident-field variation in the medium and the medium's inhomogeneity, i.e., $L \gg l_{sc}$, then the scattering proceeds as if it were in a homogeneous medium. Thus the value for the effective scattering surfaces of this body in the homogeneous medium, Σ^h, may be used in evaluating the scattered field [3.91, 187]. The inhomogeneous medium acts as a "converter" of the field. The scattered field is then customarily described by the *apparent scattering cross section* [3.91, 187] and the corresponding formula can readily be derived from the definition of the effective scattering cross section. For a point source of type (3.4.9), for example, the effective scattering cross section (3.2.15) can be recast as

$$\Sigma = 4\pi R_{23}^2 \left| \frac{A^{sc}}{A^i} \right|^2 = 4\pi R_{12}^2 R_{23}^2 \left| \frac{A}{B^0} \right|^2 , \qquad (3.8.11)$$

where $R_{12} = |r_1 - r_2|$ and $R_{23} = |r_2 - r_3|$ are the distances between points 1, 2 (transmitter, scatterer) and 2, 3 (scatterer, receiver); A^{sc} is the field at the point of reception, 3 (Fig.3.62). Considering that the source field in an inhomogeneous medium can be represented as $|A| = |B^0|/R_{eff}$, where R_{eff} is the effective distance (Sect.3.4.3), we find for a scatterer

$$|A|^2 = \left| \frac{B^0}{R_{eff,12}} \right|^2 \frac{\Sigma^h}{4\pi R_{eff,23}^2} . \qquad (3.8.12)$$

We have assumed here that the scattering proceeds by a single-ray mechanism, so that $R_{eff,12}$ and $R_{eff,23}$ are the effective distances over the routes 1-2 and 2-3, respectively (Fig.3.62). As a result, from (3.8.11, 12) we obtain for the apparent cross section of scattering

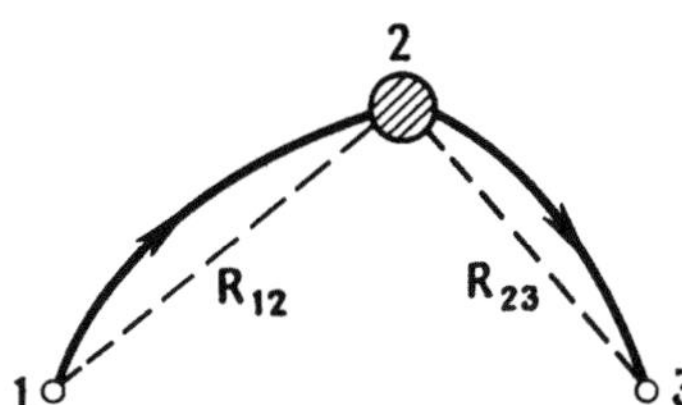

Fig.3.62. Single scattering on a body immersed in an inhomogeneous medium

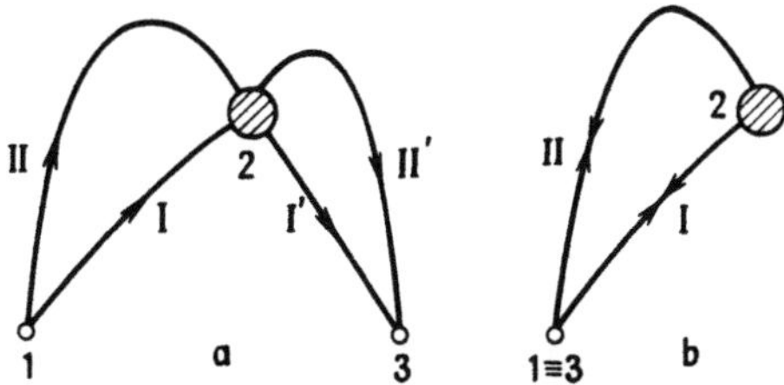

Fig.3.63. Multichannel scattering from a body embedded into an inhomogeneous medium; (a) bistatic reception and scattering; (b) combined reception and scattering

$$\Sigma = \frac{R_{12}^2}{R_{eff,12}^2} \frac{R_{23}^2}{R_{eff,23}^2} \Sigma^h = f_{12} f_{23} \Sigma^h \,, \tag{3.8.13}$$

where $f_{12} = R_{12}^2/R_{eff,12}^2$ and $f_{23} = R_{23}^2/R_{eff,23}^2$ are the field-focusing indices over the routes 1-2 and 2-3, respectively.

Owing to focusing the quantity Σ can substantially exceed Σ^h. In plane and spherically stratified media, f_{12} and f_{23} may be estimated by the formulas of Sects.3.4,5; actual computations were outlined in [3.92]. The topics received a more detailed consideration for inhomogeneous plasma media in [3.91, 187, 188]. The scattering by a body placed close to caustics was studied, particularly, in [3.189].

Note that if a scatterer is embedded in a medium capable of reflecting rays otherwise (as the ionosphere does in the decametric range), then the scattered field travels to an observation point through *several propagation paths*. Figure 3.63a sketches the situation with two incident and two scattered rays, which make *four* scattering paths II→II′, I→II′, II→I′, I→I′. In backscattering, that is when I = I′ and II = II′ (Fig.3.63b), the channels I→II and II→I turn out in phase or *coherent* (by the reciprocity theorem). The phenomenon of enhanced backscattering due to these coherent channels was considered in [3.190-182, 191].

3.8.3 Effective Scattering Surface of a Spherically Stratified Inhomogeneity

The scattering angle θ for a localized spherically stratified formation ($n^2 \to 1$ as $r \to \infty$) is determined by (3.5.21)

$$\theta = 2\theta_{tp} = 2 \int_{r_{tp}}^{\infty} \frac{\rho\, dr}{r\sqrt{r^2 n^2(r) - \rho^2}} = \theta(\rho) \,, \tag{3.8.14a}$$

where θ_{tp} and r_{tp} are the coordinates of the ray turning point (Sect.3.5).[32] Then by virtue of (3.8.7) we obtain

---

[32] Here, the scattering angle θ is measured from the z axis rather than from the incident ray, as it was done in Sects.3.8.1-2.

$$\Sigma = \frac{2\pi\rho}{\sin\theta}\left| \frac{d}{d\rho} \int_{r_{tp}}^{\infty} \frac{\rho\,dr}{r\sqrt{r^2 n^2(r) - \rho^2}} \right|^{-1} . \tag{3.8.14b}$$

Expressions (3.8.14) govern the dependence $\Sigma = \Sigma(\theta)$ parametrically: $\Sigma = \Sigma(\rho)$, $\theta = \theta(\rho)$. For a nonmonotonous dependence $\theta(\rho)$, the function $\Sigma(\theta)$ is described by (3.8.9) and may give rise to interference and focusing phenomena at caustics[33] where $d\theta/d\rho = 0$. Eq. (3.8.14b) can also be obtained from (3.5.22) directly if we pass to the limit $r \to \infty$, and then make use of the definition of the effective scattering surface (3.2.15).

For the case of an inhomogeneity bound by the sphere $r = r_{nh}$ such that $n(r) = 1$ for $r \geq r_{nh}$, from (3.8.14b) it follows

$$\Sigma = \frac{2\pi\rho}{\sin(2\theta_{tp})}\left| \frac{d\theta_{tp}}{d\rho} \right|^{-1} = \pi r_{nh}^2 \frac{\sin(2\theta_0)}{\sin(2\theta_{tp})}\left| \frac{d\theta_{tp}}{d\theta_0} \right|^{-1} , \tag{3.8.15}$$

where θ_{tp} corresponds to the turning point

$$\theta_{tp} = \theta_0 + \int_{r_{tp}}^{r_{nh}} \frac{\rho\,dr}{r\sqrt{r^2 n^2(r) - \rho^2}} ,$$

$$\rho = r_{nh}\sin\theta_0 .$$

Here θ_0 is the coordinate of the point at which the ray strikes the interface (Fig. 3.64), and the scattering angle is $\theta = 2\theta_{tp}$.

The ray with $\rho = 0$ is scattered backward from the region $n^2 = 0$ (this is possible for a *plasma* medium)[34] with

$$\Sigma^{bsc} = \pi r_{nh}^2 \left[1 + r_{nh} \int_{r_0}^{r_{nh}} \frac{dr}{r^2 n(r)} \right]^{-2} , \tag{3.8.16a}$$

and for an unbound inhomogeneity ($r_{nh} \to \infty$)

---

[33] For scattering by a spherically symmetric quantum-mechanical potential, the caustics were considered, in particular, in [3.126, 192].

[34] In backscattering ($\theta = 0$) of the ray with $\rho \neq 0$, in accord with (3.8.14, 15), $\Sigma \to \infty$, which is associated with an axial caustic being formed. The caustic causes a diffraction peak to appear, and $\Sigma(\theta)$ to oscillate (the halo scattering or glory phenomenon). A halo is observed, for instance, in spiral scattering when rays spiral around the centrum of symmetry $r = 0$ (Sect. 3.5 and [3.116, 117].

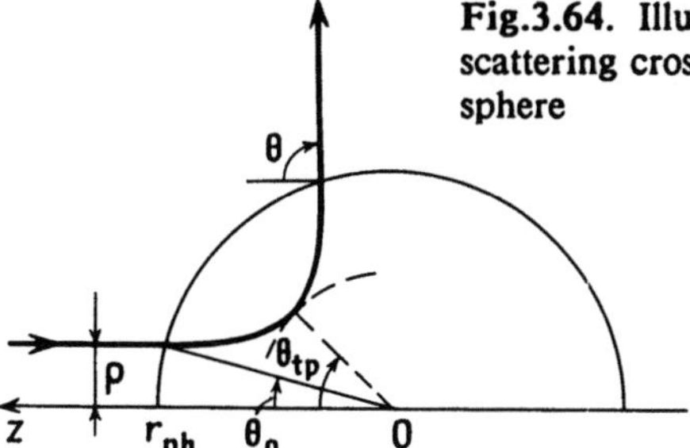

Fig.3.64. Illustrating the derivation of (3.8.15) for the scattering cross section of an inhomogeneity bounded by a sphere

$$\Sigma^{\text{bsc}} = \pi \left| \int_{r_0}^{\infty} \frac{dr}{r^2 n(r)} \right|^{-2} , \qquad (3.8.16b)$$

where r_0 is the root of $n^2(r_0) = 0$. From these equations it follows that $\Sigma^{\text{bsc}} < \pi r_0^2 < \pi r_{\text{nh}}^2$, i.e., the effective scattering surface of a plasma's spherically-symmetric inhomogeneity ($n < 1$) is smaller than that of a perfectly conducting sphere of radius r_0, due to an additional divergence of rays in the radially inhomogeneous medium with $n = n(r)$. The quantity Σ^{bsc} depends substantially upon the behavior of $n^2(r)$ close to r_0: $\Sigma^{\text{bsc}} < \infty$ for $dn(r_0)/dr \neq 0$; otherwise [i.e., when $n^2(r)$ possesses a zero of an order higher than the first], $\Sigma^{\text{bsc}} \to 0$, which corresponds to an infinite divergence of paraxial rays (close to $\rho = 0$).

For the scattering by an axially symmetric inhomogeneity, similar to (3.8.14b, 16b), we have

$$\Sigma = \pi \left| \frac{d}{d\rho} \int_{r_{tp}}^{\infty} \frac{\rho\, dr}{r\sqrt{r^2 n^2(r) - \rho^2}} \right|^{-1} , \qquad \Sigma^{\text{bsc}} = \pi \left| \int_{r_0}^{\infty} \frac{dr}{r^2 n(r)} \right|^{-1} , \qquad (3.8.17)$$

where $\theta = \theta(\rho)$ is given by (3.6.14). These expressions are valid for a bound ($r < r_{\text{nh}}$ inhomogeneity as well if we let $n^2(r) \equiv 1$ for $r \geq r_{\text{nh}}$.

The scattering cross sections for spherically and axially symmetric inhomogeneities have been the focus of numerous investigations, e.g., [3.118, 128–133, 137, 185–189].

3.8.4 Effective Scattering Surface of a Perfectly Conducting Sphere in a Spherically Stratified Medium

For a perfectly conducting sphere of the radius b placed in the region of the caustic shadow,[35] the effective scattering surface may be determined from the formulas of the previous section.

---

[35] It is implied that $r_0 - b \gg \Lambda$, where Λ is the size of the near caustic zone (Sect.2.10) and r_0 is the radius corresponding to the zero value of n^2: $n^2(r_0) = 0$.

If, on the other hand, the rays are reflect from a perfectly conducting sphere, we obtain for the scattering angle θ and the effective scattering surface the same formulas as in Sect.3.8.3, but in place of the turning-point coordinate r_{tp} the radius of reflection $r = b$ arises. Thus

$$\Sigma = \frac{2\pi\rho}{\sin\theta} \left| \frac{d}{d\rho} \int_b^\infty \frac{\rho dr}{r\sqrt{r^2 n^2(r) - \rho^2}} \right|^{-1} , \qquad (3.8.18a)$$

$$\Sigma^{bsc} = \pi \left| \int_b^\infty \frac{dr}{r^2 n(r)} \right|^{-2} ,$$

$$\theta = 2 \int_b^\infty \frac{\rho dr}{r\sqrt{r^2 n^2(r) - \rho^2}} , \qquad (3.8.18b)$$

In the general case, in (3.8.18) the integrals for Σ and θ use as the lower limit $\max(b, r_{tp})$ instead of b.

Similarly, for a perfectly conducting cylinder in a cylindrically stratified medium

$$\Sigma = \pi \left| \frac{d}{d\rho} \int_{r_{sc}}^\infty \frac{\rho dr}{r\sqrt{r^2 n^2(r) - \rho^2}} \right|^{-1} , \qquad (3.8.19)$$

$$\Sigma^{bsc} = \pi \left| \int_{r_{sc}^0}^\infty \frac{dr}{r^2 n(r)} \right|^{-1} ,$$

where $r_{sc} = \max(b, r_{tp})$, $r_{sc}^0 = \max(b, r_0)$, and $\theta = \theta(\rho)$ is given by (3.8.18b) for $b > r_{tp}$ or by (3.8.14a) for $r_{tp} > b$.

An interesting peculiarity of Σ^{bsc} is observed in a plasma of a monotonous $n^2(r)$ profile. If the size r_0 of the critical region is varied and the radii of both the ideally conducting sphere b and the inhomogeneous envelope r_{nh} are retained, then Σ^{bsc} reaches a minimum at $r_0 \simeq b$ [3.136]. This is explained by an increase, as $r_0 \to b$, in the divergence of rays, which is defined by the integral in the expression for Σ^{bsc} in (3.8.18a). For a linear dependence $n^2(r) \simeq \beta(r-r_0)$ this peculiarity was treated in depth in [3.136] and substantiated by a numerical analysis based on the exact solution.

3.8.5 Effective Scattering Surface of a Specific Two-Dimensionally Inhomogeneous Formation

Consider the axially symmetric scattering from a 2-D inhomogeneous media, (3.6.15), illuminated by a plane wave incident upon the apex ($\mathbf{l}^0 = \mathbf{e_z}$). The scattering angle θ for the rays (3.6.22) is determined from the following transcendental equation, as $r \to \infty$, i.e.,

$$\left(\int_{\theta_{tp}}^{\theta_b} \mp \int_{\theta_{tp}}^{\theta} \right) \frac{\rho \, d\theta}{\sqrt{\rho^2 - b^2 f(\theta)}} = \theta_b \; . \tag{3.8.20}$$

Then from (3.8.7, 20) we find the effective scattering surface

$$\Sigma = \frac{4\pi \rho^2}{\sin\theta \sqrt{\rho^2 - b^2 f(\theta)}} \left| \frac{d}{d\rho} \left(\int_{\theta_{tp}}^{\theta_b} \mp \int_{\theta_b}^{\theta} \right) \frac{d\theta}{\sqrt{\rho^2 - b^2 f(\theta)}} \right|^{-1} . \tag{3.8.21}$$

For the reverse incidence of the plane wave ($\mathbf{l}^0 = -\mathbf{e_z}$, Fig.3.51), we obtain for the scattering angle from (3.6.25)[36]

$$\frac{d}{d\kappa_0} \int_0^\theta \sqrt{\kappa_0^2 - b^2 f(\theta)} \; d\theta \equiv \int_0^\theta \frac{\kappa_0 \, d\theta}{(\kappa_0^2 - b^2 f(\theta))^{1/2}} = \pi \; , \qquad \text{or} \tag{3.8.22a}$$

$$\int_0^\theta \sqrt{\kappa_0^2 - b^2 f(\theta)} \; d\theta = \pi(\kappa_0 - b) \; , \tag{3.8.22b}$$

where $\kappa_0 = \sqrt{b^2 + \rho^2}$. By analogy with (3.8.21) this yields

$$\Sigma = \frac{4\pi \kappa_0^2}{\sin\theta \sqrt{\kappa_0^2 - b^2 f(\theta)}} \left| \frac{d}{d\kappa_0} \int_0^\theta \frac{\kappa_0 \, d\theta}{\sqrt{\kappa_0^2 - b^2 f(\theta)}} \right|^{-1} . \tag{3.8.23}$$

In computing Σ^{bsc} one should observe that at $\rho = 0$ the integrals of (3.8.22, 23) diverge as $\theta \to 0$:

$$\kappa_0^2 - b^2 f(\theta) = \rho^2 - \tfrac{1}{2} b^2 f''(0)\theta^2 + \dots \; . \tag{3.8.24}$$

---

[36] In this case, as in Sect.3.8.3, the scattering angle is more conveniently counted from the z axis rather than from the incident ray.

We recall that by the condition imposed in Sect.3.6 $f'(0) = 0$ and $f(0) = 1$. Therefore, in order to obtain $d\theta/d\rho$ for $\rho = 0$, we should recognize that the scattering angle θ is small at $\rho \simeq 0$, and take refuge to (3.8.24). Substituting (3.8.24) into (3.8.22) yields for $f''(0) < 0$

$$\theta \simeq \frac{\rho}{b\sqrt{\alpha}}\,\sinh\frac{\pi b\sqrt{\alpha}}{\sqrt{b^2 + \rho^2}}\,,$$

where $\alpha = |f''(0)|/2$. As a result, from (3.8.7b) we find

$$\Sigma^{\text{bsc}} = \frac{4\pi b^2 \cdot \alpha}{\sinh^2(\pi\sqrt{\alpha})}\,. \tag{3.8.25a}$$

Similar to (3.8.25a) for $f''(0) > 0$ from (3.8.22,24) and (3.8.7b) [3.143] we obtain

$$\Sigma^{\text{bsc}} = \frac{4\pi b^2 \cdot \alpha}{\sin^2(\pi\sqrt{\alpha})}\,. \tag{3.8.25b}$$

At $\alpha = 0$ (3.8.25a, b) describe the effective scattering surface $\Sigma^{\text{bsc}} = 4b^2/\pi$ corresponding to the spherically symmetric inhomogeneity of $n^2(r) = 1 - b^2/r^2$, which certainly coincides with the result of (3.8.16). At $\alpha \neq 0$, (3.8.25) provides an estimate of the effect of an angular inhomogeneity upon the effective scattering surface. A more detailed analysis of the effective scattering surface for two-dimensionally inhomogeneous formations was presented in [3.141-143]. Following [3.143] we note that the quantity Σ^{bsc} given by (3.8.25) can be both lower [for $f''(0) < 0$] and higher [for $f''(0) > 0$] than the effective scattering surface Σ^{bsc} associated with the perfectly conducting surface of revolution $r = bf(\theta)$, coinciding with the surface of $n^2(r, \theta) = 0$. The effective scattering cross section can readily be obtained from (3.2.16) to give $\Sigma_0^{\text{bsc}} = \pi b^2/[1 - f''(0)]^2$. The difference of the quantities (3.8.25a and b) is interpreted as being due to the different nature of the wave refraction for $f''(0) > 0$ and $f''(0) < 0$. This difference is most pronounced at $f''(0) = 2m^2 > 0$, $m = 1, 2$, etc. when, according to (3.8.25b), the scattering cross section becomes infinite; i.e., ideal focusing is observed for the scattered rays as $\rho \to 0$ [3.143]. The calculation of the effective scattering surface has been conducted for an inhomogeneous paraboloid and toroid with a separable eikonal equation [3.185, 186].

3.8.6 Scattering of a Spherical Wave by a Localized Inhomogeneity

By analogy with the plane-wave case (Sect.3.8.1), we have for the far field- (scattered field) amplitude

$$A = \frac{1}{r}B^0\sqrt{\frac{n^0 d\Omega^0}{d\Omega}} = \frac{1}{r}B^0\sqrt{\frac{n^0 G^0}{G}} \equiv \frac{1}{r}F\,,$$

where $d\Omega^0 = G^0 d\xi d\eta$, $d\Omega = G d\xi d\eta$ are the elementary solid angles for incident and scattered rays; ξ and η are arbitrary ray coordinates;[37] $n^0 \equiv n(r^0)$; and r^0 is the radius vector of the source. The radiation pattern of the source is then

$$F = B^0 \left| n^0 \left(\frac{\partial l^0}{\partial \xi} \times \frac{\partial l^0}{\partial \eta} \right) \right|^{1/2} \left| \frac{\partial l}{\partial \xi} \times \frac{\partial l}{\partial \eta} \right|^{-1/2}. \tag{3.8.26}$$

where l^0 is the unit vector along the ray radiated by the source, and l specifies the direction of scattering. For a directional source, $B^0 = B^0(l^0)$ is its radiation pattern in a homogeneous medium of constant $n = n^0$.

Invoking spherical scattering the (observation) angles θ and ϕ defined by (3.8.4) and similar angles of departure, θ^0 and ϕ^0, yields

$$F = B^0(\theta^0,\phi^0)\sqrt{n^0} \sqrt{ \frac{\sin\theta^0}{\sin\theta} \frac{\partial(\theta^0,\phi^0)}{\partial(\xi,\eta)} \left[\frac{\partial(\theta,\phi)}{\partial(\xi,\eta)} \right]^{-1} }$$

$$= B^0(\theta^0,\phi^0)\sqrt{n^0} \sqrt{ \frac{\sin\theta^0}{\sin\theta} \frac{\partial(\theta^0,\phi^0)}{\partial(\theta,\phi)} }. \tag{3.8.27}$$

For the axially symmetric situation

$$F = B^0 \sqrt{n^0} \sqrt{ \frac{\sin\theta^0}{\sin\theta} \frac{d\theta^0}{d\theta} }, \tag{3.8.28}$$

where $B^0 = B^0(\theta^0,\phi^0)$. Similarly, for the two-dimensional case

$$A = \frac{1}{\sqrt{r}} B^0(\theta^0)\sqrt{n^0} \sqrt{ \frac{d\theta^0}{d\theta} }. \tag{3.8.29}$$

These formulas also apply in the presence of reflecting interfaces in an inhomogeneous medium.

Consider, by way of example, the radiation field of a source in a spherically stratified medium. The partial radiation pattern associated with one scattered ray is defined by (3.8.28), and the scattering angle θ is given by (3.5.15) where one should let $r \to \infty$. The same result follows from (3.5.17) if we observe that in far field $r^2 \gg \rho^2$ and $n^2(r) \to 1$, viz.

---

[37] With the source outside the finite inhomogeneity, it would be convenient to choose ξ and η as the curvilinear coordinates of the point where the ray strikes the boundary.

$$A = \frac{1}{r} B^0 \sqrt{\frac{n^0 \sin\alpha^0}{\sin\theta \, d\theta/d\alpha^0}}$$

(in this case $\alpha^0 \equiv \theta^0$).

3.8.7 Scattering by Weak Localized Inhomogeneities

The permittivity $\epsilon(\mathbf{r}) = n^2(\mathbf{r})$ of a medium containing *weak* inhomogeneities deviates only a little from the mean, "background" values $\bar{\epsilon}(\mathbf{r}) = \bar{n}^2(\mathbf{r})$; i.e., the difference $\nu(\mathbf{r}) = \epsilon(\mathbf{r}) - \bar{\epsilon}(\mathbf{r})$ is small compared with $\bar{\epsilon}$. As a consequence, the *perturbation technique* may be invoked to describe the scattering effect (Sect. 2.9).

A salient feature of weak localized inhomogeneities is that inside them rays undergo a rather small lateral shift: if l_{nh} is the characteristic scale of the inhomogeneity, then estimates by (2.9.17, 18) indicate that $|\Delta r_{lat}| \simeq l_{nh}|\nu_{max}| \ll l_{nh}$. Therefore, we may safely replace the inhomogeneity by a phase screen, which changes only the phase of an incident beam leaving its amplitude unchanged. The pertinent variation of the eikonal $\Delta\psi$ can be derived by (2.9.8), in which the integration limits may be extended to infinity by taking advantage of the disturbance $\nu(\mathbf{r})$ which rapidly falls off as $\sigma \to \pm\infty$,

$$\Delta\psi \simeq \psi_1 = (\bar{\epsilon})^{-1/2} \int_{-\infty}^{+\infty} \nu(\sigma')d\sigma' \tag{3.8.30}$$

(here $\bar{\epsilon}$ refers to the center of the inhomogeneity).

Considering (3.8.30) as the primary eikonal in the plane of an undisturbed phase front $\psi_0 = $ constant, we may evaluate all the parameters of the ray field both in the homogeneous and in the inhomogeneous media. Specifically, (3.8.30) yields focal points and caustics arising behind the localized inhomogeneity. According to (2.9.17) the momentum increment $\Delta\mathbf{p} \sim \nabla\psi_1$ is comparable in value with the disturbance ν, so that having passed an inhomogeneity a wave forms a caustic at a distance $z_c \simeq l_{nh}/|\Delta\mathbf{p}| \sim l_{nh}/|\nu|$ [3.193].

Inhomogeneous media are interesting in that an inhomogeneity providing focusing, if placed in a homogeneous medium, may yield defocusing when embedded in an inhomogeneous medium. This phenomenon found out in [3.150] is explained by the combined effect of the local inhomogeneity and the smoothly inhomogeneous medium acting as a thick lens.

Using the equivalence of a weak localized inhomogeneity to a phase screen, we are able readily to estimate the Fresnel radius a_f for an observation point at a distance σ from the inhomogeneity

$$a_f(\sigma) = \sqrt{\lambda|1/R_0 + 1/\sigma|^{-1}}, \quad a_f(\infty) = \sqrt{\lambda|R_0|},$$

where R_0 is the radius of curvature of the phase front immediately behind the inhomogeneity, in the order of magnitude, $|R_0| \sim l_{nh}/|\nu_{max}|$. Therefore the condition of validity of geometrical optics for an infinitely distant point, $a_f(\infty) \ll l_{nh}$ leads to the inequality

$$\lambda \ll |\nu_{max}| l_{nh} \, , \tag{3.8.31}$$

which implies that the difference of the optical paths inside the inhomogeneity and beside it is equal to $|\nu_{max}| l_{nh}$ must exceed the wavelength. As the inhomogeneity becomes less dense, that is, $|\nu_{max}|$ falls, geometrical optics can no longer be applied. Under conditions inverse to (3.8.31), i.e., $|\nu_{max}| l_{nh} \ll \lambda$, we should make use of the Born approximation of scattering theory.

3.9 Pulse Propagation

3.9.1 General Relations for the Plasma (Guided) Dispersion Law

In this section we consider a number of examples in which space-time geometrical optics is employed for evaluating the fields of high-frequency pulses propagating in media with frequency dispersion.

The dispersion law

$$n(\omega, r) = \sqrt{1 - \omega_p^2(r)/\omega^2} \, , \tag{3.9.1}$$

where ω_p is the cutoff (plasma) frequency of the medium, is characteristic for waves in waveguides (then ω_p is the cutoff frequency of the waveguide) and for electromagnetic waves in a cold isotropic plasma (then ω_p is the plasma frequency). The wave field $u(r, t)$ in the medium (3.9.1) is described by a wave equation simpler than (2.7.1):

$$\left[\Delta - \frac{1}{c^2} \frac{\partial^2}{\partial t^2} - \frac{\omega_p^2(r)}{c^2} \right] u(r, t) = 0 \, , \tag{3.9.2}$$

so that the relations of geometrical optics become somewhat simpler [3, 61, 194, 195].

From (2.7.23) it follows that the group velocity is $g(\omega, r) = cn(\omega, r)$ so that the space-time ray equations become

$$\frac{dr}{dt} = \frac{c^2}{\omega} k \, , \quad \frac{(dk)}{dt} = -\frac{1}{2\omega} \nabla \omega_p^2(r) \, , \tag{3.9.3a}$$

$$\frac{d^2 r}{dt^2} = -\frac{c^2}{2\omega^2} \nabla \omega_p^2 \equiv \frac{1}{2} c^2 \nabla n^2(\omega, r) \, , \tag{3.9.3b}$$

205

where $\omega = \omega^0 = \omega^0(\xi)$ is the initial frequency, being constant on the ray.

With the identification $d\tau = cdt = cd\sigma/g = d\sigma/n$ for the group path of a pulse in the medium (3.9.1) equation (3.9.3) converts into the ordinary ray equation (2.2.19), $d^2r/d\tau^2 = \nabla n^2/2$.

The wave eikonal (2.7.35) for the plasma dispersion law (3.9.1) is given by

$$\phi = \phi^0 - \frac{1}{\omega} \int_{t^0}^{t} \omega_p^2 \, dt = \phi^0 - \omega(t - t^0) + \frac{\omega}{c} \int_{\tau^0}^{\tau} n^2 d\tau \, , \tag{3.9.4}$$

where the integration is taken along the rays (3.9.3).

3.9.2 A Homogeneous Medium with an Arbitrary Dispersion Law

In a homogeneous medium ($\nabla n \equiv 0$) the ray equations (2.7.22) take the form

$$dr/dt = g \, , \quad dk/dt = 0 \, .$$

It follows then that

$$r = r^0 + g^0(t - t^0) \equiv r(t, \xi) \, , \tag{3.9.5}$$

and for the initial conditions imposed at $r = r^0(\xi)$ and $t = t^0(\xi)$ the group velocity g is defined by (2.7.23) where $k = k^0$ and $\omega = \omega^0$ are found from (2.7.32). In accordance with (3.9.5) the space-time rays in a homogeneous medium constitute a family of straight lines whose inclination is determined by the initial group velocity g^0.

The variation of instantaneous frequency of the wave along its propagation path is determined by the initial frequency $\omega = \omega^0(\xi)$ and (3.9.5) for the ray family $r = r(t, \xi)$.

In a homogeneous medium the eikonal (2.7.35) is

$$\phi = \phi^0 + (k^0 g^0 - \omega)(t - t^0) = \phi^0 - \frac{\omega^2}{c} g^0 \frac{\partial n}{\partial \omega}(t - t^0) \, , \tag{3.9.6}$$

where $\omega = \omega^0(\xi)$. In order to evaluate the wave amplitude we find the Jacobian

$$j(t) = \frac{\partial(x, y, z)}{\partial(\xi_1, \xi_2, \xi_3)} = \prod_{m=1}^{3} \frac{\partial}{\partial \xi_m} r(t, \xi)$$

$$= a_3(t - t^0)^3 + a_2(t - t^0)^2 + a_1(t - t^0) + a_0$$

$$= a_3(t - t_1)(t - t_2)(t - t_3) \, , \tag{3.9.7}$$

where the coefficients a_i are expressed in a simple way through the derivatives $\partial r^0/\partial \xi_m$, $\partial t^0/\partial \xi_m$, and $\partial g^0/\partial \xi_m$; and t_{1-3} are the roots of the cubic equations $j(t_i) = 0$. As a result we obtain for the amplitude

$$A(t) = A(t^0)\sqrt{\frac{j(t^0)}{j(t)}} = A(t^0)\frac{\sqrt{(t^0-t_1)(t^0-t_2)(t^0-t_3)}}{\sqrt{(t-t_1)(t-t_2)(t-t_3)}} . \qquad (3.9.8)$$

The three roots of $j(t) = 0$ in (3.9.7) imply that in the general case a space-time ray can possess three *caustic points* at $t = t_{1,2,3}$ in a homogeneous medium; hence it can touch three space-time caustics. Their equations are defined by (3.9.5) at $t = t_i$, and have the form $r = r(t_i, \xi)$, $t = t_i(\xi)$. The nomenclature of the structurally stable types of the caustics formed is principally the same as in the spatial case (Sects.2.4, 3.1), though one should certainly observe the larger dimension of the physical space, the irreversibility of the time coordinate [3.194] and the bounded variation of group velocity $(g \le c)$.

3.9.3 A Plane, Frequency-Modulated Pulse in a Homogeneous Medium

In an initial plane $z = 0$ we assume the initial field of a pulse to be given in complex form:[38]

$$u(z=0, t) = A^0(t)e^{i\phi^0(t)} \equiv u^0(t) , \qquad (3.9.9a)$$

where $A^0(t)$ specifies the shape (envelope) of the pulse. The frequency-modulation law is defined by

$$\omega(0, t) = -\frac{\partial}{\partial t}\phi^0(t) \equiv \omega^0(t) . \qquad (3.9.9b)$$

The space-time rays (3.9.5) are described as

$$z = g(t - \xi) = g(\omega^0)(t - \xi) \equiv z(t, \xi) , \qquad (3.9.10)$$

where $t^0 \equiv \xi$ is the time the ray leaves the plane $z = 0$, and $\omega^0 = \omega^0(\xi)$. By the principle of casuality $g(\omega) < c$, so the rays (3.9.10) on the z, t plane may travel within the sector $z < ct$ only.

The divergence of the rays (3.9.10) is

$$\mathcal{J} = \frac{\partial z}{\partial \xi}\left(\frac{dz}{d\xi}\right)^{-1}\bigg|_{t=\xi} = 1 - \frac{z}{g^2}\frac{dg}{d\xi} \equiv 1 - \frac{z}{z_c} , \qquad (3.9.11a)$$

---

[38] The initial field which we specify as the real function $\mathrm{Re}\{u^0(t)\}$ should generally be represented in complex form using the concept of analytic signal [3.196], which does not require that the signal be narrow band. The amplitude is then introduced as $|u^0(t)|$ and the phase $\phi^0(t)$ as $\arg u^0(t)$.

so that for the pulse amplitude at $z > 0$ we obtain

$$A(z,t) = A^0(\xi)\,\mathcal{J}^{-1/2} = A^0(\xi) \left[1 - \frac{z}{z_c}\right]^{-1/2} , \qquad (3.9.11b)$$

where

$$z_c = g^2 \left[\frac{dg}{d\xi}\right]^{-1} = g^2 \left[\frac{dg}{d\omega^0}\frac{d\omega^0}{d\xi}\right]^{-1} \equiv z_c(\xi) \qquad (3.9.12a)$$

is the spatial coordinate of the caustic. Its time coordinate is derived from (3.9.10) as

$$t = \xi + g\left[\frac{dg}{d\xi}\right]^{-1} = \xi + g\left[\frac{dg}{d\omega^0}\frac{d\omega^0}{d\xi}\right]^{-1} \equiv t_c(\xi) . \qquad (3.9.12b)$$

Observing (3.9.10) we represent the eikonal (3.9.6) as

$$\phi = \sigma^0 + k^0 g(t - \xi) - \omega^0(t - \xi) = \phi^0 + k^0 z - \omega^0(t - \xi) , \qquad (3.9.13)$$

where $k^0 = (\omega^0/c)n(\omega^0)$ is the wave number at $\omega = \omega^0$.

As a result, for the field of a frequency-modulated (FM) pulse in a homogeneous medium we obtain

$$u(z,t) = u^0(\xi)\,(1 - z/z_c)^{-1/2}\,\exp[ik^0 z - i\omega^0(t - \xi)] . \qquad (3.9.14)$$

In mulitpath propagation we should, as usual, sum up the fields (3.9.14) over all the ray coordinates $\xi_j = \xi_j(z,t)$ to be deduced from the ray family equation (3.9.10). Upon a caustic (for $z > z_c > 0$), the field (3.9.14) undergoes the caustic phase shift of $-\pi/2$ if $dg/d\omega > 0$ [3.195].

In the absence of frequency modulation, when $\omega^0 = $ constant, and $g = $ constant $= g(\omega^0)$, we find from (3.9.10) that $\xi = t - z/g$, $z_c \to \infty$ and

$$u(z,t) = A^0(t - z/g)\exp(ik^0 z - i\omega^0 t) . \qquad (3.9.15)$$

According to this geometrical-optics approximation, the initial envelope of a pulse $A^0(t)$ is transferred along parallel rays (3.9.10) without its shape and width being disturbed, at a group velocity $g = g(\omega^0)$, as shown in Fig.3.65.

The FM pulse envelope changes along the propagation distance z due to a cross-sectional variation (at $t = $ constant) of the ray tubes (Fig.3.66), and in the vicinity of caustics (Fig.3.67) due to interference of different rays (3.9.10). The variation of the instantaneous carrier frequency $\omega = \omega(z,t)$ and pulse width $T_p = T_p(z)$ is attributed to the same reasoning.

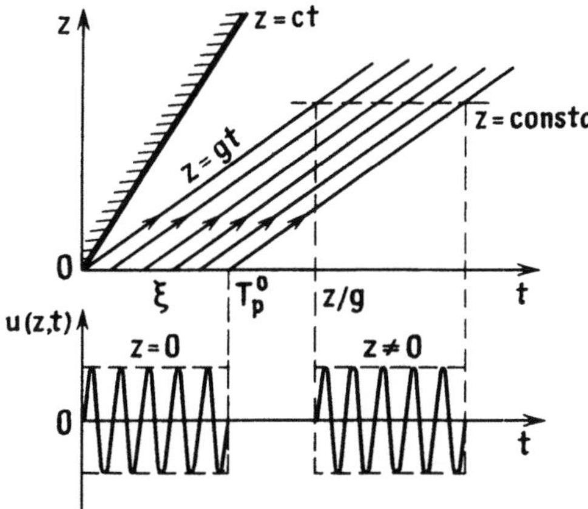

Fig.3.65. In the framework of ray theory, a constant frequency pulse propagates at a group velocity without distortion of its shape

If the inital pulse width is T_p^0, i.e., $A^0(t) \simeq 0$ for $t < 0$ and $t > T_p^0$, then the rays (3.9.10) corresponding to $\xi = 0$ and $\xi = T_p^0$ describe in the z, t plane the motion of the leading and back edges of the pulse. The time difference between their arrivals yields the pulse width T_p in the geometrical-optics approximation

$$\frac{T_p}{T_p^0} = \left| 1 + \frac{z}{T_p^0}\left[\frac{1}{g(T_p^0)} - \frac{1}{g(0)} \right] \right| ,$$

where $g(\xi) \equiv g[\omega^0(\xi)]$.

As the pulse moves closer to the caustic (3.9.12) its width T_p shortens and its height and the rate of variation of the instantaneous frequency $\omega(t)$ increase correspondingly (Fig.3.67). In the vicinity of the caustic the rays corresponding to $\xi = 0$ and $\xi = T_p^0$ cross, causing a *reversal of the signal* wherein its trailing edge overturns the leading edge so that the frequency dependency, $\omega(t)$, becomes inverted, for example, an increasing $\omega(t)$ becomes decreasing, as shown in Fig.3.67 [3.197]. It is apparent form this figure that as the ray bundle spreads behind the caustic, the pusle increases in length while its amplitude and frequency dependency $\omega(t)$ decrease. When the caustic (3.9.2) fails to form, then only a bundle spreading with all the associated effects is observed (Fig.3.68). Note that Figs.3.67,68 plot

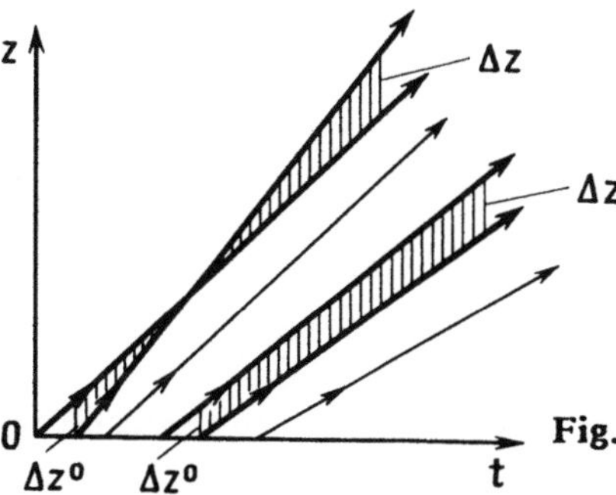

Fig.3.66. FM pulse length Δz varies in propagation

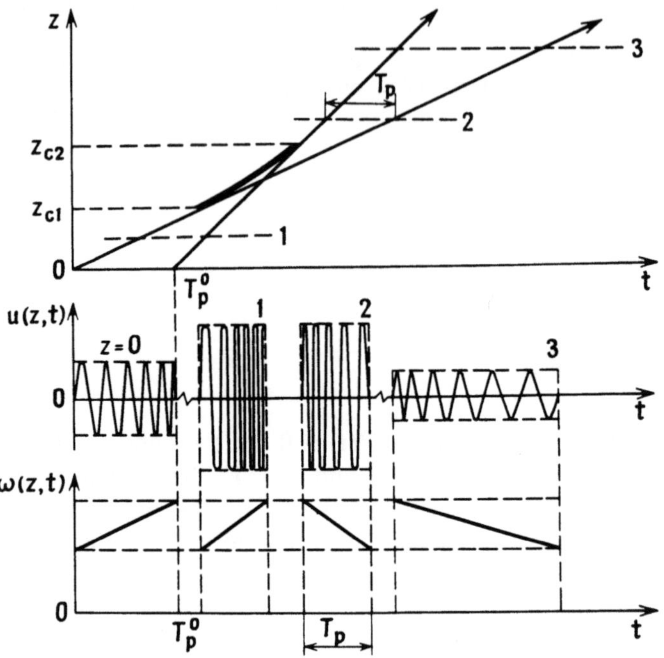

Fig.3.67. FM pulse evolution in a dispersive medium in the presence of a space-time caustic: as the signal propagates its amplitude, duration, and frequency variation law undergo changes

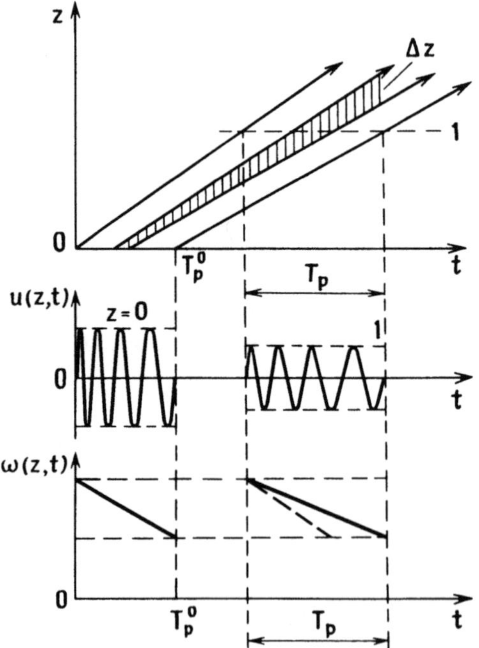

Fig.3.68. FM signal evolution in the conditions where no space-time caustic occurs

the effects qualitatively and do not illustrate how the pulse shape and modulation law change in response to a variation in ray-tube cross section.

Section 3.9.6 below presents a discussion of numerical analyses illustrating the validity limits of space-time geometrical optics.

3.9.4 Dispersive Compression of FM Pulses in Homogeneous Media

As is evident from Fig.3.67, near a caustic the FM pulse undergoes a dispersive *compression*.[39] The condition for the caustic (3.9.12) to form is

$$\frac{dg}{d\xi} = \frac{dg}{d\omega^0} \frac{d\omega^0}{d\xi} > 0 \; ;$$

at $dg/d\xi = 0$ the caustic produces an offshoot asymptotically tending to infinity. The shape of the caustic is determined by the law of dispersion $n(\omega)$ and that of pulse frequency modulation $\omega^0(\xi)$; a detailed study of its singularities and properties has been conducted in [3.195].

It is worth noting that for pulses of finite width only the terminal segments of caustics are physically realistic, corresponding to $0 \leq \xi \leq T_p^0$. With a smooth amplitude envelope $A^0(t)$ of an initial pulse (3.9.9a) these segments do not exhibit clear-cut ends, whereas the pulses of sharp leading and trailing edges, say rectangular,[40] are characteristic of *broken caustics* being formed with the discontinuity points at the pulse edges (Fig.3.67).

If $n(\omega)$ and $\omega^0(\xi)$ are chosen such that the rays (3.9.10) are brought to a *space-time focus* (Fig.3.69), rather than a caustic, the geometrical-optics approximation suggests that *optimum* (ideal) pulse compression should be expected. According to the ray-family equations (3.9.10) we have $z_f = g(t_f - \xi)$, where t_f and z_f are the focus coordinates. This yields, for the ideal compression, the condition

$$\xi = t_f - \frac{z_f}{c} \left[n(\omega^0) + (\omega^0) \frac{d}{d\omega^0} n(\omega^0) \right] \tag{3.9.16}$$

that defines the function $\xi = \xi(\omega^0)$ inverse to $\omega^0(\xi)$ specifying the required frequency dependence. With small deviations of the frequency ω^0 from a certain reference level ω_0^0 we may safely retain in the Taylor series for the function of (3.9.16) the linear term only, so this kind of frequency modulation is optimum for compression in this approximation [3.197]. Condition

---

[39] This phenomena is analogous to the FM pulse compression effected by means of appropriate signal processing facilities, say, by matched filters.

[40] Although the broken envelopes do not meet the analytic signal requirements [3.196], they are often used for pulse approximation. From the analytic signal viewpoint, such an approximation may be deemed satisfactory if in the correspondign real-valued pulse $\mathrm{Re}\{u^{0}(t)\}$ the spectrum bands clustered around $+\omega_0$ and $-\omega_0$ do not overlap practically; i.e., the condition of spectral separability [3.196] is approximately fulfilled.

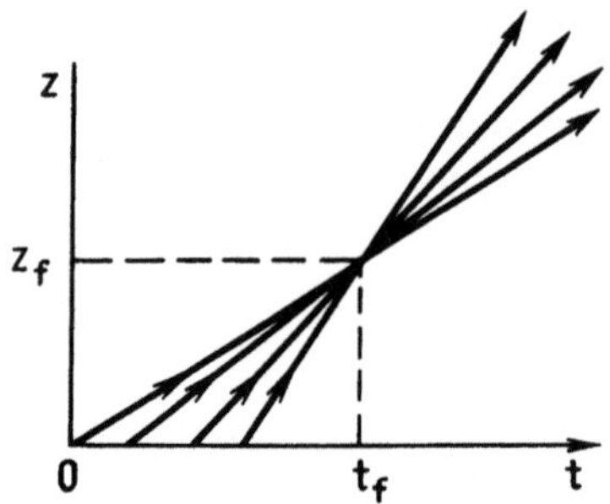

Fig.3.69. Formation of a space–time focus in a dispersive medium

(3.9.16) may readily be recast into a form allowing the dispersion law $n(\omega)$ to be determined for the ideal compression of a FM pulse having the given $\omega^0(\xi)$ [3.195].

In the particular case of the plasma dispersion law (3.9.1) we find from (3.9.16) that at ω_p = const. ideal focusing of a FM pulse is possible for the following frequency modulation [3.194, 198]:

$$\frac{\omega(0,t)}{\omega_p} = c(t_f - t)[c^2(t_f - t)^2 - z_f^2]^{-1/2} \ . \tag{3.9.17}$$

For relations $\omega^0(\xi)$ different from (3.9.17) the medium (3.9.1) of constant ω_p no longer provides an ideal space–time lens, rather it forms a caustic

$$z = \frac{c}{\omega_p^2}(\omega^{0^2} - \omega_p^2)^{3/2} \, (d\omega^0/d\xi)^{-1} \ ,$$

$$\tag{3.9.18}$$

$$t = \xi + \frac{\omega^0}{\omega_p^2}(\omega^{0^2} - \omega_p^2)(d\omega^0/d\xi)^{-1} \ .$$

The caustic exists in the considered region $z > 0$ only on the condition that $d\omega^0/d\xi > 0$, i.e., only when the frequency increases. Its asymptotes arise subject to $d\omega^0/d\xi = 0$, and the cusp points are the roots of

$$3\omega^0(d\omega^0/d\xi)^2 - (\omega^{0^2} - \omega_p^2)d^2\omega^0/d\xi^2 = 0 \ .$$

Figure 3.67 plots the caustic (3.9.18) for a pulse with linear frequency dependence on time, and Fig.3.70 that for a pulse with the quadratic dependence

$$\omega^0(t) = \omega_0[1 + \alpha(t - T_p^0/2)^2] \ .$$

Pulse compression has been discussed in depth in [3.194, 198, 199]. *Anutin* and *Orlov* [3.200] have investigated caustics (3.9.18) and computed

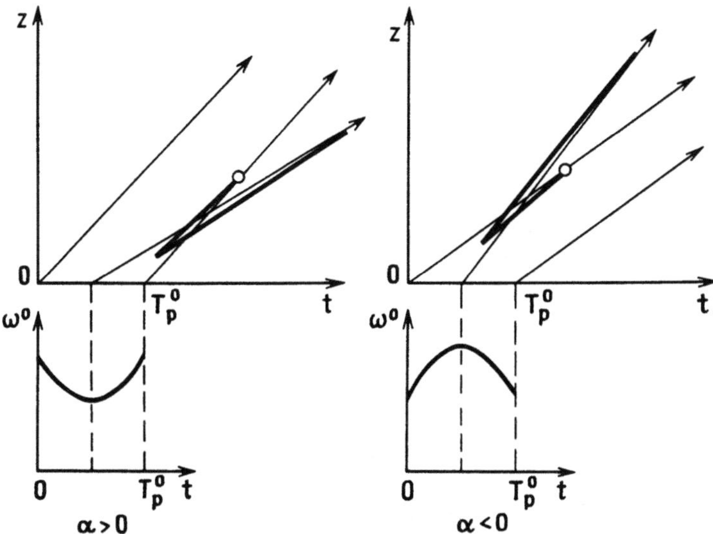

Fig.3.70. Broken caustics being formed as FM pulses propagate in a dispersive medium

the pulse field with rigorous expressions. A more complete review of the field can be found in [3.194, 195].

3.9.5 Plane-Stratified Dispersive Media

The equations of space-time rays in a medium of $n = n(z,\omega)$ may be derived in the easiest way, according to Sect.2.7, by augmenting the ray equations (2.3.37) by an equation for the pulse-propagation time, i.e.,

$$dt = \frac{d\sigma}{g} = \pm \frac{dz}{g}\sqrt{1 + \left(\frac{dx}{dz}\right)^2 + \left(\frac{dy}{dz}\right)^2} = \pm \frac{ndz}{g\sqrt{n^2(z) - (n^0\sin\theta^0)^2}},$$

so that

$$t = t^0 + \left(\int_{\tilde{z}_<}^{\tilde{z}_>} \mp \int_{z_<}^{z_>}\right)\frac{n(z,\omega)dz}{g(z,\omega)\sqrt{n^2(z,\omega) - (n^0\sin\theta^0)^2}} \equiv t(z,\xi) , \qquad (3.9.19)$$

where $\omega = \omega^0(\xi)$ and the nomenclature is the same as in Sect.3.3. Now the ray family is described by (2.3.37) and (3.9.18), which can be represented as $x = x(z,\xi)$, $y = y(z,\xi)$, and $t = t(z,\xi)$, where $\xi = (\xi_1,\xi_2,\xi_3)$ are the curvilinear coordinates on the initial hypersurface (Sect.2.7).

The Jacobian $j(t)$ required to compute the field amplitude (2.7.44) is to be found from the ray family set (2.3.37), (3.9.18) and, similar to (3.3.10), it is

213

$$j(t) = (g/n)p_z \frac{\partial(x,y,t)}{\partial(\xi_1,\xi_2,\xi_3)}$$

$$= \pm (g/n)\sqrt{n^2 - (n^0\sin\theta^0)^2} \; \frac{\partial(x,y,t)}{\partial(\xi_1,\xi_2,\xi_3)} \; , \tag{3.9.20}$$

where the derivatives of x, y, and t are to be found at z = constant. For the space-time eikonal we obtain from (2.7.35) and (3.9.9)

$$\phi = \phi^0 - \omega(t - t^0) + \frac{\omega}{c}\left(\int_{\tilde{z}_<}^{\tilde{z}_>} \mp \int_{z_<}^{z_>}\right) \frac{n^2(z,\omega).dz}{\sqrt{n^2(z,\omega) - (n^0\sin\theta^0)^2}}$$

$$= \phi^0 - \omega(t - t^0) + \frac{\omega}{c}\left\{ p_x^0(x - x^0) + p_y^0(y - y^0) \right.$$

$$\left. + \left(\int_{\tilde{z}_<}^{\tilde{z}_>} \mp \int_{z_<}^{z_>}\right) \sqrt{n^2(z,\omega) - (n^0\sin\theta^0)^2} \; dz \right\} . \tag{3.9.21}$$

A discussion of (3.9.18, 19, 21) and numerous examples of these equations as applied to various pulse-propagation problems in plane-stratified media were presented in [3.194, 195, 201]. The geometry of caustics and the structure of radio-wave pulses in a linear plasma layer have been studied in [3.195, 200] to find out that the shape of the relevant space-time caustics is very much like the spatial case (Sect. 3.3). Problems of ideal compression of FM pulses in inhomogeneous media have been discussed in [3.194, 202]

3.9.6 Near and Far Fields of a Pulse

The applicability domain of space-time geometrical optics was discussed in Sect. 2.10.5 with an example of a plane FM pulse (3.9.9a) propagation in a homogeneous dispersive medium. In accordance with (2.10.30) in the specific case of an unmodulated pulse (ω^0 = const.) the Fresnel interval at $z = 0$ is $\tau_f^0 = \sqrt{2\pi|k''0|z}$. Therefore the condition for the geometrical-optics solution (3.9.15) to be applicable $2\tau_f^0 \ll T$ (here T is the time scale of field variation at $z = 0$) may be represented as

$$z \ll \frac{(T_p^0)^2}{8\pi|k''_0|} \simeq \frac{1}{8\pi|k''_0|}\left|\frac{1}{A^0}\frac{dA^0}{dt}\right|^{-2} , \tag{3.9.22}$$

where T_p^0 is the pulse width and A^0 its amplitude envelope. The inequality (3.9.22) defines the *near field* of a pulse and it is there that geometrical-

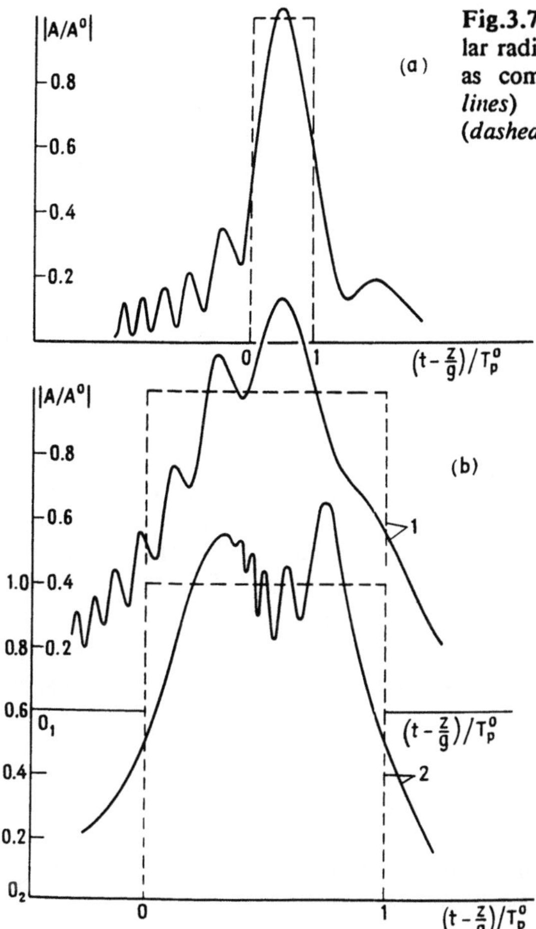

Fig.3.71. Envelopes of initially rectangular radio signals in a homogeneous plasma, as computed by exact equations (*solid lines*) and by ray theory equations (*dashed*)

optics expression (3.9.15) is valid. The ray treatment of the field can also be used in the *far zone* of the pulse, as in the far field of an antenna (Sect.3.1.7), but it will be related not with the pulse field at $z = 0$, but with the Fourier transform of this field (i.e., with the spectrum of the initial pulse). The features pertinent to the pulse field in the far zone were discussed in [3.15, 196, 203].

Figures 3.71 and 72 show the radio-signal envelopes in a homogeneous medium with the plasma (waveguided) dispersion law $n^2(\omega) = 1 - \omega_p^2/\omega^2$, computed with the exact equations [3.195, 203] (solid lines) and with the expressions of space-time geometrical optics (dashed lines). The computatiosn have been conducted for pulses with constant carrier frequency ω_0, relative pulse width $\omega_0 T_p^0 = 350$ and $\omega_0/\omega_p = 1.1$.

Figure 3.71 plots the amplitude envelope of a rectangular pulse at various distances z, corresponding to the values of $z/cT_p^0 = 4.25$ (Fig.3.71a), 1.27 (curve *1* in Fig.3.71b), and 0.53 (curve *2* in Fig.3.71b). As is apparent from the figure, (3.9.15) well describes the pulse group-delay time $t_g =$

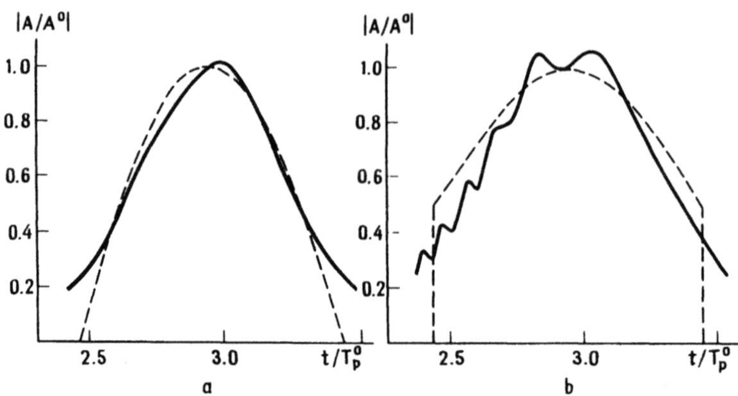

Fig.3.72. Same as in Fig.3.71 but for a cosine pulse profile

$z/g(\omega_0)$. As the observation point moves from the pulse near field into the intermediate zone distoritons show up due to dispersion (space-time diffraction of the pulses [3.195]). At large distances (Fig.3.71a) where the pulse width approximately equals the Fresnel interval ($T_p^0 \simeq \tau_f^0$) the envelope acquires the properties characterisitc of the far field of the pulse.

The evolution of the amplitude envelope in Fig.3.71 has much in common with the shape conversion of a wave bundle (Sect.3.1.7). This, in particular, is a manifestation of the *space-time analogy* between pulse and wave-bundle propagation [3.195,204].

Figure 3.72 depicts the amplitude envelope of unbiased (a) and biased (b) cosine pulses computed for the same relative distance $z/cT_p^0 = 1.02$. Figures 3.71,72 suggest that geometrical optics works more accurately for pulses with smooth envelopes exhibiting smaller transverse diffusion of the amplitude (specifically, at the beginning and end of the pulse). A similar situation occurs with wave bundles, too (Sect.3.1.7).

Note in conclusion that by virtue of (2.10.30) for the Fresnel interval τ_f^0, one may define the limits of the invalidity domain of space-time geometrical optics for different situations, say, near the front of pulse, in the vicinity of caustics and foci, etc., and estimate the field in these domains (Sect.2.10.10, [3.15]). In [3.15], for instance, we used (2.10.30) to derive for τ_f^0 an estimate of FM-pulse compression (focusing) in a homogeneous medium, see also [3.194, 197–199].

3.10 Numerical Methods in the Geometrical Optics of Inhomogeneous Media

3.10.1 The Ray-Tracing Analysis

For an arbitrary two- or three-dimensionally inhomogeneous medium, the equations of rays cannot be solved analytically. This fact suggests certain computational difficulties in applying geometrical optics directly to such real media. An alternative way has been found in the numerical methods

216

of geometrical optics, including those of ray-path computation, which have intensely been stimulated by the latest advance in computational possibilities. These methods proved to be expecially efficient for ray tracing in seismology [3.93], underwater acoustics [3.57] and radio-wave propagation in the ionosphere [3.84,119], to name but a few. Very helpful potential is also the experience gained in mechanics, in particle path synthesis, particularly in connection with the problems of celestial mechanics [3.114]. Some of the afore-mentioned ray-tracing techniques deserve to be discussed here in more detail.

The media of relatively simple structure, say, plane-stratified, radially inhomogeneous, or similar, which we considered in previous sections of this chapter, would not normally pose any computational difficulties (the respective problems were covered in [3.84,114,205]). Standard numerical techniques may be applied to compute the integrals, though sometimes the ray equations in differential form are preferred to the quadrature approach. Numerous computational examples of the integral method can be found in works on mechanics [3.114], acoustics [3.16,57,69], seismology [3.93], and radio-wave propagation in the ionosphere [3.80-83].

The ray paths in inhomogeneous media of more complex structure have to be derived by solving the differential equations of rays numerically. In order to save computational effort and time the medium can be approximated by various piecewise homogeneous or inhomogeneous models in which a simple representation leading to an explicit ray-tracing procedure is substituted at a short segment for the inhomogeneity profile under study. For the sake of illustration, a piecewise linear approximation transforms a ray path into a set of linear segments; computationally it implies a finite-difference approach to the ray equations. For the pertinent discussion of some computing algorithms, refer to [3.205-207].

Ray tracing may also be performed with the help of the Fermat principle (2.2.37) which converts the ray problem into a numerical search for the extremum of the Fermat functional (2.2.36). An illustration of how this method applies to the problem of reflection of rays from a curved interface between two homogeneous media was given in [3.205].

A direct numerical approach to ray-tracing in an arbitrary two- or three-dimensionally inhomogeneous medium means numerical solution of the Hamilton differential equations (2.2.7 or 12). In 1955, *Haselgrove* [3.116] introduced a computational technique to handle wave propagation in inhomogeneous ionosphere that became later widely recognized in other fields of ray-theory application [3.57,69,93,208-211]. Ray paths may be sought in Cartesian coordinates with (2.2.12) or in sutiable curvilinear coordinates with (2.2.39,41). The numerical integration of (2.2.12 or 39) is performed subject to the initial conditions (Sect.2.4) by some of the well-known techniques, say, of the Runge-Kutta family.

One of the components of momentum can be eliminated from the six-equation system of (2.2.12 or 39) by invoking the eikonal equation so that the computer input would consist of five rather than six equations. Another way to such a five-equation set is to use in (2.2.12) spherical angles (refraction angles) θ and ϕ which define the direction of momentum

p (3.3.6a). Moreover, in purely trajectorial analysis, when only the ray geometry is of interest, the number of equations can be reduced to four by eliminating the parameter τ, In particular, if $p_3 \neq 0$ in Cartesian coordinates ($x \equiv x_1$, $y \equiv x_2$, $z \equiv x_3$), the six-equation system (2.2.12) readily transforms into

$$\frac{dx_i}{dx_3} = \frac{p_i}{p_3} \,, \quad \frac{dp_i}{dx_3} = \frac{1}{p_3}\frac{\partial}{\partial x_i}\ln(n) \,; \quad i = 1, 2 \,, \tag{3.10.1}$$

where $p_3^2 = n^2 - p_1^2 - p_2^2 \neq 0$. In a two-dimensional problem, the system of four equations (2.2.12) can be reduced to two equations of the type (3.10.1) in a similar manner; (2.2.46-48) furnish an example of such a conversion. Thus the numerical approach to a solution yields not only the coordinates of the ray in physical space but also, in fact, more general information in the phase space $\{r, p\}$ (Sect.2.2).

The numerical integration algorithms for ray-equation systems may involve as many reflecting and refracting surfaces as practicable; the relevant computational procedure was elaborated upon in [3.93]. Under certain conditions it is rather advantageous for computation economy to apply various perturbation techniques to the equations of some slowly varying ray parameters, say, the ray oscillation period in a quasi-stratified nonuniform waveguide [3.212-215].

Numerous examples of numerical integration of the ray equations derived to solve various problems, both direct and inverse, in acoustics, seismology, and radio-wave physics, can be found in [3.57, 69, 84, 90, 116, 163, 216-219].

3.10.2 Computing the Eikonal and Wave Amplitude

The eikonal of a wave ψ can be readily obtained by numerically integrating the equations $d\psi/d\tau = n^2$ simultaneously with the equations of rays (2.2.12 or 39). Eliminating τ we get five equations which in Cartesian coordinates take the form

$$\frac{dx_i}{d\psi} = \frac{1}{n^2}p_i \,, \quad \frac{dp_i}{d\psi} = \frac{\partial}{\partial x_j}\ln(n) \,, \quad i = 1,2,3 \,; \quad j = 1,2 \tag{3.10.2}$$

with $p_3^2 = n^2 - p_1^2 - p_2^2$.

The path toward finding the intensity on a ray is more difficult because, according to (2.3.9), to do so one has to determine the divergence of the rays (2.3.10) expressed in terms of the derivatives of the ray-equation solutions. An approximation to the divergence

$$\mathscr{J} = \frac{n^0 da^0}{n\,da} = \frac{\mathscr{D}(\tau^0)}{\mathscr{D}(\tau)}$$

is based on numerical differentiating. With a set of ray equations of the type (3.10.2), one may approximately determine the specific cross section of a ray tube as

$$\left| \frac{da}{d\xi d\eta} \right| \simeq \left| \frac{\Delta a}{\Delta\xi\Delta\eta} \right| = \left| \frac{\Delta r_1}{\Delta\xi} \times \frac{\Delta r_2}{\Delta\eta} \right|$$

$$= \left| \left(\frac{\Delta r_1}{\Delta\xi} \right)^2 \left(\frac{\Delta r_2}{\Delta\eta} \right)^2 - \left(\frac{\Delta r_1 \Delta r_2}{\Delta\xi\Delta\eta} \right)^2 \right|^{1/2} . \qquad (3.10.3)$$

Here Δa is the area cut from a wave front ψ = const. by the tube of rays emanated at (ξ,η), $(\xi+\Delta\xi,\eta)$, $(\xi,\eta+\Delta\eta)$, and $(\xi+\Delta\xi,\eta+\Delta\eta)$; and Δr_1 and Δr_2 are the increments of the ray paths (3.10.2) evaluated, respectively, at constant η and ψ, or at constant ξ and ψ. According to (3.10.3), to determine the ray-tube cross section Δa one has to determine the paths for three adjoining rays with close coordinates $(\xi,\eta),\xi+\Delta\xi,\eta)$, and $(\xi,\eta+\Delta\eta)$, and then the increments of the ray paths Δr_1 and Δr_2. Obviously, this procedure extends markedly the amount of computation, the accuracy of the numerical differentiation being usually low.

The Jacobian $\mathscr{D}(\tau)$ and the field amplitude can be found directly by numerically integrating the adjoint to (2.2.12 or 39) set of equations for the derivatives $\partial r/\partial\xi$ and $\partial r/\partial\eta$ of the ray equation solution [3.93,216,219-222]. The set results from (2.2.12 and 39) differentiated with respect to ξ and η[41] to become in Cartesian coordinates

$$\frac{d}{d\tau}\left(\frac{\partial x_i}{d\xi_j} \right) = \frac{\partial p_i}{\partial\xi_j} ,$$

$$\frac{d}{d\tau}\left(\frac{\partial p_i}{\partial\xi_j} \right) = \sum_{k=1}^{3} \frac{\partial}{\partial x_k}\left(n\frac{\partial n}{\partial x_i} \right)\frac{\partial x_k}{\partial\xi_j} , \qquad (3.10.4a)$$

where $i = 1,2,3$; and $j = 1,2$. To have the set in compact form we have denoted $\xi = \xi_1$ and $\eta = \xi_2$. In vector form (3.10.4a) is as follows:

$$\frac{d}{d\tau}\left(\frac{\partial r}{d\xi_j} \right) = \frac{\partial p}{\partial\xi_j} , \quad \frac{d}{d\tau}\left(\frac{\partial p}{\partial\xi_j} \right) = \frac{d}{dr}\left(n\frac{dn}{dr} \right)\frac{\partial r}{\partial\xi_j} . \qquad (3.10.4b)$$

Numerically solving (3.10.4) simultaneously with the ray equations (2.2.12) paves the way for evaluating

---

[41] This operation is justified by the existence of the derivatives in initial values of the solutions of a set of ordinary differential equations [3.223].

$$\mathscr{D}(\tau) = \left[\frac{\partial \mathbf{r}}{\partial \xi} \times \frac{\partial \mathbf{r}}{\partial \eta}\right] \cdot \mathbf{p}$$

and the field amplitude (2.3.9).

In a similar manner we obtain for $\mathscr{D}(\tau)$ in an arbitrary system of orthogonal curvilinear coordinates q_i

$$\mathscr{D}(\tau) = \frac{\partial(x_1, x_2, x_3)}{\partial(\xi, \eta, \tau)} = \frac{\partial(x_1, x_2, x_3)}{\partial(q_1, q_2, q_3)} \frac{\partial(q_1, q_2, q_3)}{\partial(\xi, \eta, \tau)}$$

$$= h_1 h_2 h_3 \frac{\partial(q_1, q_2, q_3)}{\partial(\xi, \eta, \tau)} , \tag{3.10.5}$$

where h_i is the Lame coefficient of the coordinate q_i. Thus, to get $\mathscr{D}(\tau)$ we have to calculate numerically $\partial q_i / \partial \xi_j$. The adjoint system for $\partial q_i / \partial \xi_j$ is obtained from (2.2.41) in a way similar to that for (3.10.4):

$$\frac{d}{d\tau}\left(\frac{\partial q_i}{\partial \xi_j}\right) = \frac{1}{h_i} \frac{\partial \tilde{p}_i}{\partial \xi_j} - \frac{1}{h_j^2} \sum_{k=1}^{3} \frac{\partial h_i}{\partial q_k} \frac{\partial q_k}{\partial \xi_j} ,$$

$$\frac{d}{d\tau}\left(\frac{\partial \tilde{p}_i}{\partial \xi_j}\right) = \sum_{k=1}^{3} \left(\frac{\partial f_i}{\partial q_k} \frac{\partial q_k}{\partial \xi_j} + \frac{\partial f_i}{\partial \tilde{p}_k} \frac{\partial \tilde{p}_k}{\partial \xi_j}\right) , \tag{3.10.6}$$

where $i = 1,2,3$; $j = 1,2$. Here $\tilde{p}_i = (1/h_i)\partial \psi / \partial q_i$ are the components of momentum in the curvilinear coordinates, and f_i is the function on the right-hand side of (2.2.41):

$$f_i = \frac{1}{h_i} n \frac{\partial n}{\partial q_i} + \frac{1}{h_i} \sum_{m=1}^{3} \frac{\tilde{p}_m}{h_m}\left(\tilde{p}_m \frac{\partial h_m}{\partial q_i} - \tilde{p}_i \frac{\partial h_i}{\partial q_m}\right) \equiv f_i(q_k, \tilde{p}_k) . \tag{3.10.7}$$

Explicit expressions for $\partial f_i / \partial q_k$, $\partial f_i / \partial p_k$ may be deduced from (3.10.7) and are not given here.

The adjoint system (3.10.4 or 6) integrated along with the ray equations makes the total count of equations to be integrated equal to eighteen (six if in vector form). Eliminating one of the momentum components, say p_3, between the eikonal equation and the system leaves fifteen of them. Various ways to reduce the number of equations, thereby speeding up the computation of the field amplitude, were discussed in [3.93]. The examples of actual physical problems in seismology, acoustics, and radio-wave physics solved by the discussed numerical approach can be found in [3.57, 93, 116, 216].

3.10.3 Problems of Numerical Analysis

The application of computers to solving equations of geometrical optics has brought about some specific computational problems [3.93, 205]. Among them we would like to stress those of specifying (approximating) the parameters of the medium and initial data in general, handling large data arrays in ray-tracing, and aiming the rays onto a given point of observation.

When the distribution of the medium parameters is given by an analytic function, no specific computational difficulties would be expected. In actual physical problems, one has to seek an approximation to the parameters often specified graphically (by some curves or maps) or by data arrays. Choosing an approximating model one should compromise its simplicity, to speed up computation, and the accuracy of the approximation which, in general, depends on the objectives of the analysis. Thus, striving for the field amplitude, one should avoid modeling the medium by a function which has discontinuous derivatives for the refractive index n(r), since it may lead to imaginary effects and, hence, to large errors (Sect. 3.3 and [3.73]). This recommendation should be taken into account when working out an interpolation for n(r). Most convenient for the purpose are splines [3.224], for example, cubic splines which approximate n(r), warranting two continuous derivatives. Examples of spline application for ray tracing are contained in [3.93, 225], relevant problems of n(r) approximation were discussed in [3.93, 205].

Of particular practical interest is the problem of numerical evaluation of the rays passing through a given observation point. Its solution is required to compute the complete field (2.3.31) as a function of the observation point. So far any general-solution algorithm is unavailable. The aforementioned numerical techniques are invoked to solve the initial-value (Cauchy) problem which defines the field variation along the ray. In order to calculate the field in a given point of observation r one needs to develop a program to seek for initial data, ray coordinates ξ, η, and τ.[42] In principle, this program may be based on a known numerical algorithm for solving transcendential equations. Practically, however, devising a more or less general algorithm is a formidable task in the presence of multiple rays when the transcendental equation has several roots, $\xi_\nu(r)$, $\eta_\nu(r)$, $\tau_\nu(r)$, which exhibit fairly complex behavior on changing the observation point r, complicating primarily the choice of an adequate initial approximation. In some actual computations, an exhaustive search of initial data may prove helpful, and in situations when a suitable initial approximation is known one may use to advantage various iterative techniques [3.93] and perturbation techniques [3.226] (see also [3.227]).

It should be stressed that the methods discussed may be employed to compute both harmonic and nonharmonic fields, and the fields in aniso-

---

[42] In the most recurrent problem on the field due to a point source, one has as known the coordinates of two points, the radiation source and the receiver, and seeks the directions of rays issued by the source and defined by the initial components of momentum $p_i{}^0$. This is said to be a two-point ray-tracing problem.

tropic and nonstationary media, to name only a few. Numerical methods may also be used for evaluating various characteristics of the field associated with rays (pulse propagation time, Doppler-frequency shifts, polarization characteristics, etc. [3.57, 69, 93, 116, 228]).

Geometrical-optics field computations can be augmented by evaluating Fresnel volumes so as to control the validity of the method and to estimate numerically the fields in the regions of ray-optics inapplicability. Notably, Fresnel-volume evaluations can, in fact, be carried out by the standard ray-tracing procedure.

3.11 Inverse Problems of Geometrical Optics

The inverse problem is to deduce features of sources or scattering objects from the emitted or scattered radiation that has propagated to a detector. Inverse problems of geometrical optics are of great interest for many physical and technical studies such as radar, sonar, the synthesis of radiation sources with specified characteristics, the wave methods to study the structure and parameters of media (plasma diagnostics, geophysical survey, and structural geology problems), evaluation of the potentials of quantum-mechanical interaction, etc. The interpretation of experimental data in terms of ray optics will, of course, be valid if the diffraction effects are small. This condition is met in many situations of practical interest; nevertheless, the diffraction errors should not be discounted in assessing the potential accuracy of solution techniques for inverse problems based on geometrical optics.

3.11.1 Reflection and Refraction at Interfaces

The simplest inverse problem of geometrical optics concerns the evaluation of the characteristics of an interface between two homogeneous media by the known (measured or specified) characteristics of reflected field [3.229-231]. In detection and ranging (scattering) problems, say, from the scattered plane-wave field in the far zone (3.2.5), i.e., from the effective scattering surface Σ, one can derive the functions for the reflection coefficient Γ and Gaussian curvature $\tilde{K}$ of the boundary at the point of reflection, namely, $\tilde{K}/|\Gamma|^2 = \pi/\Sigma$. This yields for perfect reflection $\tilde{K} = \pi/\Sigma$. The solution of the inverse reflection problem derived in the physical-optics approximation [3.232] may be used as a generalization of the above result.

The *synthesis* of a reflector antenna poses the problem of selecting a reflector shape or a geometry of the radiator so as to provide for the specified radiation pattern.[43] For the case of a point source, the reflector profile is to be found in the geometrical-optics approximation with (3.2.20) [3.42, 230, 233]; the solution can be refined with the generalizations of geometrical optics that allow for the diffraction phenomena at the reflector

---

[43] The synthesis of a lens antenna calls for the required parameters of the refracting boundary to be determined [3.120].

222

edges. Of a more complicated variety is the geometrical-optics synthesis for two-reflector antennas, where one should have to account for the reflection effects from two surfaces.[44] The synthesis of several reflecting and refracting boundaries represents an important problem of instrumental geometrical optics dealing with the design of optical instruments to meet the performance specifications [3.1, 3, 29-34].

3.11.2 Inverse Problems for Given Models of the Inhomogeneous Medium

When some structural information on the studied medium is available beforehand, the solution of the inverse problem is reduced to evaluating the profile, or even a discrete set of parameters, of the inhomogeneity in the framework of an *assumed model* of the medium, say, a plane-stratified or radially inhomogeneous model. Any possible deviations of the medium parameters from the adopted model should be accounted for, so as to keep the accuracy of the inverse problem below a tolerable level.

This problem formulation was utilized to handle the synthesis of inhomogeneous lenses and lens antennas [3.1, 120] and numerous problems with the remote sensing of media, specifically a laboratory plasma, the ionosphere, an interstellar plasma, the earth's core, and a thermal boundary layer to name but a few [3.66, 80-83, 147, 228, 234-241][45] The problem arising in the synthesis of two-dimensioanlly inhomogeneous lens antennas and plane-stratified and radially inhomogeneous lens antennas were discussed in [3.120] (also [3.14] and Sects. 3.5, 6).

Remote sensing of a medium allows for several problem statements aiming at the reconstruction of the inhomogeneity profile $n(z)$ or $n(r)$ by the following measured field characteristics: the phase (eikonal), amplitude (say, in terms of the effective scattering surface), and ray path (using the features of ray refraction).[46] The solution of inverse problems is based upon Abel's inversion formula for the Volterra integral equation [3.243]

$$g(x) = \int_0^x \frac{f(y)}{\sqrt{x-y}}\, dy \,, \tag{3.10.8}$$

---

[44] One of the first problems with synthesizing the reflector profile has been handled by Kinber [3.229] to demonstrate that in the two-dimensional problem two reflections from the curvilinear boundaries would suffice to derive from an arbitrary ray field the wave with the given phase front and specified distribution of amplitude on the front. The ideas suggested in this work were utilized in [3.180] to design a nonconverting coupler for a waveguide.

[45] The quasi-classical approximation was invoked for a detailed study of the inverse quantum scattering problems aimed at retrieving the potentials of nuclear or molecular interaction [3.109, 117, 242-244].

[46] On some occasions, for example, the optical monitoring of a medium, the tracking of caustics may prove helpful. Of related monitoring techniques we can point out those based on interference effects.

which for g(0) = f(0) = 0 has the form

$$f(y) = \frac{1}{\pi} \int_0^y \frac{dg}{dx} \frac{dx}{\sqrt{y-x}} \, . \tag{3.10.9}$$

By way of illustration, in the problem of plane-wave scattering by a radially inhomogeneous cylinder, one may take as the measured function g(x) the phase shift $\Delta\Phi$ as a function of the aiming distance ρ (Sect.3.5). Then (3.10.9) defines $n^2(r)$ through the dependence of scattering angle θ upon ρ, as $\theta = d\Delta\Phi/d\rho$. In the diagnostics of a plasma discharge, the role of the function g(x) may also be plyed by the dependence of phase shift on the wave frequency. Elsewhere [3.109, 243] a similar solution was given for a specified dependence of the effective scattering surface on the scattering angle. Other examples of solutions to (3.10.9) applied to plane-stratified and radially inhomogeneous media can be found in [3.238, 245, 246], and to quantum scattering in [3.117, 242-244]. In a number of application where the inhomogeneity profile is a well-known function, the inverse problem is reduced to determining the parameters of this function, say, the parameters of a parabolic law $n^2(z)$. In fact, the diagnostics in this case makes use of the solution of the direct problem [3.224, 229].

Various generalizations of geometrical optics can be employed to refine the inverse-problem solution in the framework of a chosen inhomogeneity model (Sect.5 and [3.117, 134, 242-244, 247]). Formulating an inverse problem one should not overlook more sophisticated models of multi-dimensionally inhomogeneous media allowing for a separation of variables (Sect.2.8). Of great interest are also the inverse problems concerned with the reflection of waves from one or several interfaces in an inhomogeneous medium.

3.11.3 Multidimensional Inverse Problems

In a general (multidimensional) form, inverse problems of geometrical optics can be formulated for any geometrical optics equation, whether it be the eikonal, transfer, or ray equation. Of course, the multidimensional inverse problems can be handled by numerical methods only. A special difficulty inherent in these problems is their ill-definition in the classical sense; therefore they call for special study [3.237, 238, 248-250].

Despite a great interest in the multidimensional problems, the actual results available have been so far only a few [3.237, 238, 250, 251]. In a number of cases the perturbation technique proves helpful (Sect.2.9) by providing a multidimensional perturbation for an already known law of (say one-dimensional) inhomogeneity of the medium. The resulting inverse problem turns out to be closely related to a problem of integral geometry [3.237, 238].

3.11.4 Nonstationary Inverse Problems

Space-time geometrical optics allow for inverse problem formulations similar to those of Sects.3.11.1-3. The relevant information on pulse propagation time must somewhat widen the possibilities of solution of these problems.

The nonstationary problem of the reflection from interfaces draws great attention since in some cases it leads to separate centers of reflection. In nondispersive media, the inverse problems of deriving medium characteristics by wave propagation time do not practically differ from the stationary inverse problems (Sects.3.11.2,3), as, in accordance with Sect.2.7, the time is proportional to the wave eikonal. The nonstationary inverse problems for nondispersive media, especially inverse kinematic geophysical problems, have collected a vast literature [3.234-236,250,252].

A separate group of topical problems form the nonstationary inverse problems in frequency dispersive media, such as plasma (laboratory, ionospheric, free space, and other types). The most elaborated technique is that of pulse sounding of the ionosphere [3.80-83,239-241]. The analysis of ionograms (the dependences of group delay time for a reflected signal upon the carrier frequency) in geometrical optics terms may be viewed as the simplest inverse problem. A more sophisticated formulation recovers from an ionogram the vertical electronic density profile (that portion which causes the ray reflection).

Currently, attention has been drawn to the fine structure of backscattered pulse, as the information about the pulse shape can be used to deduce the characteristics of interfaces and to retrieve the parameters of both dispersive and nondispersive media. Among the studies in this direction we should primarily point out those dealing with seismogram analysis in geophysics [3.234-236]; for other applications, see also [3.253-256].

The applicability of inverse problem solutions is obviously limited by the domain where the ray method is valid (Sect.2.10). Therefore, in remote sensing problems, for instance, a parameter derivation is feasible only from the regions lighted by rays. Otherwise, as we have already noted, the inverse problems may be approached by different generalizations of geometrical optics [3.237,238,251] (see also Chap.5).[47]

---

[47] In addition to the methods of this section, for practical applications, we would point to holographic methods [3.257-259] and other wave methods of remote sounding and nondestructive testing of materials [3.260,261].

4. Vector Wave Fields

In this chapter we devote our attention to features that differentiate vector fields from scalar fields, namely, polarization and co-existance of several modes. For the purpose of illustration, we shall consider examples from electromagnetic theory, which are of major interest for applications in optics and radio engineering. For some vector problems of elastic-wave theory and quantum mechanics, we shall refer the reader to certain reports and publications. Of new results discussed here we should mention the theory of normal wave interaction in weakly anisotropic media (the quasi-isotropic approximation, Sect.4.3), the general geometrical-optics theory of modulated wave propagation in dispersive media, and the conditions defining the existence of the adiabatic invariant (Sect.4.4).

4.1 Transverse Electromagnetic Waves in Isotropic Media

4.1.1 Maxwell Equations for Monochromatic Waves

The propagation of monochromatic electromagnetic waves ($\mathbf{E}$, $\mathbf{H}$, and $\mathbf{D} \propto \exp(-i\omega t)$) in stationary media is governed by the Maxwell equations (written here in Gaussian units with $k_0 = \omega/c$)

$$\text{curl}\mathbf{H} + ik_0\mathbf{D} = 0 \; ,$$

$$\text{curl}\mathbf{E} - ik_0\mathbf{B} = 0 \; . \tag{4.1.1}$$

We assume that the electric- and magnetic-flux densities are related to the respective field intensities by linear constitutive equations of the form

$$D_\alpha(\mathbf{r}) = \epsilon_{\alpha\beta}(\omega,\mathbf{r}) \, E_\beta(\mathbf{r}) \; ,$$

$$\mathbf{B}(\mathbf{r}) = \mathbf{H}(\mathbf{r}) \; , \tag{4.1.2}$$

where $\epsilon_{\alpha\beta}$ is the dielectric permittivity tensor in the general case of an anisotropic medium (repeated indices imply summation throughout this text). The magnetic permeability tensor $\mu_{\alpha\beta}$ is supposed to be the unit tensor $\delta_{\alpha\beta}$, which is true for the majority of media (ferrites form a possible exclusion) of interest for short-wavelength propagation. We shall defer the discussion of electromagnetic waves accounting for spatial dispersion, non-stationarity of the medium, and field nonmonochromaticity until Sect.4.4.

4.1.2 The Debye Expansion and the Iterative Equations[1]

In the isotropic medium the permittivity tensor is proportional to the unit tensor, $\epsilon_{\alpha\beta} = \epsilon\delta_{\alpha\beta}$, so $\mathbf{D} = \epsilon\mathbf{E}$ and Maxwell's equations (4.1.1) take the form

$$\text{curl}\mathbf{H} + ik_0\epsilon\mathbf{E} = 0 \ ,$$

$$\text{curl}\mathbf{E} - ik_0\mathbf{H} = 0 \ . \tag{4.1.3}$$

Using the procedure introduced by P. Debye[2] we represent the electric and magnetic fields as the expansions

$$\mathbf{E} = \sum_{m=0}^{\infty} \frac{\mathbf{E}_m}{(ik_0)^m} e^{ik_0\psi} \ , \quad \mathbf{H} = \sum_{m=0}^{\infty} \frac{\mathbf{H}_m}{(ik_0)^m} e^{ik_0\psi} \ . \tag{4.1.4}$$

Substituting (4.1.4) into (4.1.3) and collecting the coefficients of like powers of k_0 yields, in a zero-order approximation,

$$\mathbf{p}\times\mathbf{H}_0 + \epsilon\mathbf{E}_0 = 0 \ , \quad \mathbf{p}\times\mathbf{E}_0 - \mathbf{H}_0 = 0 \ , \tag{4.1.5}$$

where $\mathbf{p} \equiv \nabla\psi$. In a first approximation we have

$$\mathbf{p}\times\mathbf{H}_1 + \epsilon\mathbf{E}_1 = - \text{curl}\mathbf{H}_0 \equiv \mathbf{X} \ ,$$

$$\mathbf{p}\times\mathbf{E}_1 - \mathbf{H}_1 = - \text{curl}\mathbf{E}_0 \equiv \mathbf{Y} \ . \tag{4.1.6}$$

The equations of higher-order approximations result from (4.1.6) by substituting m-1 for 0 and m for 1, respectively.

4.1.3 The Eikonal Equation

Eliminating the magnetic field $\mathbf{H}_0 = \mathbf{p}\times\mathbf{E}_0$ from the zero-order approximation equations (4.1.5) we obtain for the electric field

$$\mathbf{p}\times(\mathbf{p}\times\mathbf{E}_0) + \epsilon\mathbf{E}_0 = 0$$

or

$$(p^2 - \epsilon)\mathbf{E}_0 - \mathbf{p}(\mathbf{p}\mathbf{E}_0) = 0 \ . \tag{4.1.7}$$

---

[1] *Ignatovsky* [4.1] was the first who applied the ray approach to describe electromagnetic waves, but he failed to consider field polarization. A complete analysis of the ray field in the isotropic medium is due to *Rytov* [4.2,3]. Later the results of the analysis have been reiterated in the literature [4.4,5].

[2] Here, as in the scalar problem, the most systematic approach is based, of course, on expanding the fields in powers of the small dimensionless parameter $1/kL \ll 1$ (the Rytov expansion); it is this approach that was used in [4.2].

With

$$q_{\alpha\beta} = (p^2 - \epsilon)\delta_{\alpha\beta} - p_\alpha p_\beta \qquad (4.1.8)$$

(4.1.7) yields

$$q_{\alpha\beta} E_\beta = 0 \qquad (4.1.9)$$

(henceforth we drop the subscript $_0$ of the zero-approximation amplitudes).

In order for the set of linear equations (4.1.9) to possess a nontrivial solution its determinant must be zero

$$\mathscr{H} = \det q_{\alpha\beta} = -\epsilon(p^2 - \epsilon)^2 = 0 , \qquad (4.1.10)$$

which holds if either

$$p^2 - \epsilon = (\nabla\psi)^2 - \epsilon = 0 \quad \text{or} \quad \epsilon = 0 . \qquad (4.1.11)$$

The condition $\epsilon = 0$ reflects the possibility for longitudinal waves to exist (Sect. 4.1.4), whereas (4.1.11) coincides with the eikonal equation (2.1.11) in the scalar formulations of the problem. Consequently, all the results of the scalar theory allied with the eikonal are valid for electromagnetic waves in isotropic media as well.

4.1.4 Transverse Nature of Zero Approximation Waves. Polarization Degeneracy

The dot product of (4.1.5) by the unit vector $\mathbf{l} = \mathbf{p}/p$ tangent to the ray results in $\mathbf{E}\cdot\mathbf{l} = \mathbf{H}\cdot\mathbf{l} = 0$. Hence, $\mathbf{E}$ and $\mathbf{H}$ are perpendicular to the ray; i.e., in a zero approximation the field has a *transverse structure*. From (4.1.5) it also follows that $\mathbf{E}$ and $\mathbf{H}$ are orthogonal.

At $p^2 = \epsilon$ the elements of the matrix $q_{\alpha\beta}$ are $p_\alpha p_\beta$, thus

$$\mathscr{H} = \det \begin{bmatrix} p_1{}^2 & p_1 p_2 & p_1 p_3 \\ p_1 p_2 & p_2{}^2 & p_2 p_3 \\ p_1 p_3 & p_2 p_3 & p_3{}^2 \end{bmatrix} = -p_1 p_2 p_3 \det \begin{bmatrix} p_1 & p_2 & p_3 \\ p_1 & p_2 & p_3 \\ p_1 & p_2 & p_3 \end{bmatrix} . \qquad (4.1.12)$$

It is readily evident that the rank of $q_{\alpha\beta}$ equals unity, so two of the three components of $\mathbf{E}$ may be chosen arbitrarily [4.6]. The transverse structure of the field suggests, as a natural choice for the two arbitrary quantities, the projections of the field $\mathbf{E}$ onto the principal normal to the ray $\boldsymbol{\nu}$ and onto the binormal $\mathbf{b}$:

$$\mathbf{E} = \Phi_\nu \boldsymbol{\nu} + \Phi_b \mathbf{b} ; \qquad (4.1.13)$$

then

$$\mathbf{H} = \mathbf{p} \times \mathbf{E} = \sqrt{\epsilon}\,(\mathbf{l} \times \mathbf{E}) = \sqrt{\epsilon}\,(\Phi_\nu \mathbf{b} - \Phi_b \boldsymbol{\nu}) . \qquad (4.1.14)$$

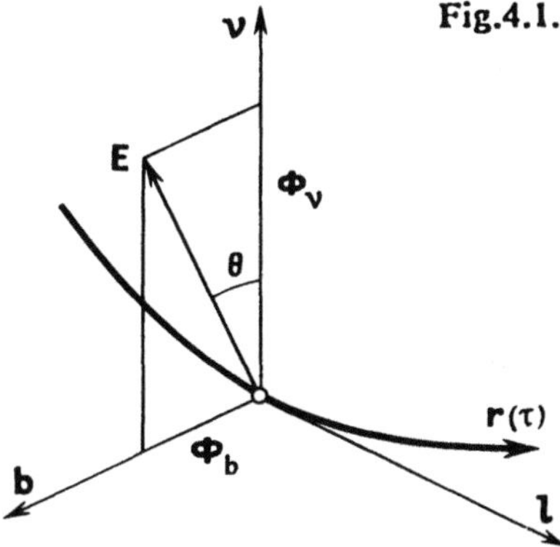

Fig.4.1. Natural reference frame associated with the ray

To the real-valued Φ_ν and Φ_b there correspond linearly polarized waves (Fig.4.1). In the general case of complex Φ_ν and Φ_b the electromagnetic wave is elliptically polarized.

Thus the zero-approximation equations cannot completely specify the field polarization, as the orientation of the field vectors in the plane transverse to the ray remains to be defined. This gives rise to the phrase the *polarization degeneracy of waves* in isotropic media. The phenomenon is related to multiple eigenvalues of the matrix $q_{\alpha\beta}$ (Sects.4.2.13 and 4.3).

4.1.5 Consistency of the First-Approximation Equations

Eliminating H_1 from (4.1.6) yields the equation for E_1:

$$p \times (p \times E_1) + E_1 = X + p \times Y \equiv - \text{curl}H - p \times \text{curl}E \equiv Z . \tag{4.1.15}$$

For $p^2 = \epsilon$ the determinant of the left-hand side along with all its second-order minors is zero [this immediately follows from (4.1.12)].

In accord with the Fredholm theorem [4.6], for the inhomogeneous equations (4.1.15) to be consistent, the vector Z on the right-hand side must be orthogonal to all solutions E_j of the transposed homogeneous system of equations; i.e., it must be $E_j \cdot Z = 0$, $j = 1,2$. In this case, the solutions of the transposed system coincide with those of the original system (since the matrix $q_{\alpha\beta}$ is symmetric). Therefore the solutions E_j may be the normal ν and the binormal b, so that the conditions of consistency assume the form

$$\nu \cdot Z = 0 , \quad b \cdot Z = 0 . \tag{4.1.16}$$

Substituting into (4.1.16) the zero-approximation fields (4.1.13,14) and observing that $p = \sqrt{\epsilon}l$, we get

$$\nu \text{curl}[\sqrt{\epsilon}(\Phi_\nu b - \Phi_b \nu)] - \sqrt{\epsilon}b\text{curl}(\Phi_\nu \nu + \Phi_b b) = 0 ,$$

$$\tag{4.1.17}$$

$$b\text{curl}[\sqrt{\epsilon}(\Phi_\nu b - \Phi_b \nu)] - \sqrt{\epsilon}\nu\text{curl}(\Phi_\nu \nu + \Phi_b b) = 0 .$$

Making use of the relationship $\mathrm{curl}\, a\phi = \nabla\phi \times \mathbf{a} + \phi\,\mathrm{curl}\, a$, (2.1.49) for the torsion κ, and the relation

$$\mathrm{div}\, \mathbf{l} = \mathrm{div}(\boldsymbol{\nu}\times\mathbf{b}) = \mathbf{b}\,\mathrm{curl}\boldsymbol{\nu} - \boldsymbol{\nu}\,\mathrm{curl}\mathbf{b} ,$$

we transform (4.1.17) into the form

$$\mathbf{l}(2\sqrt{\epsilon}\,\nabla\Phi_\nu + \Phi_\nu\nabla\sqrt{\epsilon}) + \Phi_\nu\sqrt{\epsilon}\,\mathrm{div}\mathbf{l} + 2\sqrt{\epsilon}\kappa\Phi_b = 0 , \tag{4.1.18}$$

$$\mathbf{l}(2\sqrt{\epsilon}\,\nabla\Phi_b + \Phi_b\nabla\sqrt{\epsilon}) + \Phi_b\sqrt{\epsilon}\,\mathrm{div}\mathbf{l} - 2\sqrt{\epsilon}\kappa\Phi_\nu = 0 . \tag{4.1.19}$$

These equations enable us to find the complex amplitudes Φ_ν and Φ_b, and thereby completely define the polarization state of the field. We shall consider this process in more detail.

4.1.6 Conserving Energy Flow in a Ray Tube

On multiplying (4.1.18) by Φ_ν^* and (4.1.19) by Φ_b^* and summing up the real parts of the resulting expressions, we get

$$\mathrm{div}(p\mathbf{A}^2) = 0 , \quad \mathbf{A}^2 \equiv |\Phi_\nu|^2 + |\Phi_b|^2 = |\mathbf{E}|^2 . \tag{4.1.20}$$

Observing that Poynting vector $\mathbf{S} = (c/8\pi)\mathrm{Re}\{\mathbf{E}\times\mathbf{H}^*\}$ in an isotropic medium becomes $\mathbf{S} = (c/8\pi)p\mathbf{A}^2$ by virtue of (4.1.13,14), Eq.(4.1.20) may be interpreted as an energy conservation law, $\mathrm{div}\mathbf{S} = 0$. The amplitude $\mathbf{A} = |\mathbf{E}|$ is found by (4.1.20) in the same manner as in the scalar problem (Sect.2.3).

4.1.7 Preserving the Polarization Ellipse

The set of (4.1.18,19) allows for another conservation law: adding up (4.1.18) multiplied by Φ_ν and (4.1.19) multiplied by Φ_b leads to

$$\mathrm{div}(p\Phi^2) = 0 , \quad \Phi^2 = \Phi_\nu^2 + \Phi_b^2 = \mathbf{E}^2 . \tag{4.1.21}$$

The *squared intensity* of the electrical field $\Phi^2 = \mathbf{E}^2$ generally does not coincide with $\mathbf{A}^2 = |\mathbf{E}|^2$, the *squared modulus* of the electric field, but both these quantities vary along the ray in the same manner, i.e.,

$$\mathbf{A} = \mathbf{A}^0 \sqrt{\mathscr{D}(\tau^0)/\mathscr{D}(\tau)} , \quad \Phi = \Phi^0 \sqrt{\mathscr{D}(\tau^0)/\mathscr{D}(\tau)} . \tag{4.1.22}$$

Therefore $\chi = \Phi/\mathbf{A}$ is constant along the ray and is equal to the initial value

$$\chi = \Phi/\mathbf{A} = \mathrm{const} = \Phi^0/\mathbf{A}^0 = \chi^0 . \tag{4.1.23}$$

In the case of linear polarization (both Φ_ν and Φ_b are real) $\chi = 1$, and for circular polarization ($\Phi_b = \pm i\Phi_\nu$) we have $\chi = 0$. In the general case of an elliptically polarized wave, $|\chi|$ defines the *eccentricity* of the polariza-

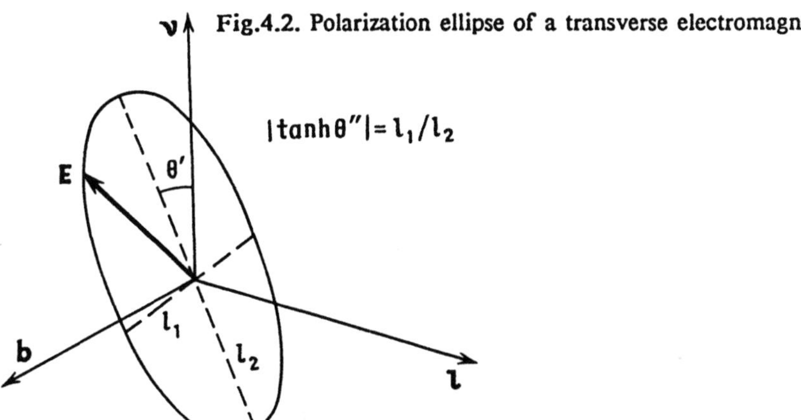

tion ellipse, and $\arg\chi$ the position of the field vectors on the ellipse. Hence, preserving of χ indicates that the *polarization ellipse retains its shape* along the ray.

4.1.8 Rotation of Field Vectors (Rytov's Law)

Let us trace out how the polarization ellipse changes its orientation along the ray. For this purpose we introduce the angle θ as

$$\tan\theta = \Phi_\nu/\Phi_b \ . \tag{4.1.24}$$

This angle is compex-valued in the general case; its real part $\theta' = \text{Re}\{\theta\}$ subtends the angle between the major axis of the polarization ellipse and the normal to the ray, ν, as shown in Fig.4.2, and the imaginary part $\theta'' = \text{Im}\{\theta\}$ defines the ratio of the semiminor axis l_1, to the semimjor axis l_2 so that $l_1/l_2 = |\tanh\theta''|$. The sign of θ'' specifies the sense of rotation of the field vector $\mathbf{E}$ over the polarization ellipse, so positive values of θ'' correspond to clockwise rotation, as viewed in the ray direction. For linear polarization (all quantities Φ, θ, Φ_ν, and Φ_b are real) the angle θ is simply the angle between the field vector $\mathbf{E}$ and the principal normal to the ray ν (Fig.4.1).

The functions $\Phi = (\Phi_\nu^2 + \Phi_b^2)^{1/2}$ and $\theta = \arctan(\Phi_b/\Phi_\nu)$ can be used to describe the field polarization instead of Φ_ν and Φ_b. The latter can be expressed via Φ and θ as

$$\Phi_\nu = \Phi\cos\theta \ , \quad \Phi_b = \Phi\sin\theta \ . \tag{4.1.25}$$

Substituting (4.1.25) into (4.1.18) and observing (4.1.21) we obtain

$$\mathbf{l}\cdot\nabla\theta - \kappa = 0 \quad \text{or} \quad d\theta/d\sigma = \kappa \ . \tag{4.1.26}$$

This equation was derived by *Rytov* [4.2,3] and is referred to as Rytov's field-vectors rotation law with respect to a moving trihedral $(\mathbf{l},\nu,\mathbf{b})$.

Integrating (4.1.26) along the ray yields

$$\theta = \theta^0 + \int_{\sigma^0}^{\sigma} \kappa d\sigma \quad \text{or} \quad \theta = \theta^0 + \int_{\tau^0}^{\tau} \kappa\sqrt{\epsilon}\,d\tau \;. \qquad (4.1.27)$$

As torsion κ is real-valued, from (4.1.27) it follows that

$$\theta' = (\theta^0)' + \int_{\sigma^0}^{\sigma} \kappa d\sigma \;, \quad \theta'' = (\theta^0)'' = \text{const} \;; \qquad (4.1.28)$$

that is, only the orientation of the polarization ellipse, characterized by θ', varies along the ray, rather than its shape (as we have already mentioned in Sect.4.1.7). For *plane* trajectories with the ray torsion equal to zero, the direction of the major axis of the polarization ellipse also remains unchanged with respect to $(\mathbf{l},\boldsymbol{\nu},\mathbf{b})$. This is characteristic of plane, cylindrically- and spherically-stratified media.

For *linearly polarized* waves (θ real-valued), the variation of θ signifies the rotation of the field polarization plane with respect to the trihedral $(\mathbf{l},\boldsymbol{\nu},\mathbf{b})$. Having introduced the unit vectors

$$\hat{\mathbf{e}} = \frac{\mathbf{E}}{E} = \boldsymbol{\nu}\cos\theta + \mathbf{b}\sin\theta \;, \quad \hat{\mathbf{h}} = \frac{\mathbf{H}}{H} = \mathbf{b}\cos\theta - \boldsymbol{\nu}\sin\theta \;, \qquad (4.1.29)$$

we may state that at sufficiently large (i.e., comparable with the torsion radius $\rho_{\text{tors}} = 1/\kappa$) distances the vectors $\hat{\mathbf{e}}$ and $\hat{\mathbf{h}}$ generally do not return to the initial positions $\hat{\mathbf{e}}^0$ and $\hat{\mathbf{h}}^0$ even if at the end of the trajectory, $\mathbf{l}$ coincides with its initial position $\mathbf{l}^0$ [4.7].[3]

At comparatively small distances, on the other hand, where $\sigma \ll 1/\kappa$ the field vectors practically hold their positions with respect to some unmoving coordinate system, since as the field vectors rotate with respect to the moving trihedral at the rate of $d\theta/d\sigma = \kappa$ the trihedral itself revolves around $\mathbf{l}$ at the same velocity but in the opposite direction [by virtue of one of the Frenet-Serret formulas (2.2.20) $d\mathbf{b}/d\sigma = \kappa\boldsymbol{\nu}$]. To convince ourselves we differentiate (4.1.29) with respect to σ, observing (4.1.26 and 2.2.20), to obtain [4.5, 7]

$$\frac{d\hat{\mathbf{e}}}{d\sigma} = - K\mathbf{l}\cos\theta \;, \quad \frac{d\hat{\mathbf{h}}}{d\sigma} = K\mathbf{l}\sin\theta \;. \qquad (4.1.30)$$

These formulas indicate that the elementary increments $d\hat{\mathbf{e}}$ and $d\hat{\mathbf{h}}$ are directed along the tangent to the ray $\mathbf{l}$ and have no components in either $\boldsymbol{\nu}$

---

[3] Note that in the non-euclidean space of metric $ds^2 = \epsilon(x,y,z)(dx^2 + dy^2 + dz^2)$, Rytov's equation (4.1.26) describes the so-called pseudoparallel transfer of the field vectors [4.4,5,7].

232

or **b**. Rytov's equation (4.1.26) thus reflects a certain *inertia* of the field vectors which fail locally to follow the ray torsion.[4] Rytov's law may be considered a particular case of a more general phenomenon called *Berry's* "topological phase" [4.8].

4.1.9 Polarization of Transverse Waves

For a linearly polarized wave, the wave amplitude Φ coincides with A and can be found from (4.1.22) with the real initial value $\Phi^0 = A^0$. The angle θ^0 in (4.1.27) is also real, and the field is defined by

$$\mathbf{E} = \Phi_\nu \boldsymbol{\nu} + \Phi_b \mathbf{b} = \Phi(\boldsymbol{\nu}\cos\theta + \mathbf{b}\sin\theta) \ . \tag{4.1.31}$$

The same result can be derived by directly integrating (4.1.30): $\mathbf{E} = \Phi\hat{\mathbf{e}}$ where $\hat{\mathbf{e}} = \hat{\mathbf{e}}^0 - \int \mathbf{K} \mathbf{l}\cos\theta d\sigma$.

In the general case of elliptically polarized waves, Φ and θ in (4.1.31) are determined by (4.1.22, 27) with the complex initial values Φ^0 and θ^0. Of course, we can arrive at the same result by resolving the elliptically polarized wave into *a sum of two* linearly polarized waves and employing more straightforward procedures to the linearly polarized waves.

In the majority of applications the trajectories of rays deviate from the plane curves only slightly. Therefore, in general, the instantaneous angle θ only insignificantly differs from the respective initial values θ^0. Nevertheless, this phenomenon should be taken into account in a number of situations. The deviations of θ form θ^0 reveal the so-called geometrical depolarization of light waves in inhomogeneous random media. This phenomenon was estimated quantitatively in [4.9] by a solution derived with the perturbation technique. The depolarization of waves reflected from the bodies embedded into an inhomogeneous medium is also related to Rytov's rotation of the polarization plane [4.10, 11]. We note also that Rytov's rotation should be accounted for in the problems of radiation transfer in inhomogeneous media [4.12]

4.1.10 Longitudinal Components of the Field

Projecting (4.1.6) onto the vector **l** tangent to the ray yields

$$E_{1\parallel} = \mathbf{l}\cdot\mathbf{E}_1 = - \frac{1}{\epsilon}\mathbf{l}\operatorname{curl}\mathbf{H}_0 \ ,$$

$$\tag{4.1.32}$$

$$H_{1\parallel} = \mathbf{l}\cdot\mathbf{H}_1 = \mathbf{l}\operatorname{curl}\mathbf{E}_0 \ .$$

By the order of magnitude $|\operatorname{curl}\mathbf{H}_0| \propto |\mathbf{H}_0|/L$, where L is the characteristic scale of amplitude variation. Recognizing that $\mathbf{E}_1$ is a coefficient of $1/ik_0$ we can estimate the "actual" longitudinal field $E_{1\parallel}/ik_0$ as

$$\left| \frac{E_{1\parallel}}{ik_0} \right| \propto \frac{1}{\epsilon} \frac{|\mathbf{H}_0|}{k_0 L}$$

---

[4] This lag is connected with the conservation of field momentum.

(similarly $|H_{1\parallel}/ik_0| \propto |E_0|/k_0 L$). This estimate indicates that the longitudinal component is small everywhere with the exclusion of the region for $\epsilon \le 1/k_0 L$. In a plasma, it is the region of the plasma resonance $\epsilon \to 0$ where transverse waves can excite intensive longitudinal (plasma) oscillations [4.13]. The analysis of higher-order approximations was outlined in [4.4].

4.1.11 Reflection of Transverse Electromagnetic Waves from Interfaces

The reflection (Γ) and transmission (T) coefficients for electromagnetic waves depend on the polarization of the incident wave. The computations similar to those performed in Sect.2.5 but with the electrodynamic boundary conditions imposed on the tangential field components

$$E_{1\tan} = E_{2\tan} , \quad H_{1\tan} = H_{2\tan} , \tag{4.1.33}$$

lead to the local Fresnel expressions in the zero approximation of the ray method. The relevant phase relationships and ray-reflection laws remain the same as in the scalar problem.

Let the wave incident on the interface be represented as $E_i = E_\parallel + E_\perp$, where $E_\parallel$ is the component in the plane of incidence and $E_\perp$ the component perpendicular to the plane. Then we have for the reflected and refracted (transmitted) amplitudes

$$E_r = \Gamma_\parallel E_\parallel + \Gamma_\perp E_\perp , \quad E_t = T_\parallel E_\parallel + T_\perp E_\perp , \tag{4.1.34}$$

where [4.14]

$$\Gamma_\parallel = \frac{n_1 \cos\theta_t - n_2 \cos\theta}{n_1 \cos\theta_t + n_2 \cos\theta} , \quad T_\parallel = \frac{2n_1 \cos\theta_t}{n_1 \cos\theta_t + n_2 \cos\theta} ,$$

$$\Gamma_\perp = \frac{n_1 \cos\theta - n_2 \cos\theta_t}{n_1 \cos\theta + n_2 \cos\theta_t} , \quad T_\perp = \frac{2n_1 \cos\theta}{n_1 \cos\theta + n_2 \cos\theta_t} . \tag{4.1.35}$$

The angle of incidence, θ, and the angle of refraction, θ_t, are related by Snell's law (2.5.6), and $n_{1,2}$ are the refractive indices of two media. The amplitude reflection coefficient $\Gamma_\parallel$ falls off to zero at the Brewster angle $\theta_B = \arctan(n_2/n_1)$. The expression for field amplitudes at weak interfaces can be deduced in the same way as in Sect.2.5.

4.1.12 Polarization Degeneracy in Problems of Quantum Mechanics and Theory of Elasticity

Quantum-mechanical equations assume the vectorial form when the particles involved possess spin. The first attempt to handle the Dirac equations by a quasi-classical (geometrical-optics) approach was due to *Pauli* [4.15], who interpreted only one of the consistency conditions of the first approximation, leading to the conservation of the probability density flow.

The rotation of the electron spin traveling along a classic path was predicted by *Rytov* [4.2, 3] and computationally verified by *Galanin* [4.16], who derived an equation similar to (4.1.36) for the spin orientation. In this case the polarization degeneracy is lifted by an external magnetic field.

The polarization plane rotates also when transverse elastic waves propagate in an isotropic inhomogeneous medium. For this case an equation coinciding with (4.1.26) was derived in [4.17] and [4.18, 19]. Anisotropy would lift the polarization degeneracy. The elastic media are characteristic in that they are able to propagate, besides transverse waves, longitudinal waves as well. At sharp interfaces these wave types convert one into another. An elaborated treatment of this topic and an exhaustive list of references on geometrical optics of elastic waves are to be found in [4.20, 21].

4.2 Independent Normal Waves in an Anisotropic Medium

4.2.1 Equation of the Eikonal

Without dwelling on the properties of plane waves in homogeneous anisotropic media - they have been treated in many books, e.g. [4.13, 14, 22-24] - we come consider directly waves in inhomogeneous media. The geometrical optics of one-dimensionally inhomogeneous anisotropic media is at present a fairly well-developed discipline [4.13, 25-29] (an exhaustive bibliography on waves in an inhomogeneous magnetoactive plasma was collected in [4.13]). The three-dimensional problems are far less studied [4.30-35]. In our discussion we shall follow the basic ideas developed in [4.31, 36, 37].

The equations of zero and first approximations for waves in anisotropic media differ from (4.1.5-6) only in that in place of the scalar ϵ they contain the permittivity tensor $\hat{\epsilon} = \{\epsilon_{\alpha\beta}\}$, which we assume at first to be Hermitian,[5] i.e., $\epsilon_{\beta\alpha}{}^* = \epsilon_{\alpha\beta}$. Eliminating the magnetic field from the zero-approximation equations, we obtain for the vector $\mathbf{E}$ the system of homogeneous equations allied to (4.1.7):

$$\mathrm{p}^2 \mathbf{E} - \hat{\epsilon}\mathbf{E} - \mathbf{p}(\mathbf{pE}) = 0 \qquad \text{or} \tag{4.2.1}$$

$$q_{\alpha\beta} E_\beta = 0 , \quad \alpha,\beta = 1,2,3 . \tag{4.2.2}$$

In contrast to Sect. 3.11, now we have

$$q_{\alpha\beta} = \mathrm{p}^2 \delta_{\alpha\beta} - \mathrm{p}_\alpha \mathrm{p}_\beta - \epsilon_{\alpha\beta} , \quad \mathrm{p}_\alpha = \frac{\partial \psi}{\partial x_\alpha} . \tag{4.2.3}$$

Equating the determinant of (4.2.1) to zero we get the eikonal equation for waves in an anisotropic medium

$$\mathcal{H} = \det[q_{\alpha\beta}] = \det[\mathrm{p}^2 \delta_{\alpha\beta} - \mathrm{p}_\alpha \mathrm{p}_\beta - \epsilon_{\alpha\beta}] = 0 . \tag{4.2.4}$$

---

[5] For lossless anisotropic dielectrics, the tensor $\epsilon_{\alpha\beta}$ is real-valued, while in gyrotropic media, say in a magnetoactive plasma, $\epsilon_{\alpha\beta}$ is complex-valued even in lossless media.

It has the same form as the equation of dispersion for electromagnetic waves in a homogeneous anisotropic medium. Therefore, it may be viewed as a local dispersion equation for quasi-plane waves in a smoothly inhomogeneous anisotropic medium.

Substituting

$$\mathbf{p} = p\mathbf{l} \quad \text{and} \quad \mathbf{l}^2 = 1 \tag{4.2.5}$$

into (4.2.4) we arrive at the biquadratic equation in $p = |\mathbf{p}|$ (the sixth-degree terms cancel out each other):

$$\mathcal{H} = Ap^4 + 2Bp^2 + C = 0 . \tag{4.2.6}$$

In the general case, A, B, and C are complex functions of the components of tensor $\epsilon_{\alpha\beta}$ and the unit vector $\mathbf{l}$. These functions, however, are drastically simplified in suitable coordinates where the expression for $\epsilon_{\alpha\beta}$ is the simplest. These are the coordinate systems related to specific directions in the medium such as the optical axes, the direction of an external magnetic field, etc. [4.13,14,22,24]. Note that in isotropic media $A = -\epsilon$, $B = \epsilon^2$, and $C = -\epsilon^2$, so that (4.2.6) takes the form of (4.1.10).

For a Hermitian permittivity tensor, the coefficients of (4.2.6) are real, and its roots

$$p_{1,2}^2 = \frac{-B \pm \sqrt{B^2 - AC}}{A} = n_{1,2}^2(\mathbf{r}, \mathbf{l}) , \tag{4.2.7}$$

are either real or complex conjugates. The quantities $n_{1,2}$ are the refractive indices of the anisotropic medium. We shall confine ourselves to considering the real-valued roots associated with the propagating waves. We factor the Hamiltonian (4.2.6) subject to (4.2.7) to obtain

$$\mathcal{H} = A(p^2 - n_1{}^2)(p^2 - n_2{}^2) = 0 . \tag{4.2.8}$$

This equation eventually yields two independent equations which, for convenience, we represents in the form

$$\mathcal{H}_j(p,\mathbf{r}) = \frac{1}{2}[p^2 - n_j{}^2(\mathbf{r}, \mathbf{l})] = 0 , \quad j = 1,2 . \tag{4.2.9}$$

4.2.2 Independent Normal Mode

Each of the factors, $\mathcal{H}_1$ or $\mathcal{H}_2$, in (4.2.9) approaching to zero is associated with a certain type of electromagnetic *normal* wave or *mode*. Each such mode possesses its own refractive index n_j, phase speed $v_j = c/n_j$, and a definite polarization.

Recognizing the possibility that two normal modes may exist, we should somewhat modify the geometrical-optics expansion (4.1.4) by representing the field $\mathbf{E}$ as the sum of two series

$$E = \sum_{m=0}^{\infty} \left[\frac{E_{m1}}{(ik_0)^m} e^{ik_0 \psi_1} + \frac{E_{m2}}{(ik_0)^m} e^{ik_0 \psi_2} \right] . \qquad (4.2.10)$$

Substituting (4.2.10) into the Maxwell equation (4.1.1) we have now to set to zero the coefficients not only of like powers of k_0, but also of rapidly oscillating functions $\exp(ik_0\psi_1)$ and $\exp(ik_0\psi_2)$. By this we assume that normal modes are *independent*.

Strictly speaking, the considered modes are independent in homogeneous media only. In an inhomogeneous medium we may assume independent propagation of normal modes only to some approximation. A criterion of independence can be deduced from the following consideration. The superposition of waves (4.2.10) has three characteristic space scales: the wavelength $\bar{\lambda}_0/n$ (scale of the exponential variation), the space/beating scale $l_b = 1/k_0|n_1-n_2|$ between two oscillating exponents [at a distance Δr of order l_b the phase difference $k_0(\psi_1-\psi_2) \simeq l_0(p_1-p_2)\Delta r$ varies by 1 radian], and the scale of the field amplitude variation L.

Normal modes may be deemed independent when $l_b \ll L$ or

$$\Delta n \equiv |n_1 - n_2| \gg 1/k_0 L \sim \mu . \qquad (4.2.11)$$

Whenever this condition is satisfied, the difference in the refractive indices (phase speeds) of two waves can be detected by physical devices. When, to the contrary, the condition is valid

$$\Delta n \le \mu , \qquad (4.2.12)$$

the variations of the resultant field (4.2.10) due to the difference in refractive indices can no longer be separated from those associated with the variation of amplitudes. Under these circumstances we can no longer speak of an independence of normal waves because in the framework of the representation of modes, passing to the limit, $\epsilon_{\alpha\beta} \rightarrow \epsilon\delta_{\alpha\beta}$ or $\Delta n \rightarrow 0$, from an anisotropic to isotropic medium, is impossible, in principle. This passage is enabled, though, by the approximation accounting for the *interaction* of normal waves (Sect.4.3).

4.2.3 Ray Equations

To each normal wave there corresponds its own family of rays, that is, a family of characteristics of the eikonal equation (4.2.9). The ray equations in an anisotropic medium have the form

$$\frac{d\mathbf{r}}{d\tau} = \frac{\partial \mathscr{H}}{\partial \mathbf{p}} = \mathbf{p} - \frac{1}{2}\frac{\partial n^2}{\partial \mathbf{p}} \equiv \mathbf{s}(\mathbf{p}, \mathbf{r}),$$

$$\qquad (4.2.13)$$

$$\frac{d\mathbf{p}}{d\tau} = -\frac{\partial \mathscr{H}}{\partial \mathbf{r}} = \frac{1}{2}\frac{\partial n^2}{\partial \mathbf{r}}$$

(the subscript j = 1 or 2 referring to the type of wave is dropped for compactness). The parameter τ is related to the arclength σ by

$$d\sigma = \sqrt{(d\mathbf{r})^2} = \left| \frac{\partial \mathcal{H}}{\partial \mathbf{p}} \right| d\tau = |\mathbf{s}| d\tau \; . \tag{4.2.14}$$

The initial conditions for the trajectory $\mathbf{r}(\tau)$ and momentum $\mathbf{p}(\tau)$ are supplied in the same manner as with the scalar problem (Sect.2.2).

In an anisotropic medium, the direction of the rays, (4.2.13), is defined by the *ray vector* $\mathbf{s} = d\mathbf{r}/d\tau$, which in the general case does not coincide with the direction of the normal $\mathbf{p}$ to the phase front.[6] The difference between $\mathbf{p}$ and $\mathbf{s}$ is given by

$$\mathbf{q} = \mathbf{s} - \mathbf{p} = -\frac{1}{2} \frac{\partial n^2}{\partial \mathbf{p}} = -n \frac{\partial n}{\partial \mathbf{p}} \tag{4.2.15}$$

(in an isotropic medium, $\mathbf{q} = 0$). Since the refractive index n depends on the momentum $\mathbf{p}$ via direction $\mathbf{l} = \mathbf{p}/p$ only, we have

$$\frac{\partial n}{\partial p_i} = \frac{\partial n}{\partial l_j} \frac{\partial l_j}{\partial p_i} = \frac{\partial n}{\partial l_j} \frac{1}{p} (\delta_{ij} - l_i l_j) \; .$$

Then by virtue of (4.2.15) and the identity $p = n$, we get

$$q_i = -\frac{\partial n}{\partial l_j}(\delta_{ij} - l_i l_j) \quad \text{or} \quad \mathbf{q} = \mathbf{l} \times \left(\mathbf{l} \times \frac{\partial n}{\partial \mathbf{l}} \right) . \tag{4.2.16}$$

From (4.2.16) it immediately follows that $\mathbf{q}$ and $\mathbf{p}$ are orthogonal (Fig.4.3) and $s = \sqrt{(p^2 + q^2)}$. The angle α between $\mathbf{s}$ and $\mathbf{p}$ can be found from

$$\cos\alpha = \frac{p}{s} = \frac{p}{\sqrt{p^2 + q^2}} = (1 + (q/n)^2)^{-1/2} = (1 + (\partial n/\partial p)^2)^{-1/2} \; . \tag{4.2.17}$$

The vector $\mathbf{s}$ tangent to the ray is parallel to the group-velocity vector $\mathbf{g}$. To prove it, we may put in correspondence to the Hamiltonian (4.2.9), depending on $\mathbf{p}$, $\mathbf{r}$, and (implicitly) ω, another Hamiltonian [(4.4.1) below]:

$$\mathcal{H}(\mathbf{k},\omega,\mathbf{r}) = \frac{1}{2}\left[k^2 - \frac{\omega^2}{c^2} n^2 \left(\mathbf{r},\omega,\frac{\mathbf{k}}{k} \right) \right] = 0 \; ,$$

depending on $\mathbf{k} = (\omega/c)\mathbf{p}$, $\mathbf{r}$, and ω. Then by the formulas of Sect.2.7 [also (4.4.2)] for the group velocity $\mathbf{g}$ we find

---

[6] In other words, in an anisotropic medium the rays can no longer be identified with the *phase trajectories*, i.e., trajectories perpendicular to the phase front.

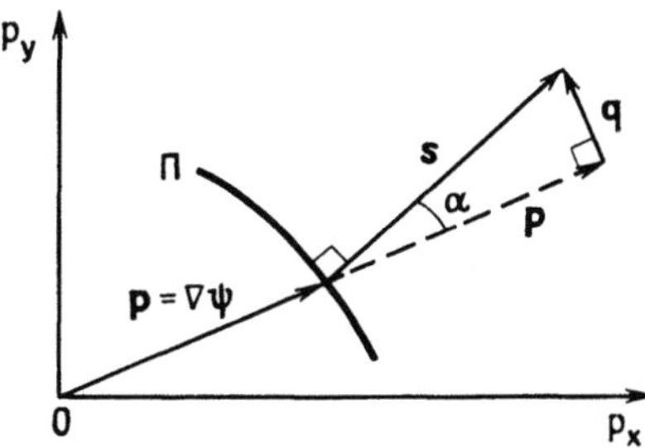

Fig.4.3. Refractive index surface π, momentum p and ray vector s

$$g = -\frac{\partial\mathscr{H}}{\partial k}\left(\frac{\partial\mathscr{H}}{\partial\omega}\right)^{-1} = \frac{k - \dfrac{\omega^2}{2c^2}\dfrac{\partial n^2}{\partial k}}{\dfrac{1}{2c^2}\dfrac{\partial(\omega^2 n^2)}{\partial\omega}} = \frac{cs}{n\,\dfrac{\partial(\omega n)}{\partial\omega}}\,. \tag{4.2.18a}$$

Since $s = p/\cos\alpha = n/\cos\alpha$ (Fig.4.3), we obtain for the group-velocity modulus

$$g = \frac{c|s|}{n}\left(\frac{\partial\omega n}{\partial\omega}\right)^{-1} = \frac{c}{\cos\alpha}\left(\frac{\partial\omega n}{\partial\omega}\right)^{-1}\,. \tag{4.2.18b}$$

In an isotropic medium $\alpha = 0$, and (4.2.18) reduces to (2.7.23) for g.

The surface Π produced in the momentum space by the tip of the vector $p = \ln(1)$ as a consequence of the unit vector 1 changing its direction, is referred to as the *surface of refractive index*. On this surface, $\mathscr{H}(p,r) = 0$ so that the ray vector $s = \partial\mathscr{H}/\partial p \equiv \nabla_p\mathscr{H}$ is normal to the surface, as Fig.4.3 illustrates. This leads to a simple graphic procedure for ray tracing whenever the direction of the normal p to the phase front $\psi =$ const., is known.

In p space the surface of refractive index has two envelopes (more correctly, a bivalued surface) associated with two types of normal waves, say, with ordinary and extraordinary waves in a crystal or in magnetoactive plasma. Along with the surface of refractive index $p = \ln(1)$ use is often made of the relative surfaces of phase speed (the surface of normals) $v = c\mathbf{1}/n(1)$ and of group velocities (ray surface) $g = g(1)$. The general features of these surfaces have been discussed in [4.14, 22, 24] as applied to crystals, and in [4.23, 25] as applied to magnetoactive plasma.

4.2.4 Solving the Eikonal Equation

When the solutions of the ray equations (4.2.13) are known, the eikonal can be obtained by (2.2.9). In this case, for the mode of type j we have

$$\left(p\cdot\frac{\partial\mathscr{H}}{\partial p}\right)_j = (p\cdot s)_j = p_j^2 = n_j^2$$

and therefore

$$\psi_j = \psi_j{}^0 + \int_{\tau_0}^{\tau} n_j{}^2 d\tau \ . \qquad\qquad (4.2.19)$$

We recast this expression, considering that by virtue of (4.2.14, 17) $nd\tau = |s|d\tau(n/|s|) = d\sigma\cos\alpha$, to obtain

$$\psi_j = \psi_j{}^0 + \int_{\sigma^0}^{\sigma} n_j \cos\alpha_j \, d\sigma \ . \qquad\qquad (4.2.20)$$

The quantity $n_r = n\cos\alpha$ describes the variation of the eikonal along the ray and therefore is referred to as the *ray* refractive index.

4.2.5 Definition of Mode Polarization Vectors

Considering one type of wave, let

$$\mathbf{E} = \Phi\mathbf{f} \ , \quad \mathbf{H} = \mathbf{p}\times\mathbf{E} = \Phi(\mathbf{p}\times\mathbf{f}) \ , \qquad\qquad (4.2.21a)$$

where Φ is the complex field amplitude, and $\mathbf{f}$ is the polarization vector (generally speaking, also complex). It is convenient to work with the vectors normalized to unity, $\mathbf{f}\cdot\mathbf{f}^* = 1$, so that

$$|\mathbf{E}|^2 = |\Phi|^2 \ , \qquad\qquad (4.2.21b)$$

though other normalizations are also possible.

The components of the polarization vector $\mathbf{f}$ as well as the components of the zeroth-approximation field itself, $\mathbf{E}$, satisfy the system of homogeneous equations

$$q_{\alpha\beta}f_\beta = p^2\delta_{\alpha\beta} - p_\alpha p_\beta - \epsilon_{\alpha\beta}f_\beta = 0 \ , \qquad\qquad (4.2.22)$$

from which not only the ratios $f_1:f_2:f_3$, but also the magnitudes of f_β can be defined. The above normalization to unity eliminates the ambiguity in choosing the moduli $|f_\beta|$; however, the uncertainty in choosing the phase factor $\exp(i\gamma)$ defining the position of $\mathbf{f}$ in the polarization ellipse is left. Accurate to this factor, (4.2.22) yields $\mathbf{f}$ for any point $\mathbf{r}$ and any direction $\mathbf{l} = \mathbf{p}/p$. As we shall see below the uncertainty in chossing $\exp(i\gamma)$ has no effect on the magnitude of the field $\mathbf{E} = \Phi\mathbf{f}$.

4.2.6 Consistency of Equations of the First Approximation

The equation for the field $\mathbf{E}_1$,

$$\mathbf{p}\times(\mathbf{p}\times\mathbf{E}_1) + \hat{\epsilon}\mathbf{E}_1 = \mathbf{Z} = -\operatorname{curl}\mathbf{H}_0 - \mathbf{p}\times\operatorname{curl}\mathbf{E}_0 \ , \qquad\qquad (4.2.23)$$

is similar to (4.1.15) but in contrast to that for the isotropic medium, now the determinant $\mathcal{H} = \det[q_{\alpha\beta}]$ has no multiple roots (polarization degeneracy is absent), and the rank of the matrix $[q_{\alpha\beta}]$ equals two for $\mathbf{p} = \mathbf{p}_1$ or $\mathbf{p} = \mathbf{p}_2$. Thus, in accordance with the Fredholm theorem, we can write down only *one* consistency condition of the form $\mathbf{E}' \cdot \mathbf{Z} = 0$.

In our case the tensor $\hat{\epsilon}$ is assumed to be Hermitian; therefore as a solution $\mathbf{E}'$ of the transposed homogeneous system may be taken the complex-conjugate field of zero approximation $\mathbf{E}^*$, so that the consistency condition $\mathbf{E}' \cdot \mathbf{Z} = 0$ takes the form

$$\mathbf{E}^* \cdot \mathbf{Z} = 0 . \tag{4.2.24}$$

Substituting $\mathbf{E}$ and $\mathbf{H}$ from (4.2.21) recasts this condition into

$$\mathbf{E}^* \mathrm{curl}\mathbf{H} - \mathbf{H}^* \mathrm{curl}\mathbf{E} = 0 . \tag{4.2.25}$$

This complex equation leads us to the modulus and argument of the amplitude factor $\Phi = |\phi|\exp(i\delta)$.

4.2.7 The Transfer Equation

Setting the real-valued part of (4.2.25) to zero yields

$$(\mathbf{E}^* \mathrm{curl}\mathbf{H} - \mathbf{H}^* \mathrm{curl}\mathbf{E}) + (\mathrm{cc}) \triangleq \mathrm{div}(\mathbf{E}\times\mathbf{H}^* + \mathbf{E}^*\times\mathbf{H}) = 0 , \tag{4.2.26}$$

where cc stands for the complex conjugate. Equation (4.2.26) can be rewritten in the form of an energy-flow conservation law [4.30, 35, 36]

$$\mathrm{div}\mathbf{S} = 0 , \quad \mathbf{S} = \frac{c}{16\pi}(\mathbf{E}\times\mathbf{H}^* + \mathbf{E}^*\times\mathbf{H}) , \qquad \text{or} \tag{4.2.27}$$

$$\mathrm{div}(\boldsymbol{\sigma}|\Phi|^2) = 0 , \tag{4.2.28}$$

where $\boldsymbol{\sigma} = 16\pi\mathbf{S}/c|\Phi|^2$ is the vector parallel to the Poynting vector

$$\boldsymbol{\sigma} = \mathbf{f}\times(\mathbf{p}\times\mathbf{f}^*) + \mathbf{f}^*\times(\mathbf{p}\times\mathbf{f}) = 2\mathbf{p} - \mathbf{f}(\mathbf{p}\mathbf{f}^*) - \mathbf{f}^*(\mathbf{p}\cdot\mathbf{f}) . \tag{4.2.29}$$

The vector $\boldsymbol{\sigma}$ is parallel to the ray vector $\mathbf{s} = d\mathbf{r}/d\tau$ and the vector of group velocity $\mathbf{g}$. The proportionality coefficient ζ between $\boldsymbol{\sigma}$ and $\mathbf{s}$ can be found from the relation $\mathbf{S} = w\mathbf{g}$ between the Poynting vector and the density of electromagnetic energy

$$w = \frac{1}{16\pi} \frac{\partial(\omega^2 \epsilon_{\alpha\beta})}{\omega\partial\omega} f_\alpha f_\beta^* |\Phi|^2 , \tag{4.2.30}$$

which we shall derive in Sect. 4.4. Using (4.2.18, 30) leads to

$$\boldsymbol{\sigma} = \frac{16\pi\mathbf{g}w}{c|\Phi|^2} = 2\mathbf{s} \frac{\partial(\omega^2\epsilon_{\alpha\beta})}{\partial\omega} f_\alpha f_\beta^* \left[\frac{\partial\omega^2 n^2}{\partial\omega}\right]^{-1} \equiv \zeta\mathbf{s} . \tag{4.2.31}$$

Integrating (4.2.28) over the ray-tube volume (Fig.2.5), we obtain

$$|\Phi| = |\Phi^0| \sqrt{\frac{|\sigma^0|\,\mathrm{d}a^0}{|\sigma|\,\mathrm{d}a}} = |\Phi^0| \sqrt{\frac{\varsigma^0|s^0|\,\mathrm{d}a^0}{\varsigma|s|\,\mathrm{d}a}} \, , \qquad (4.2.32)$$

which differs from (2.3.18) by the refractive index n replaced by $|\sigma| = \varsigma|s|$.

By virtue of (4.2.61) we can arrive at an alternative solution for the amplitude. Substituting (4.2.31) into (4.2.28) yields

$$s\nabla(\varsigma|\Phi|^2) + \varsigma|\Phi|^2\,\mathrm{divs} = 0 \, . \qquad (4.2.33)$$

Since $s = \mathrm{d}r/\mathrm{d}\tau$, the first term in (4.2.33) is $\mathrm{d}(\varsigma|\Phi|^2)/\mathrm{d}\tau$, while the second term can be expressed by Liouville's formula through the Jacobian $\mathscr{D}(\tau)$ as

$$\mathrm{divs} = \frac{\mathrm{d}}{\mathrm{d}\tau}\ln\mathscr{D}(\tau) \, , \quad \mathscr{D}(\tau) = \frac{\partial(x,y,z)}{\partial(\xi,\eta,\tau)} \, .$$

As a result, from (4.2.30) we obtain the equation

$$|\Phi| = |\Phi^0| \sqrt{\frac{\varsigma^0\mathscr{D}(\tau^0)}{\varsigma\mathscr{D}(\tau)}} \, , \qquad (4.2.34)$$

equivalent to (4.2.30) because $|\sigma|\,\mathrm{d}a = \varsigma|s|\,\mathrm{d}a = \varsigma\mathscr{D}(\tau)\mathrm{d}\xi\mathrm{d}\eta$.

4.2.8 Equation for the Argument of a Complex Amplitude

An equation for $\delta = \arg\Phi$ results from equating the imaginary part of (4.2.25) to zero:

$$\sigma\nabla\delta = M \, , \quad M = \mathrm{Im}\{f^*\,\mathrm{curl}(p\times f) + (p\times f)\mathrm{curl}f^*\} \, . \qquad (4.2.35)$$

In view of (4.2.31) the left-hand side of (4.2.35) is $\varsigma s\nabla\delta = \varsigma\mathrm{d}\delta/\mathrm{d}\tau$, whence

$$\delta = \delta^0 + \int_{\tau^0}^{\tau} \frac{M}{\varsigma}\,\mathrm{d}\tau \, . \qquad (4.2.36)$$

A closer examination indicates that M equals zero in inhomogeneous crystals (by virtue of $\epsilon_{\alpha\beta}$ and f being real) and in homogeneous media (it suffices to show that p and f are constant along the ray), and is different from zero in inhomogeneous gyrotropic media only.

Lewis was the first to find the argument δ variation along the ray [4.31]. He also noted the invariance of the field with respect to the phase factor of vector f. Should we change the argument of the polarization vector f by letting $\tilde{f} = f\exp(i\gamma)$, the respective $\tilde{M}$ becomes

242

$$\tilde{M} = M - \varsigma \frac{d\gamma}{d\tau} \; .$$

From (4.2.35) it follows that $\tilde{\delta} = \delta - \gamma$ and

$$\tilde{\Phi} = |\tilde{\Phi}| \exp(i\tilde{\delta}) = |\tilde{\phi}| \exp[i(\delta - \gamma)] = \Phi \exp(-i\gamma) \; .$$

We have taken into account here that $|\tilde{\Phi}| = |\Phi|$ since $|\tilde{\sigma}| = |\sigma|$. Hence,

$$\tilde{E} = \tilde{\Phi}\tilde{f} = |\Phi| \exp(i\delta - i\gamma) f \exp(i\gamma) = \Phi f = E \; ;$$

that is, the resultant field $E = \Phi f$ is indeed independent of the phase γ of the polarization vector.

We eventually arrive at the following scheme for evaluating the ray field in an anisotropic medium:

 i) Determine the rays form (4.2.13) for each mode.
 ii) Calculate the eikonal by (4.2.19) or (4.2.20).
 iii) Find the polarization vector f by (4.2.22).
 iv) Determine the amplitude moduli of the modes by (4.2.32 or 34), and the arguments of these amplitudes by (4.2.36).

The resultant zero approximation field is a sum of two waves, namely

$$E = \Phi_1 f \exp(ik_0 \psi_1) + \Phi_2 f_2 \exp(ik_0 \psi_2) \; . \tag{4.2.37}$$

This computational procedure supposes that initial conditions for the amplitudes $\Phi_{1,2}$ and the eikonals $\psi_{1,2}$ are supplied on the starting surface Q. When caustics arise, (4.2.37) should be augmented by the caustic phase shifts as in Sect.2.4, and the summation should be taken over all rays passing through the observation point.

4.2.9 Rays and Energy Paths.
The Fresnel Volumes in Anisotropic Media

From Sect. 4.2.7 it follows that although in anisotropic media the rays can no longer be treated as *phase trajectories* - i.e., lines perpendicular to the phase fronts - they retain functions of the *energy paths*, as the energy flow S is directed along the ray vector $s = dr/d\tau$. This is intimately connected with the interference nature of the wave field forming in anisotropic media [4.22, 23, 39].

Consider a bundle of plane wave (a wave packet) whose momenta lie within the cone $(p, p+\Delta p)$ shown in Fig.4.4. If in the initial moment of time all the plane waves possess equal phases, then over a time t their phase fronts $\psi_p = pr = ct$ and $\psi_{p+\Delta p} = (p+\Delta p)r = ct$ reach the positions indicated in Fig.4.4 by thin lines. It turns out that the phase fronts cross at the point P lying in the extension of the ray vector s [4.22, 23, 39]. Therefore, the addition of interfering waves and, consequently, the propagation of electromagnetic energy do proceed along the vector s, tangentially to the ray.

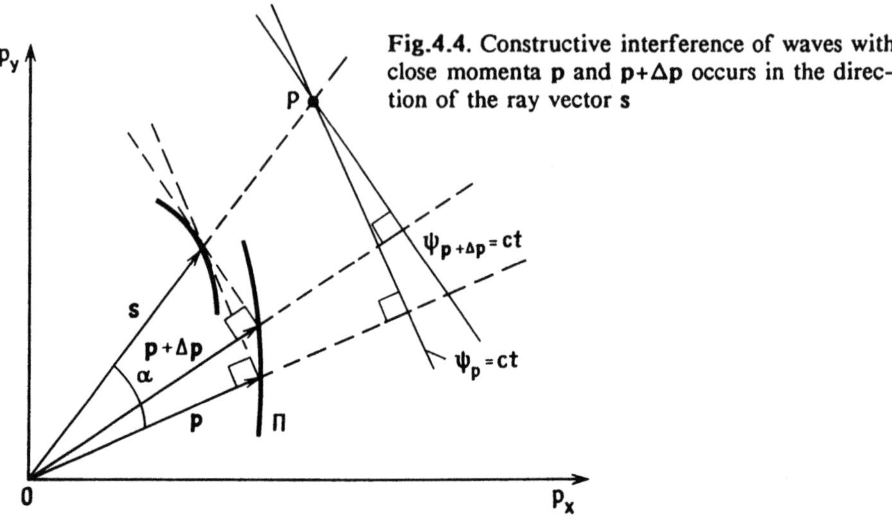

Fig.4.4. Constructive interference of waves with close momenta **p** and **p**+Δ**p** occurs in the direction of the ray vector **s**

Deviations of the direction of **s** break down the phase coincidence of the waves composing the wave packet; hence the ray is the axis of the Fresnel volume responsible for the formation of the field in a given point. The boundary of this volume can be defined, similar to (2.10.6), by

$$\left| \psi_{\text{virt}}(\mathbf{r}') - \psi_{\text{ref}} \right| - \lambda/2 = 0. \tag{4.2.38}$$

Figure 4.5 illustrates the shape of the Fresnel volume for a plane wave in a homogeneous medium.

4.2.10 An Account of Weak Absorption

In a weakly absorbing medium, the tensor $\epsilon_{\alpha\beta}$ acquires an anti-Hermitian term $\epsilon_{\alpha\beta}{}^{a}$, which is small compared with the Hermitian component $\epsilon_{\alpha\beta}^{h}$,

$$\epsilon_{\alpha\beta} = \epsilon_{\alpha\beta}^{h} + \epsilon_{\alpha\beta}^{a} \, , \quad \left| \epsilon_{\alpha\beta}^{a} \right| \ll \left| \epsilon_{\alpha\beta}^{h} \right| . \tag{4.2.39}$$

Similar to ϵ'' as in Sect.2.3.7, we refer the small term $ik_0\hat{\epsilon}^{a}\mathbf{E}$, arising in the Maxwellian equations, to the first approximation equations; as a conse-

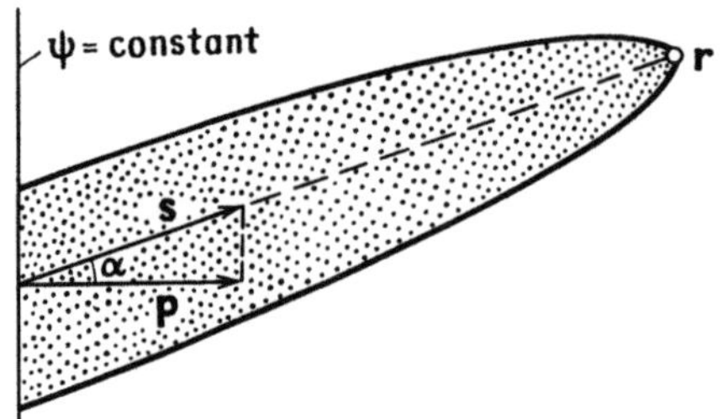

Fig.4.5. Fresnel volume is oriented along the ray vector **s** in an anisotropic medium

244

quence, we include $\mathrm{curl}H_0 + ik_0\hat{\epsilon}^a E$ instead of $\mathrm{curl}H_0$, and the consistency condition (4.2.24) takes the form

$$E^* \mathrm{curl}H - H^* \mathrm{curl}E - ik_0\epsilon^a_{\alpha\beta}E_\alpha E_\beta^* = 0 \ . \tag{4.2.40}$$

The new term in $\epsilon^a_{\alpha\beta}$ is real-valued as $(\epsilon^a_{\alpha\beta})^* = -\epsilon^a_{\beta\alpha}$, and therefore in a first approximation in $\epsilon^a_{\alpha\beta}$ absorption will affect the modulus of the amplitude Φ only. Taking the real part of (4.2.40) we arrive at the equation

$$\mathrm{div}(\sigma|\Phi|^2) - ik_0\epsilon^a_{\alpha\beta}f_\alpha f_\beta^*|\Phi|^2 = 0 \ ,$$

whose solution differs from that of (4.2.34) by an exponential factor responsible for damping

$$|\Phi| = |\Phi^0| \sqrt{\frac{\varsigma^0\mathscr{D}(\tau^0)}{\varsigma\mathscr{D}(\tau)}} \exp\left\{ik_0 \int_{\tau^0}^{\tau} \varsigma^{-1}\epsilon^a_{\alpha\beta}f_\alpha f_\beta^* \mathrm{d}\tau\right\} \ . \tag{4.2.41}$$

4.2.11 Reflection and Refraction at the Boundaries of Anisotropic Media

Consider an electromagnetic wave indicent from an isotropic medium onto a sharp interface with an anisotropic medium. In the general case this launches in the anisotropic medium, *two refracted (transmitted) waves* (birefringence) so that the boudnary conditions (4.1.33) become

$$(E_i + E_r)_{tan} = (E_{t1} + E_{t2})_{tan} \ ,$$
$$(H_i + H_r)_{tan} = (H_{t1} + H_{t2})_{tan} \ . \tag{4.2.42}$$

Seeking solutions of Maxwell's equations in geometrical form, we can satisfy (4.2.42) by only letting the eikonals be equal for all waves at the interface Q:

$$\psi_i = \psi_r = \psi_{t1} = \psi_{t2} \ . \tag{4.2.43}$$

These relationships play the role of the initial conditions for eikonals of reflected and refracted (transmitted) waves.

initial conditions for the momenta of all waves leaving the interface can be derived from (4.2.43). Differentiating (4.2.43) we obtain the equality for the tangential components of momentum

$$(p_i)_{tan} = (p_r)_{tan} = (p_{t1})_{tan} = (p_{t2})_{tan} \ , \tag{4.2.44}$$

where $p_{tan} \equiv p - N(N\cdot p)$, and N is the normal to Q (Fig.4.6a). Introducing the anlges θ_i, θ_r, θ_{t1}, and θ_{t2} we can recast (4.2.44) into

$$n\sin\theta = n\sin\theta_r = n_{1,2}(\theta_{t1,2})\sin\theta_{t1,2} \ , \tag{4.2.45}$$

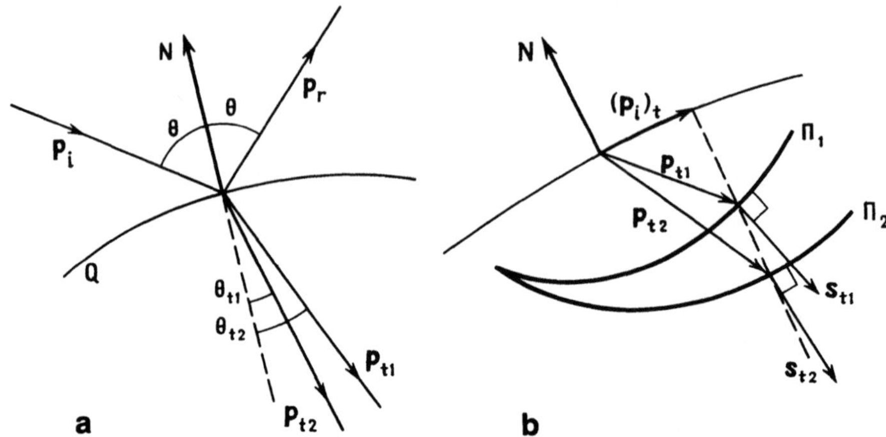

Fig.4.6a,b. Incidence of a wave on the interface between isotropic and anisotropic media: (a) two refracted waves occur; (b) graphical evaluation of the ray vectors of the refracted waves

where n is the refractive index of the isotropic medium, and n_1 and n_2 are the refractive indices of the two normal waves in the anisotropic medium.

Relations (4.2.45) generalize Snell's law over the case of anisotropic media. In contrast to isotropic media, the refractive indices n_1 and n_2 depend on the direction of $l = p/p$; therefore the ratios $\sin\theta_{t1,2}/\sin\theta$ are now functions of the angle of incidence. The angles $\theta_{t1,2}$ can be found either from the transcendental relations (4.2.45) or by a graphic approach, using the surfaces of refractive index (Fig.4.6b). The graphic approach also yields the directions of the rays s_{t1} and s_{t2} which, according to Sect.4.2.3, are perpendicular to the surfaces of refraction indices, Π_1 and Π_2. It is essential that the directions of the rays, in the general case, do not belong to the plane of incidence.

The polarization vectors of the refracted waves $f_{t1,2}$ are found from (4.2.22), whereas the two amplitudes Φ_{t1} and Φ_{t2} are determined, simultaneously with the amplitudes Φ_ν and Φ_b of the reflected wave, from the set of two vector equations (4.2.42). The respective generalized Fresnel equations – fairly bulky and therefore dropped – can be found in [4.40].

The number of refracted waves can be less than two if one or even both normal waves suffer total reflection, or the axes of symmetry of the anisotropic medium appear to be suitable oriented, or the incident-wave polarization is appropriately chosen. Various specific situations of reflection and refractions of normal waves were elaborated in [4.23,41]. A specific kind of focusing of normal waves for a point source placed near the interface was described in [4.42]; this phenomenon resulted from the dependence of the refractive index upon the direction of propagation.

4.2.12 Some Specific Results

Here we dwell on some topics of potential application, which have only been touched on, among numerous other topics, as in [4.13,14,22-29].

a) The Field of a Point Source in an Anisotropic Medium

The field in the immediate vicinity of the source may be described assuming that the medium around it is homogeneous. In an homogeneous medium the momentum $\mathbf{p}$ and the ray vector $\mathbf{s}$ are constant along the ray, $\mathbf{p} = \mathbf{p}^0$ and $\mathbf{s} = \mathbf{s}^0(\mathbf{p}^0)$, so that the rays appear as straight lines, $\mathbf{r} = \mathbf{r}^0 + \mathbf{s}^0\tau$, and the eikonal increases linearly with the distance of the source;

$$\psi = \psi^0 + n^2\tau = \psi^0 + n^2 r/s = \psi^0 + nr\cos\alpha .$$

Taking polar θ and azimuth ϕ angles as the ray coordinates, we define the initial amplitude Φ_j^0 of the type-j wave by an expression similar to (2.3.25):

$$\Phi_j^0 = B_j G_j(\theta,\phi)/r_0 , \quad j = 1,2 , \tag{4.2.46}$$

where r_0 is the radius of a sphere, small compared with the scale L of the medium's inhomogeneity, yet sufficiently large for the zone of $r \geq r_0$ to lie in the far (Fraunhofer) field of the source. In contrast to (2.3.25), the factor $G_j(\theta,\phi)$ in (4.2.46) is defined not only by the radiation pattern of the source in a homogeneous medium but also by the *type* of the wave. This factor can be found from the asymptotics of the exact solution of the problem of radiation in a given mode in a homogeneous anisotropic medium.

Using (4.2.46) we have from (4.2.34)

$$\Phi_j = \Phi_j^0 \sqrt{\frac{\varsigma^0 \mathscr{D}(\tau^0)}{\varsigma \mathscr{D}(\tau)}} = B_j G_j(\theta,\phi) \sqrt{\frac{\varsigma^0 W(\theta,\phi)}{\varsigma \mathscr{D}(\tau)}} , \tag{4.2.47}$$

where $W(\theta,\phi) = \mathscr{D}(\tau_0)/(r_0)^2$ is a quantity independent of the choice of the radius r_0. The connection of the field amplitude with the radii of curvature of the refractive index surface was discussed in [4.23].

Specific conditions arise when the refractive index surface Π contain points of inflection (Fig.4.7a). In this case an observation point may be simultaneously reached by three rays having coaxial ray vectors $\mathbf{s}$ but different eikonals and, hence, different momenta $\mathbf{p}$ (Fig.4.7b shows the surface of a constant ψ corresponding to Fig.4.7a). In singled-out directions corresponding to the inflection points of surface Π (dashed lines in Fig.4.7) two of the three rays merge, and the Jacobian $\mathscr{D}(\tau)$ in (4.2.45) goes to zero while the amplitude Φ_j goes to infinity [4.43]. This results in a specific caustic being formed by two straight rays going out into singular directions. As is evident from Fig.4.7b, the phase fronts then correspond to cross sections of a swallowtail catastrophe.

b) Waves in Plane-Stratified Anisotropic Media

Let $\epsilon_{\alpha\beta} = \epsilon_{\alpha\beta}(z)$. From (4.2.13) we have then $p_x = \text{constant} = p_x^0$ and $p_y = \text{const.} = p_y^0$. Without loss of generality, p_y^0 may be put equal to zero, and the constancy of p_x recast into a kind of Snell's law (4.2.45):

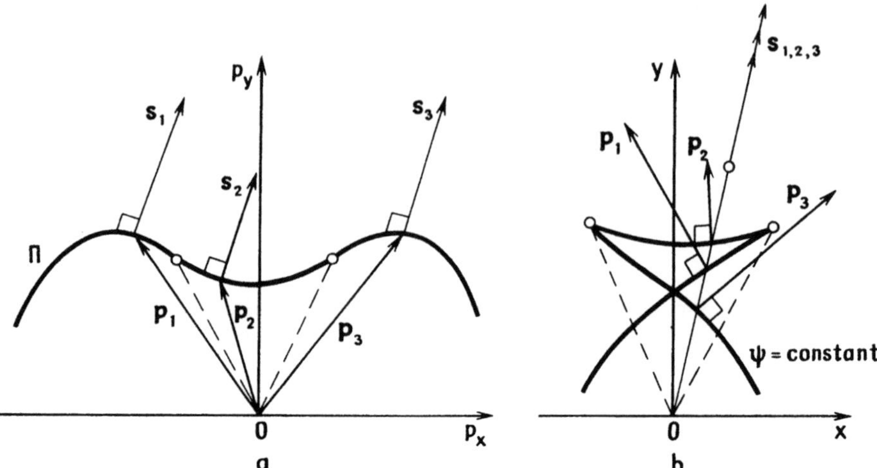

Fig.4.7 (a) Colinearity of ray vectors in case when the ray surface has points of reflection. (b) Orientation of the momenta and ray vectors relative to the phase front ψ = const.

$$p_x = \text{const.} = p\sin\theta = n(\theta)\sin\theta \ , \tag{4.2.48}$$

where θ is the angle of incidence, i.e., the angle between **p** and the z axis.

Substituting (4.2.48) and $p_y = 0$ into the eikonal equation $\mathcal{H} = \det[q_{\alpha\beta}] = 0$, we obtain a fourth-degree polynomial (Booker's quartic) in $p_z = \partial\psi/\partial z$:

$$a_0 p_z^4 + a_1 p_z^3 + a_2 p_z^2 + a_3 p_z^1 + a_4 p_z^0 = 0 \ ,$$

where $a_m = a_m(z,\theta)$. Solving this equation for p_z yields for the eikonal

$$\psi = \psi^0 + p_x^0(x - \xi) + \int_{z^0}^{z} p_z\,dz \ .$$

Two of the four Booker's quartic roots correspond to two types of incident wave, and the other two correspond to two types of reflected wave.

Although the momentum **p** lies in the plane of incidence y = 0, the ray paths $\mathbf{r} = \mathbf{r}(\tau)$ are, generally speaking, not plane curves, as Fig.4.8 schematizes for the case of an electromagnetic wave reflected from an inhomogeneous magnetoactive ionosphere. Numerical analysis for ray tracing in a plane stratified magnetoactive plasma was treated, e.g., in [4.13, 25-28]. Interesting features of phase trajectories forming cusps and pockets were also discussed there.

248

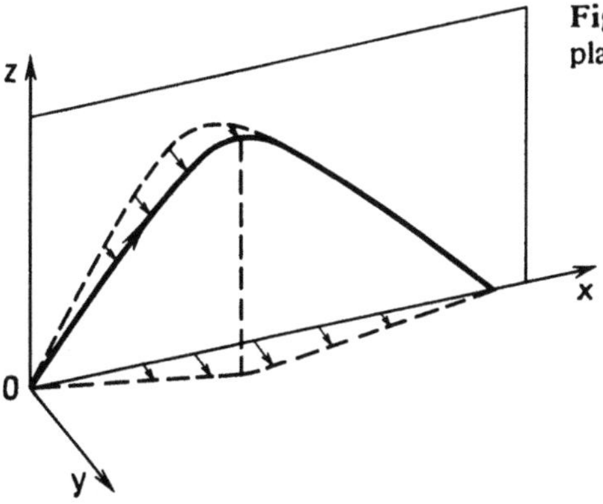

Fig.4.8. Lateral deviation of a ray in a plane-stratified magnetoactive ionosphere

c) Separation of Variables in the Equation of Eikonal in the General Case

The conditions that render the eikonal equation separable for an inhomogeneous anisotropic medium have been studied comparatively little. This topic was studied in [4.44] as applied to a two-dimensionally inhomogeneous magnetoactive plasma. As an example, [4.44] considered the propagation of a wave in a plane perpendicular to the static magnetic field. Of other works on the subject, we should point out [4.45].

d) Perturbation Theory

A perturbation theory for the eikonal in an anisotropic medium[7] has been suggested in [4.46, 47]; the perturbation method has been used for ray-tracing computations in an inhomogeneous magnetoactive plasma. The perturbation in [4.47] was not the magnetic field $\mathbf{H}_0$ itself but a certain expression that falls off to zero simultaneously with $\mathbf{H}_0$. Already the first iteration described a small displacement of rays from the plane of incidence, like that shown in Fig.4.8.

4.2.13 Divergence of First-Approximation Fields at Polarization Degeneracy

Let λ_j be the eigenvalues, and $\mathbf{f}_j$ be the eigenvectors of the matrix $\hat{q} = [q_{\alpha\beta}]$,

$$\hat{q}\mathbf{f}_j = \lambda_j \mathbf{f}_j . \tag{4.2.49}$$

To the normal waves (modes) that interest us there correspond, according to (4.2.22), the zero eigenvalues $\lambda_j = 0$ (the respective vectors $\mathbf{f}_j$, which we

---

[7] In crystals, the difference between the principal values of the permittivity tensor, ϵ_1, ϵ_2, and ϵ_3, is normally small. Let $\bar{\epsilon} = (\epsilon_1 + \epsilon_2 + \epsilon_3)/3$. Treating $\Delta\epsilon_i = \epsilon_i - \bar{\epsilon}$ as small perturbations one might compute the proportional corrections to the eikonal and other wave parameters, simplifying thereby wave analysis in uniaxial and, especially, in biaxial crystalline media, including liquid crystals. This approach, however, has not been given due recognition.

treat as the polarization vectors, are referred to as *zero vectors* [4.30,31]). The determinant $\mathcal{K}$ of the matrix $\hat{\mathbf{q}}$ is equal to the product of its eigenvalues, $\mathcal{K} = \lambda_1 \lambda_2 \lambda_3$, so that the eikonal equation $\mathcal{K} = 0$ may be written in the form

$$\mathcal{K} = \lambda_1 \lambda_2 \lambda_3 = 0 \; . \tag{4.2.50}$$

In an anisotropic medium $\lambda_1 = \lambda_2 = -(p^2 - \epsilon)$ and $\lambda_3 = -\epsilon$ so that (4.2.50) becomes equivalent to (4.1.10). In an anisotropic medium, the eigenvalues $\lambda_{1,2,3}$ are generally nonidentical, and $\lambda_j(\mathbf{p})$ vanishes as the Hamiltonian (4.2.9) does.

Consider one of the normal modes labeled with the subscript 1. For this mode the eikonal equation (4.2.50) reduces to $\lambda_1(\mathbf{r},\mathbf{p}) = 0$. Let $\psi_1(\mathbf{r})$ be the solution of this equation, $\mathbf{p}_1 = \nabla \psi_1$ the respective normal to the phase front, and $\mathbf{f}_1(\mathbf{r})$ the polarization vector of the first normal mode corresponding to $\lambda_1 = 0$. In a zero approximation, the field is proportional to $\mathbf{f}_1$, $\mathbf{E}_0 = \Phi \mathbf{f}_1$, whereas in the first approximation the field also contains other components. Making use of the orthogonality of the Hermitian-matrix eigenvectors, $\mathbf{f}_k^* \cdot \mathbf{f}_j = \delta_{kj}$, we represent the first approximation field as the sum

$$\mathbf{E}_1 = \sum_{j=1}^{3} \Phi_j^1 \mathbf{f}_j \; , \tag{4.2.51}$$

and derive the components of (4.2.23) in the eigenvectors $\mathbf{f}_k$. For this purpose we premultiply (4.2.23) by $\mathbf{f}_k^*$ and observe (4.2.49) to become

$$\lambda_j \, \Phi_j^1 = \mathbf{f}_j^* \cdot \mathbf{Z}|_{\mathbf{p}=\mathbf{p}_1} \tag{4.2.52}$$

From (4.2.52) we are able to determine all the amplitude coefficients Φ_j^1 for which[8] $\lambda_j \neq 0$, i.e., Φ_2^1 and Φ_3^1 in this case (not Φ_1^1 as $\lambda_1 = 0$):

$$\Phi_j^1 = \mathbf{f}_j^* \cdot \mathbf{Z}/(\lambda_j)|_{\mathbf{p}=\mathbf{p}_1} \; , \quad j = 2,3 \; . \tag{4.2.53}$$

The value Φ_1^1 left nondefined is then found from the consistency of the equations of the second approximation, and so on.

Equation (4.2.53) is valuable in that it leads to the *divergence of the first approximation field* $\mathbf{E}_1$ for the case of polarization degeneracy, i.e., as $\lambda_2(\mathbf{p}_1)$ decreases, because $\Phi_2^1 \to \infty$, as $\lambda_2(\mathbf{p}_1) \to \lambda_1(\mathbf{p}_1) = 0$ [4.30].

---

[8] At $j = 1$ the left-hand side of (4.2.52) equals zero (as $\lambda_1 = 0$), thus entailing a zero equality for the right side, $\mathbf{f}_1^* \cdot \mathbf{Z} = 0$; that is, we have independently deduced the consistency condition (4.2.24). If two eigenvalues, say λ_1 and λ_2, vanish simultaneously, then (4.2.52) yields the equations $\mathbf{f}_1^* \cdot \mathbf{Z} = \mathbf{f}_2^* \cdot \mathbf{Z} = 0$, the specific case of which is the consistency conditions (4.1.16) since in an isotropic medium we may take the normal $\boldsymbol{\nu}$ and the binormal $\mathbf{b}$ as $\mathbf{f}_1$ and $\mathbf{f}_2$.

Now we find the condition on the relative smallness of E_1. Letting $p_1 = p_2 + \Delta p$, with p_2 being the solution of the eikonal equation $\lambda_2(p_2) = 0$ for the second mode, we get

$$\lambda_2(p_1) = \lambda_2(p_2 + \Delta p) \simeq \frac{\partial \lambda_2}{\partial p_2} \Delta p \ , \qquad (4.2.54)$$

from which it follows that Φ_2^1 increases as $1/\Delta p \sim 1/\Delta n$, where Δn is the difference of the refractive indices. Recall that $Z \sim \Phi/L$ and $|\partial\lambda/\partial p| \sim 1$ by an order-of-magnitude consideration, and that the first approximation field E_1 enters (4.2.10) with the factor $1/ik_0$, the condition of E_1/ik_0, being small compared with E_0, can be stated as

$$\frac{|E_1|}{k_0|E_0|} \sim \frac{|\Phi_2^1|}{k_0|\Phi|} = \frac{1}{k_0|\Phi|} \left| \frac{(f_2^{*}Z)}{\lambda_2} \right| \sim \frac{1}{k_0 L \Delta n} \ll 1 \ . \qquad (4.2.55)$$

It is an easy matter to see that this inequality is equivalent to the condition (4.2.11) of the independence of normal modes deduced from the qualitative considerations.

The modifications of the geometrical-optics method, which are valid in the polarization-degeneracy region $k_0 L \Delta n \leq 1$ and describe the interaction of normal waves there, will be discussed in Sect.4.3 below.

4.2.14 Other Vector Problems

A systematic ray theory of waves in inhomogeneous elastic media has been developed by *Babich* [4.48]. The theory of elasticity equations allows, in the general case, for three types of normal modes to exist, two of which stay in their parameters close to transverse waves, and the third to the longitudinal wave. Various aspects of the ray theory of waves in anisotropic elastic media have been considered in [4.21, 49].

Situations equivalent to the propagation of normal modes in anisotropic media arise in quantum theory when considering the behavior of spin-possessing particles in a magnetic field. The quasi-classic description of electron spin precessing in an inhomogeneous magnetic field was treated in [4.50] (see also [4.51, 52]). In all the situations mentioned the change to polarization degeneracy, i.e., the appearance of multiple roots λ_j, calls for special methods to describe the interaction of the normal modes.

4.3 Interaction of Normal Modes
in Inhomogeneous Anisotropic Media

4.3.1 Waves in Weakly Anisotropic Media.
The Quasi-Isotropic Approximation

As we pointed out in Sect.4.2, when the condition (4.2.11) is violated [or, which is the same, (4.2.55) fails], i.e., for $k_0 L \Delta n \leq 1$, the normal modes can no longer be deemed independent and in the inhomogeneous anisotropic medium the geometrical-optics method has to be modified. Such a modification is to describe the linear interaction of normal modes in the region of polarization degeneracy $k_0 L \Delta n \leq 1$, where the eigenvalues λ_1 and λ_2 of the matrix $\hat{q} = [q_{\alpha\beta}]$ turn out to be almost equal values (Sect. 4.2.13)

In weakly anisotropic media, closeness of the eigenvalues λ_1 and λ_2 is due to the smallness of the off-diagonal components of the tensor $\epsilon_{\alpha\beta}$ (and differences of the diagonal components) compared with the diagonal components of the tensor. For these media, the tensor $\epsilon_{\alpha\beta}$ can be divided into a large isotropic part $\epsilon\delta_{\alpha\beta}$ and a small anisotropic part $\chi_{\alpha\beta} = \epsilon_{\alpha\beta} - \epsilon\delta_{\alpha\beta}$ so that

$$\epsilon_{\alpha\beta} = \epsilon\delta_{\alpha\beta} + \chi_{\alpha\beta} . \tag{4.3.1}$$

We may take for ϵ one of the diagonal components ϵ_{11}, ϵ_{22}, or ϵ_{33}, or their mean value $(\epsilon_{11} + \epsilon_{22} + \epsilon_{33})/3 = \mathrm{Sp}\{\epsilon_{\alpha\beta}\}/3$.

In a weakly anisotropic environment the structure of the electromagnetic wave will differ only a little from the field in the isotropic medium. Assuming that in the zero approximation the medium is isotropic, $\epsilon_{\alpha\beta} = \epsilon\delta_{\alpha\beta}$, and considering the anisotropy (i.e., the anisotropic part of the permittivity tensor) as a weak disturbance, we arrive at the *quasi-isotropic approximation of geometrical optics* [4.53].

The equations for the quasi-isotropic approximation may be derived on the following grounds: Represent the constitutive equation $\mathbf{D} = \hat{\epsilon}\mathbf{E}$ as $\mathbf{D} = \epsilon\mathbf{E} + \hat{\chi}\mathbf{E}$. The refractive-index differential $\Delta n = |n_1 - n_2|$ is of order of $|\chi_{\alpha\beta}|$ so that the quantity

$$\mu_1 = \Delta n/n \sim \chi_{\alpha\beta}/n \ll 1 \tag{4.3.2}$$

may be viewed as an additional (to $\mu \sim 1/kL$) small parameter of the problem. Under these circumstances, a search for an asymptotic solution suggests that the fields should be developed simultaneously in two small parameters μ and μ_1, deemed to be the quantities of like order. The same results can be obtained in a simpler way by invoking the Debye procedure and formally ascribing, to the anisotropy, $\hat{\chi}$ the first order of smallness in $1/k_0$, i.e., by letting $\hat{\chi} = \hat{\xi}/k_0$. The term $\hat{\chi}\mathbf{E} = \hat{\xi}\mathbf{E}/k_0$ is referred to the first-order equations, and the quantities $\mathbf{X}$ and $\mathbf{Z}$ in the first approximation equations (4.1.6, 15) will obtain the increment $i\hat{\xi}\mathbf{E} = ik_0\hat{\chi}\mathbf{E}$, e.g.,

$$\mathbf{Z}_{\mathrm{new}} = -\,\mathrm{curl}\mathbf{H} + ik_0\hat{\chi}\mathbf{E} - \mathbf{p}\times\mathrm{curl}\mathbf{E} . \tag{4.3.3}$$

The equations of zero approximation, of course, remain the same as for the isotropic medium: in particular, the transverse field structure (4.1.13, 14) characteristic of an isotropic media remains unchanged. However, the amplitudes Φ_ν and Φ_b now obey equations which are more complicated than (4.1.18) and follow from the consistency conditions (4.1.16) subject to (4.3.3):

$$l(2\sqrt{\epsilon}\nabla\Phi_\nu + \Phi_\nu\nabla\sqrt{\epsilon}) + \Phi_\nu\sqrt{\epsilon}\,\mathrm{div}\,l + 2\kappa\sqrt{\epsilon}\Phi_b - ik_0(\chi_{\nu\nu}\Phi_\nu + \chi_{\nu b}\Phi_b) = 0$$

$$(4.3.4)$$

$$l(2\sqrt{\epsilon}\nabla\Phi_b + \Phi_b\nabla\sqrt{\epsilon}) + \Phi_b\sqrt{\epsilon}\,\mathrm{div}\,l - 2\kappa\sqrt{\epsilon}\Phi_\nu - ik_0(\chi_{b\nu}\Phi_\nu + \chi_{bb}\Phi_b) = 0$$

where $\chi_{\nu\nu} = (\nu\hat{\chi}\nu)$, $\chi_{b\nu} = (b\hat{\chi}\nu)$, $\chi_{\nu b} = (\nu\hat{\chi}b)$, and $\chi_{bb} = (b\hat{\chi}b)$ are the components of the anisotropy tensor in the frame of reference (l,ν,b) attached to the ray.

4.3.2 Different Forms of the Equations of the Quasi-Isotropic Approximation

Equations (4.3.4) can be stated in several equivalent forms, each of which has advantages of its own as applied to actual problems.

Let $\Phi_{\nu,b} = A\Gamma_{\nu,b}$ with $A^2 = |\Phi_\nu|^2 + |\Phi_b|^2$ obey the transfer equation (4.1.20). Then, for Γ_ν and Γ_b of (4.3.4), we obtain

$$\frac{d\Gamma_\nu}{d\sigma} = \frac{ik_0}{2\sqrt{\epsilon}}(\chi_{\nu\nu}\Gamma_\nu + \chi_{\nu b}\Gamma_b) - \kappa\Gamma_b \,,$$

$$(4.3.5)$$

$$\frac{d\Gamma_b}{d\sigma} = \frac{ik_0}{2\sqrt{\epsilon}}(\chi_{b\nu}\Gamma_\nu + \chi_{bb}\Gamma_b) + \kappa\Gamma_\nu \,.$$

These equations describe the effect of weak anisotropy of the medium and the ray torsion, with the divergence of rays being already accounted for in the amplitude A. It is a straightforward to verify that in the absence of absorption, when $\chi_{\alpha\beta} = \chi_{\alpha\beta}{}^*$, (4.3.5) leads to the conservation law $|\Gamma_\nu|^2 + |\Gamma_b|^2 = 1$.

For $\theta = \arctan(\Phi_b/\Phi_\nu) = \arctan(\Gamma_b/\Gamma_\nu)$ (being complex in the general case) and $\ln\Gamma^2 \equiv \ln(\Gamma_\nu{}^2 + \Gamma_b{}^2)$, Eq.(4.3.5) yields two equations [4.53]:

$$\frac{d\ln\Gamma^2}{d\sigma} = \frac{ik_0}{2\sqrt{\epsilon}}[(\chi_{\nu\nu} + \chi_{bb}) + (\chi_{\nu\nu} - \chi_{bb})\cos2\theta + (\chi_{\nu b} + \chi_{b\nu})\sin2\theta] \,,$$

$$(4.3.6)$$

$$\frac{d\theta}{d\sigma} = \kappa + \frac{ik_0}{4\sqrt{\epsilon}}[(\chi_{b\nu} - \chi_{\nu b}) + (\chi_{b\nu} + \chi_{\nu b})\cos2\theta - (\chi_{\nu\nu} - \chi_{bb})\sin2\theta] \,.$$

By virtue of (4.3.6) the angle θ may, in principle, be found (from the second line) independently of the amplitudes,[9] whereupon Γ^2 can be found from the first line by quadratures.

In the limit of vanishingly small anisotropy, as $\chi_{\alpha\beta} \to 0$ and consequently $\Delta n/\mu \sim \mu_1/\mu \to 0$, the equations for the quasi-isotropic approximation obviously transfer into those for transverse waves in isotropic media. On the other hand, with $\Delta n/\mu \sim \mu_1/\mu \gg 1$ when (4.2.11) is already satis-

---

[9] Note, in addition, that the equation for θ can be recast into a Riccati equation. Under certain conditions, this facilitates the derivation of approximate solutions [4.54].

fied - assuming the modes to be independent with the parameter $\mu_1 = \Delta n/n$ still sufficiently small ($\mu_1 \ll \epsilon$) - the equations for the quasi-isotropic approxiamtion should produce results in agreement with the field representation as a sum of noninteracting waves (Sect.4.2).

Analysis of (4.3.4,5) indicates that this correspondence actually takes place, but for a refinement stating that the field

$$\mathbf{E} = (\Phi_\nu \boldsymbol{\nu} + \Phi_b \mathbf{b})\, e^{ik_0\psi} = A(\Gamma_\nu \boldsymbol{\nu} + \Gamma_b \mathbf{b})\, e^{ik_0\psi} \tag{4.3.7}$$

transforms into the sum of simplified modes

$$\mathbf{E} = \tilde{\Phi}_1 \tilde{\mathbf{f}}_1 e^{ik_0\tilde{\psi}_1} + \tilde{\Phi}_2 \tilde{\mathbf{f}}_2 e^{ik_0\tilde{\psi}_2} \ , \tag{4.3.8}$$

whose amplitudes $\tilde{\Phi}_{1,2}$, polarization vectors $\tilde{\mathbf{f}}_{1,2}$, and eikonals $\psi_{1,2}$ are obtained not from the exact formulas of Sect.4.2 but from approximate expressions accounting only for the first-order terms in $\chi_{\alpha\beta}$.[10] Specifically, the simplified eikonal $\tilde{\psi}_{1,2} = \psi + \delta\psi_{1,2}$ differs from the eikonal ψ in an isotropic medium by

$$\delta\psi_{1,2} = \frac{k_0}{4\sqrt{\epsilon}} \int_{\sigma^0}^{\sigma} \left[\chi_{\nu\nu} + \chi_{bb} \pm \sqrt{(\chi_{\nu\nu} - \chi_{bb})^2 + \chi_{\nu b}\chi_{b\nu}} \right] d\sigma \ . \tag{4.3.9}$$

This expression can be obtained by perturbation theory as well [4.45].

Considering that the simplified modes (4.3.8) reconstruct the exact (unabbreviated) independent waves more simply than the solutions (4.3.7) do, the equations for the quasi-isotropic approximations can be transformed into equations for interacting normal modes by letting

$$\Phi_\nu = F_1 f_{1\nu} e^{ik_0\delta\psi_1} + F_2 f_{2\nu} e^{ik_0\delta\psi_2} \ ,$$

$$\tag{4.3.10}$$

$$\Phi_b = F_1 f_{1b} e^{ik_0\delta\psi_1} + F_2 f_{2b} e^{ik_0\delta\psi_2} \ ,$$

and considering F_1 and F_2 as unknowns [4.54]. Without displaying the equations for F_1 and F_2, we note only that they are intimately related with Budden's equations [4.13,23,24,27] and differ from the latter in two respects: they contain only the terms of first order in $\chi_{\alpha\beta}$ (Budden's

---

[10] This result is derivable by solving the equations for the quasi-isotropic approximation by Khokhlov's technique of secondary simplifaction [4.56]. While the primary shortening reduces to the conversion from the complete Maxwell equations to the equations of geometrical optics, the secondary shortening implies solving (4.3.5) asymptotically, invoking the large parameter with $\mu_1/\mu \gg 1$ because it is essentially the order of the ratio taken between the terms containing anisotropy $\chi_{\alpha\beta}$ and the terms containing the torsion $\kappa = 1/\rho_{\text{tor}} \sim 1/L$:

$$k_0 \chi_{\alpha\beta} \epsilon^{-1/2}/\kappa \ \sim k_0 \mu_1 \rho_{\text{tor}} \ \sim k_0 L \mu_1 \ \sim \mu_1/\mu \gg 1 \ .$$

equations do not employ this approximation), but instead they account for the 3-D inhomogeneity of the medium, whereas Budden's equation are stated for a one-dimensional inhomogeneous medium only.

Finally, another form of equation for the quasi-isotropic approximation deals with the vector **D** in place of **E**, expanding the fields in the anisotropic part of the tensor inverse to $\hat{\varepsilon}$. This succeeds in accounting for the longitudinal component of the vector $\mathbf{E} = \hat{\varepsilon}^{-1}\mathbf{D}$, appearing due to anisotropy. The advantagues of this approach were discussed in [4.57, 58].

4.3.3 Solution Techniques for the Quasi-Isotropic Approximation

The field evaluation in the framework of the quasi-isotropic approximation reduces to solving a system of ordinary differential equations of first order with variable coefficients. Such equations often arise in problems of mathematical physics. They are not only encountered in the theory of oscillation and wave propagation (in addition to the already-mentioned books [4.13, 25-29] we point also to review papers [4.59, 60]: they appear in quantum theory as well (for the transitions induced by adiabatic perturbations and related topics, see [4.61-63]). Analytic solutions for such equations have obtained for a few specific situations only;[11] therefore use is frequently made of approximate techniques. We list the more widely employed ones.

The **phase integral technique** provides an analytic continuation of the solution by encircling the area of interaction in the complex-variable plane. This technique is applicable to a wide class of variable coefficients, but fails to yield the field in the area of interaction itself. Applications of the phase-integral technique to the evaluation of mode conversion were exemplified in [4.13, 25, 26, 29].

The **perturbation technique** is valid only where the anisotropy $\chi_{\alpha\beta}$ is sufficiently small. It is able to describe a weak-mode conversion only [4.13, 61, 64].

The **method of canonic functions** is employed where the coefficients of the respective equation behave in qualitative similarity with the set possessing an analytic solution. The scheme can be illustrated by solutions involving the functions of a parabolic cylinder, Airy functions, etc. [4.57, 58, 62, 65-70].

Beyond these cases, the perturbation technique is applicable to (4.3.6) for a complex angle θ [4.53] or to a Riccati equation related to (4.3.5) [4.54].

4.3.4 On the Error of the Quasi-Isotropic Approximation

The accuracy of the quasi-isotropic approximation has not received a comprehensive treatment so far, although it has been attempted in a number of works [4.71, 72]. We shall confine ourselves here to some general remarks.

The main source of error in the quasi-isotropic approximation is the unaccounted term quadratic in $\mu_1 \sim \Delta n \sim \chi_{\alpha\beta}$ terms which gives rise to the accumulating error in phase

---

[11] Of the treatments related to electromagnetic theory, we note [4.64].

$$\gamma_1 \propto k_0 \int_{s_0}^{\sigma} (\Delta n)^2 d\sigma \ . \tag{4.3.11}$$

A formal application of the quasi-isotropic approximation calls for $\gamma_1 \ll 1$. But actually the inequality may often be markedly relaxed.

For one thing, the majority of applications is not concerned so much the phase corrections as the parameters of the polarization ellipse of a wave beyond the interaction area; as model computations indicate, these parameters are less sensitive to the effect of quadratic terms than the phase itself.

Furthermore, when Δn in (4.3.11) increases rapidly enough, it is the limited region, where the quasi-isotropic solution (4.3.7) converts into the sum of independent normal modes (4.3.8), that matters. A significant interaction of normal modes takes place in the interval $l_{int} = |\sigma_2 - \sigma_1|$ where these modes are practically equiphase, i.e.,

$$|\psi_1 - \psi_2| \propto k_0 \left| \int_{\sigma_1}^{\sigma_2} \Delta n d\sigma \right| \leq 1 \ . \tag{4.3.12}$$

Let $\overline{\Delta n}$ stand for the mean value of Δn across the interval (σ_1, σ_2). Then (4.3.12) leads to the following estimate for the length of interaction:

$$l_{int} \sim (k_0 |\overline{\Delta n}|)^{-1} \ .$$

Over this length the quadratic error (4.3.11) turns out to be a small quantity, namely

$$\gamma_1 \sim k_0 |\overline{\Delta n}|^2 \, l_{int} \sim |\overline{\Delta n}| \ll 1 \ .$$

It goes without saying that on coming from the quasi-isotropic field to the sum of normal modes one has to account for an error γ_2 in representing the field by (4.3.8). By virtue of (4.2.55) this error is defiend by

$$\gamma_2 \sim (k_0 L |\Delta n|)^{-1} \ . \tag{4.3.13}$$

This quantity, in contrast to γ_1, increases inversely with Δn. Therefore a problem appears for the optimum matching of two solutions so that the total error $\gamma_1 + \gamma_2$ is the least [4.71, 72]. The place of the matching is sought with a particular dependence $\Delta n(\sigma)$ in view.

4.3.5 Applications of the Quasi-Isotropic Approximation

Weak anisotropy is characteristic primarily of high-frequency electromagnetic waves in magnetoactive plasma when the gyrofrequency $\omega_e = eH_0/mc$ is small compared with ω. These conditions are not infrequently realized both in laboratory and in natural plasma (the earth's ionosphere, the solar corona, and free space plasma).

256

A weak depolarization of the wave in a magnetoactive plasma in the simultaneous presence of the Faraday and Cotton-Mouton effects[12] has been outlined in [4.53], and the role of this depolarization in ionospheric scattering of VHF radiowaves was studied in [4.73]. The strong depolarization of the field in the region of quasi-lateral propagation was investigated in [4.58,68], the results of which agree well with the studies employing the phase-integral technique [4.25,29]. Numerical results referring to the wave interaction in the region of neutral magnetic field were given in [4.74].

The quasi-isotropic approximation can be used to describe the *limiting polarization* of a wave radiated by a magnetized plasma into a vacuum [4.71,72] (the problem statement for the limiting polarization was dealt with in [4.13,25,26,28,70]), and to define field depolarization in a random, inhomogeneous, free-space plasma [4.75]. Note also that the dynamo-optic effects due to the wave polarization induced by a nonuniform motion of the plasma can also be described by the quasi-isotropic approximation [4.76-78] because the nonuniform motion of the plasma gives rise to a weak anisotropy (Sect. 4.5).

Finally, we point out that the quasi-isotropic approximation is a convenient tool for studying the weak anisotropy effects in elasticity theory [4.79] and in optics (natural optical activity and dichroism, the anisotropy of absorption).

4.3.6 The Quasi-Degenerate Approximation of Geometrical Optics

In contrast to weakly anisotropic media where the refractive-index difference Δn is small in all directions of travel, in strongly anisotropic media Δn can appear as a small quantity in the vicinity of definite (degenerate) directions. In these directions the eikonal equation (4.3.4) allows for close, or multiple, roots $p_1 = n_1 l$ and $p_2 = n_2 l$ or, which is the same, close eigenvalues $\lambda_1 \simeq \lambda_2$. It is in the vicinity of the degenerate directions that the interaction of the normal modes propagating with about the same phase speeds is possible. Such an interaction can be described by the *quasi-degenerate approximation* of geometrical optics suggested in [4.80] (a similar approach was adopted in [4.81]).

The fundamental computational procedure of the quasi-degenerate approximation is to transfer small terms of the order $\lambda_2(p_1) \simeq (\partial\lambda_2/\partial p)\Delta p$, where $\Delta p = p_1 - p_2$, form the equation of zero approximation to that of first approximation, thus transforming the almost degenerate system for E_0 into a strictly degenerate one. The procedure can be realized, for instance, in the following way.

As will be recalled, the linear form $\hat{q}E_0$ with a Hermitian matrix $[q_{\alpha\beta}]$ can be represented as [4.81]

---

[12] The Faraday effect (the rotation of the plane of polarization of a wave) is linear with respect to the magnetic field H_0. The Cotton-Mouton effect (the noncoincidence of the phase speeds of linearly polarized waves traveling transverse to H_0) is of second order in H_0. Both effects will apply simultaneously when the direction of beam travel smoothly varies from quasi-parallel to quasi-transverse.

$$\hat{q}E = \sum_{j=1}^{3} (f_j \times f_j^*) \, \lambda_j(p) E_0 = 0 \; , \tag{4.3.14}$$

where $(f_j \times f_j^*)$ stands for the direct product of the eigenvectors f_j and f_j^*.[13] For a wave of the first type, $\lambda_1(p_1) = 0$ and $\lambda_2(p_1)$ is close to zero. Replacing in the equation of zero approximation the exact matrix $\hat{q}$, possesing, generally speaking, different (though close) eigenvalues $\lambda_2(p_1) \neq \lambda_1(p_1) = 0$, e.g.,

$$\hat{q}_0 = (f_3 \times f_3^*) \, \lambda_3(p_1) E_0 \; , \quad \lambda_{20} = \lambda_1 = 0 \; . \tag{4.3.15}$$

We have to account for the small difference

$$(\hat{q} - \hat{q}_0)E_0 = (f_2 \times f_2^*)\lambda_2(p_1)E_0 \simeq f_2 \times f_2^* \frac{\partial \lambda_2}{\partial p_2} \Delta p E_0$$

in the equation of the *first* approximation.[14]

Two consistency conditions for the first approximation equations will set up a system of two first-order equations for the amplitudes of interacting waves. As with the quasi-isotropic approximation, the solution obtained within the area of interaction has to be matched with the normal modes outside the area.

On approaching weakly anisotropic media, the quasi-degenerate approximation converts into the quasi-isotropic one. Possible applications of the quasi-degenerate approximation were discussed in [4.80].

Note in conclusion that the outliend types of linear interactions do not exhaust all the possible situations; specifically, they fail to describe mode conversion in the vicinity of singularities (poles) n_j and the conversion of longitudinal modes into transverse modes. These topics were dealt with in [4.59, 60].

4.4 Equations of Geometrical Optics for Nonharmonic Electromagnetic Waves in the General Case of Inhomogeneous and Nonstationary Dispersive Media

4.4.1 The Maxwell Equations in Inhomogeneous and Nonstationary Media of Temporal and Spatial Dispersion

In Sect.2.7 we restricted our attention to scallar nonmonochromatic waves in homogeneous, stationary media with account made only of temporal dis-

---

[13] The direct product of two vectors $a \times b$ is essentially the tensor of elements $a_\alpha b_\beta$.

[14] A singular matrix $\hat{q}_0$ close to the almost singular matrix $\hat{q}$ can be composed in a variety of other ways.

persion. Sections 4.1,2 were devoted to harmonic electromagnetic fields. Now we discuss the more general case of nonmonochromatic electromagnetic waves in inhomogeneous and nonstationary media exhibiting both temporal and spatial dispersion. The simultaneous presence of all these features brings about some qualitatively novel phenomena, say, the appearance of anisotropy in an isotropic medium possessing simultaneously some inhomogeneity and spatial dispersion, or the nonconservation of electromagnetic energy because of the medium being nonstationary.

Efforts to construct a possibly general geometrical theory have been attempted in many studies, e.g., [4.31,83,84]. A systematic evaluation of the problem, taking into consideration all the aforementioned features, has been performed in [4.85], to the lines of which we will adhere in what follows. Also in the sections to come we will substantially borrow from the surveys [4.86,87].

For nonmonochromatic waves the Maxwell equations become

$$\mathrm{curl}\mathbf{H} = \frac{1}{c}\frac{\partial \mathbf{D}}{\partial t} \, ,$$

$$\mathrm{curl}\mathbf{E} = -\frac{1}{c}\frac{\partial \mathbf{B}}{\partial t} \, . \tag{4.4.1}$$

A fairly general case of relationship between the electric and magnetic flux densities and the respective field intensities is described by the constitutive equations

$$D_\alpha(t,\mathbf{r}) = \int_{-\infty}^{t} dt' \int d\mathbf{r}' \tilde{\epsilon}_{\alpha\beta}(t{-}t',\mathbf{r}{-}\mathbf{r}';t',\mathbf{r}') E_\beta(t',\mathbf{r}') \, , \quad \mathbf{B} = \mathbf{H} \, , \tag{4.4.2}$$

which assume that the electric and magnetic properties of the medium are defined by the permittivity tensor $\tilde{\epsilon}_{\alpha\beta}$.[15] The dependence of the kernel $\tilde{\epsilon}_{\alpha\beta}$ in (4.4.2a) on the time difference $\Delta t = t{-}t'$ describes the temporal dispersion, whereas that on $\Delta \mathbf{r} = \mathbf{r}{-}\mathbf{r}'$ corresponds to the spatial dispersion of the medium: the electric displacement $\mathbf{D}(t,\mathbf{r})$ at time t at point $\mathbf{r}$ depends not only upon $\mathbf{E}(t,\mathbf{r})$ but also upon the electric field intensity $\mathbf{E}(t',\mathbf{r}')$ at previous instants of time $t' < t$ taken at neighboring points $\mathbf{r}'$.

An explicit dependence of $\tilde{\epsilon}_{\alpha\beta}$ on t and $\mathbf{r}$ reflects the nonstationarity and the inhomogeneity of the medium. The permittivity tensor $\tilde{\mathrm{E}}_{\alpha\beta}$ is found, generally speaking, from considering the microprocesses ocurring in the material medium under the electromagnetic field. Henceforth, unless pointed out otherwise, we shall assume this tensor to be specified.

---

[15] Recall that for variable fields, the introduction of the magnetic permeability tensor is hampered by ambiguity in separating electric and magnetic properties of the material medium in the presence of dispersion; therefore it pays to include in the polarization $\mathbf{P} = (\mathbf{D}{-}\mathbf{E})4\pi$ all the currents induced in the medium [4.88,89].

Space-time geometrical optics has the field vectors represented as

$$E(t,r) = \tilde{E}(t,r)e^{i\phi(t,r)} \; ,$$

$$D(t,r) = \tilde{D}(t,r)e^{i\phi(t,r)} \; ,$$

(4.4.3)

etc., where $\tilde{E}$, $\tilde{D}$, etc. are slowly varying amplitudes.

Now the necessary conditions (2.7.11) of space-time geometrical-optics applicability should be augmented by another inequality $\rho_0/L \ll 1$, where ρ_0 is the radius of spatial dispersion (dissipation range), i.e., the scale over which the kernel $\tilde{\epsilon}_{\alpha\beta}(\Delta t, \Delta r, t, r)$ varies with Δr significantly. When $\rho_0 \ll L$ is met, the wave $E = \tilde{E}\exp(i\phi)$ may be deemed plane within the region $\Delta r \leq \rho_0$ essential for the integration in (4.4.2). Therefore, taking into account spatial dispersion leads to the inequality constraint

$$\mu = \max(\bar{\tau}/T, \tau_0/T, \lambda/L, \rho_0/L) \ll 1 \; ,$$

(4.4.4)

where $\bar{\tau} \sim 1/\bar{\omega}$, $\lambda \sim 1/k$ are the average period and wavelength of the field, respectively, and T, L are the scales defining the nonstationarity and non-homogeneity of the field variables $\tilde{E}$, $\omega = -\partial\phi/\partial t$, $k = \nabla\phi$, and the parameters of the material medium.

4.4.2 The Constitutive Equation in Differential Form

By analogy with (2.7.12) we develop the amplitude $E(t',r')$ and phase $\phi(t',r')$ into Taylor series in powers of $\Delta t = t - t'$ and $\Delta r = r - r'$:

$$E_\beta(t',r') = \exp[i\phi(t,r) + i\omega\Delta t - ik\Delta r]\left\{\tilde{E}_\beta(t,r) - \frac{\partial\tilde{E}_\beta}{\partial t}\Delta t - \frac{\partial\tilde{E}_\beta}{\partial x_j}\Delta x_j\right.$$

$$\left. + \frac{i\tilde{E}_\beta}{2}\left[\frac{\partial^2\phi}{\partial t^2}(\Delta t)^2 + 2\frac{\partial^2\phi}{\partial t\partial x_j}\Delta t\Delta x_j + \frac{\partial^2\phi}{\partial x_j\partial x_m}\Delta x_j\Delta x_m\right] + ...\right\}$$

(4.4.5)

(the summation here and below is taken over the repeated indices).

On substituting (4.4.5) into (4.4.2) we have to integrate the kernel $\tilde{\epsilon}_{\alpha\beta}$ with its exponential factor $\exp(i\omega\Delta t - ik\Delta r)$ and powers of Δt and Δx_j. Similar to (2.7.14), the integrals of the kind can be obtained as derivatives with respect to ω and k from the quantity

$$\epsilon_{\alpha\beta}(\omega, k, t, r) \equiv \int_{-\infty}^{t} dt \int dr' \tilde{\epsilon}_{\alpha\beta}(\Delta t, \Delta r, t, r)e^{i\omega\Delta t - ik\Delta r} \; ,$$

(4.4.6)

called the *complex permittivity tensor*.[16] For the electric displacement D [4.85] this leads to the expansion

$$D_\alpha(t,\mathbf{r}) = e^{i\phi(t,\mathbf{r})}[\epsilon_{\alpha\beta}\tilde{E}_\beta + i\hat{\Gamma}_{\alpha\beta}\tilde{E}_\beta + O(\mu^2)] \,, \qquad (4.4.7)$$

where $\hat{\Gamma}_{\alpha\beta}$ is the differential operator defined by

$$\hat{\Gamma}_{\alpha\beta}\tilde{E}_\beta = \frac{\partial\epsilon_{\alpha\beta}}{\partial\omega}\frac{\partial\tilde{E}_\beta}{\partial t} - \frac{\partial\epsilon_{\alpha\beta}}{\partial k_j}\frac{\partial\tilde{E}_\beta}{\partial x_j} \qquad (4.4.8)$$

$$- \frac{\tilde{E}_\beta}{2}\left[\frac{\partial^2\phi}{\partial t^2}\frac{\partial^2\epsilon_{\alpha\beta}}{\partial\omega^2} - 2\frac{\partial^2\phi}{\partial t\partial x_j}\frac{\partial^2\epsilon_{\alpha\beta}}{\partial\omega\partial k_j} + \frac{\partial^2\phi}{\partial x_j\partial x_m}\frac{\partial^2\epsilon_{\alpha\beta}}{\partial k_j\partial k_m}\right] \,.$$

Subject to (4.4.4), all the terms joined under $\hat{\Gamma}_{\alpha\beta}\tilde{E}_\beta$ have a first order of smallness in μ, $\epsilon_{\alpha\beta}\tilde{E}_\beta$ being the leading term.

The above procedure helps us to replace the integral relationship (4.4.2) by the approximate differential expression (4.4.7). When dispersion, both temporal and spatial, is absent, $\Gamma_{\alpha\beta}$ vanishes and the connection between $\mathbf{D}$ and $\mathbf{E}$ reduces to the algebraic form $D_\alpha = \epsilon_{\alpha\beta}(t,\mathbf{r})E_\beta$.

4.4.3 Equations for the Fields in Zeroth and First Approximations

Expanding the amplitudes $\tilde{E}$, $\tilde{H}$, and $\tilde{D}$ in the small parameter μ as, e.g.,

$$E = E^0 + iE^1 + i^2 E^2 + \dots \,,$$

we arrive, first, at the equations in zeroth approximation (henceforth we delete the tilde):

$$(\mathbf{k}\times\mathbf{H}^0)_\alpha + (\omega/c)\epsilon_{\alpha\beta}E_\beta{}^0 = 0 \,,$$
$$(4.4.9)$$
$$(\mathbf{k}\times\mathbf{E}^0)_\alpha + (\omega/c)H_\alpha^0 = 0 \,;$$

and then at the equations in first approximation:

$$(\mathbf{k}\times\mathbf{H}^1)_\alpha + (\omega/c)\epsilon_{\alpha\beta}E_\beta{}^1$$
$$= -\operatorname{curl}_\alpha H^0 + \frac{1}{c}\frac{\partial}{\partial t}(\epsilon_{\alpha\beta}E_\beta{}^0) - i\frac{\omega}{c}\epsilon_{\alpha\beta}{}^a E_\beta{}^0 - \frac{\omega}{c}\hat{\Gamma}_{\alpha\beta}E_\beta{}^0 \equiv \tilde{X}_\alpha \,,$$
$$(4.4.10)$$

$$(\mathbf{k}\times\mathbf{E}^1)_\alpha - \frac{\omega}{c}H_\alpha{}^1 = -\operatorname{curl}_\alpha E^0 - \frac{1}{c}\frac{\partial H_\alpha{}^0}{\partial t} \equiv \tilde{Y}_\alpha \,,$$

etc. As usual, $\mathbf{k} = \nabla\phi$, $\omega = -\partial\phi/\partial t$. We isolated from the tensor $\epsilon_{\alpha\beta}(\omega,\mathbf{k};t,\mathbf{r})$ its anti-Hermitian part $\epsilon_{\alpha\beta}^a = \frac{1}{2}(\epsilon_{\alpha\beta}-\epsilon_{\alpha\beta}{}^*)$, assuming that it is small compared with the Hermitian part $\epsilon_{\alpha\beta}^h = \frac{1}{2}(\epsilon_{\alpha\beta}+\epsilon_{\alpha\beta}{}^*)$. Hence, the quantities of order μ in the inequality (4.3.4) should be augmented by another parameter

---

[16] The dependence of this tensor on $\mathbf{k} = \nabla\phi$ corresponds to the spatial dispersion of the medium similar to how that on $\omega = -\partial\phi/\partial t$ is related to the temporal dispersion.

$$|\epsilon^{a}_{\alpha\beta}| \sim \mu|\epsilon^{h}_{\alpha\beta}| \ll |\epsilon^{h}_{\alpha\beta}| \ .$$

Similar to the procedure of Sects.3.1-3, it is convenient to eliminate H^0) and H^1 from (4.4.9, 10) to obtain for E^0 and E^1 the following equations extending (4.2.2, 23) to the case of non-monochromatic waves:

$$\hat{Q}E^0 = 0, \tag{4.4.11}$$

$$\hat{Q}E^1 = (\omega/c)\tilde{X} + k \times \tilde{Y} \equiv \tilde{Z} \ , \tag{4.4.12}$$

where $\hat{Q}$ is the matrix with the components

$$Q_{\alpha\beta} = k^2 \delta_{\alpha\beta} - k_\alpha k_\beta - (\omega/c)^2 \epsilon_{\alpha\beta} \ . \tag{4.4.13}$$

4.4.4 The Eikonal Equation. Space–Time Rays

Setting the determinant of the system (4.4.11) equal to zero yields the equation of the eikonal (the local dispersion equation)

$$\mathcal{H} = \det[k^2 \delta_{\alpha\beta} - k_\alpha k_\beta - (\omega/c)^2 \epsilon_{\alpha\beta}] = 0 \ , \tag{4.4.14}$$

which describes the behavior of phase $\phi = \phi(t, r)$. In stationary problems where $\epsilon_{\alpha\beta}$ does not depend explicitly on time, and L is constant, the phase is

$$\phi = \frac{\omega}{c} \psi(\mathbf{r}) - \omega t \ ,$$

where $\psi(\mathbf{r})$ is the eikonal which we dealt with in considering monochromatic waves. correspondingly,

$$k = \nabla\phi = (\omega/c)\nabla\psi = \frac{\omega}{c}\, p \ ,$$

so that (4.4.14) is equivalent to (4.2.4) in the stationary case.

Equation (4.4.14) can take a variety of forms. Specifically it can be factored like (4.2.8) or be represented as a product of the matrix $\hat{Q}$ eigenvalues, viz., $\mathcal{H} = \lambda_1 \lambda_2 \lambda_3 = 0$. All three, rather than two, of the eigenvalues can vanish on account of spatial dispersion, which corresponds to launching longitudinal waves (Sect.4.4.9).

In contrast to the formulation of the scalar and stationary problems, adopted in Sect.2.7, the Hamiltonian $\mathcal{H}(\omega, k; t, r)$ here depends on time; and, as a consequence, on a space-time ray $\{t(\varsigma), r(\varsigma)\}$ satisfying (2.7.20):

$$\frac{d\omega}{d\varsigma} = \frac{\partial\mathcal{H}}{\partial t} \neq 0 \ ; \tag{4.4.15}$$

i.e., the frequency ω does not remain constant. The associated phenomena – energy and wave-packet length variation – will be discussed in Sect.4.6.

One more difference from the scalar-problem statement is seen in that the vector field $\mathcal{H}$ becomes zero at several surfaces (envelopes) in the space $\{\omega,\mathbf{k}\}$ each associated with a different type of normal mode [these envelopes occur where one or another eigenvalue $\lambda_j(\omega,\mathbf{k})$ approaches to zero].

To each of the normal modes there corresponds its own family of rays. We write the Hamiltonian $\mathcal{H}$ for one of the modes in the form analogous to (4.2.9):

$$\mathcal{H}_j(\omega,\mathbf{k};t,\mathbf{r}) = \frac{1}{2}\left[\mathbf{k}^2 - \frac{\omega^2}{c^2}\,n_j^2(\omega,\mathbf{l};t,\mathbf{r})\right] = 0\ , \tag{4.4.16}$$

where $\mathbf{l} = \mathbf{k}/k$ points in the direction of the wave vector, and n_j is the refractive index of the jth normal mode. On substituting (4.4.16) into (4.4.15) and eliminating ζ, we arrive at the ray equations used in [4.86, 87, 90, 91]; this extends (2.7.23) over anisotropic nonstationary media:

$$\frac{d\mathbf{r}}{dt} = 2c^2\left[\mathbf{k} - \frac{1}{2}\frac{\omega^2}{c^2}\frac{\partial n^2}{\partial \mathbf{k}}\right] \Big/ \frac{\partial \omega^2 n^2}{\partial \omega} = \mathbf{g}\ ,$$

$$\frac{d\mathbf{k}}{dt} = \omega^2\frac{\partial n^2}{\partial \mathbf{r}} \Big/ \frac{\partial \omega^2 n^2}{\partial \omega} = \omega\left(\frac{\partial \omega n}{\partial \omega}\right)^{-1}\frac{\partial n}{\partial \mathbf{r}}\ , \tag{4.4.17}$$

$$\frac{d\omega}{dt} = -\omega^2\frac{\partial n^2}{\partial t} \Big/ \frac{\partial \omega^2 n^2}{\partial \omega} = -\omega\left(\frac{\partial \omega n}{\partial \omega}\right)^{-1}\frac{\partial n}{\partial t}\ .$$

In a stationary medium without spatial dispersion ($\partial n/\partial \mathbf{k} = 0$, $\partial n/\partial t = 0$) these equations reduce to (2.7.22).

4.4.5 The Transfer Equation for Independent Normal Modes in an Anisotropic Medium

As in Sect. 4.2, let $\mathbf{E}^0 = \Phi\mathbf{f}$, where $\mathbf{f}$ is the polarization vector associated with one of the normal modes. The consistency condition for the first-order approximation (4.4.7) has a form similar to (4.2.24):

$$\check{\mathbf{X}}\mathbf{E}^* - \tilde{\mathbf{Y}}\mathbf{H}^* = 0\ , \tag{4.4.18}$$

where, according to (4.4.9)

$$\mathbf{H} = (c/\omega)\mathbf{k}\times\mathbf{E} \equiv \mathbf{h}\Phi\ , \quad \mathbf{h} = (c/\omega)\,\mathbf{k}\times\mathbf{f} = \mathbf{p}\times\mathbf{f}\ ;$$

the superscripts ^{0} are dropped for shortness. Equating the real part of (4.4.18) to zero we obtain an equation for the modulus of the amplitude Φ, the imaginary part of (4.4.18) defining the behavior of its argument $\delta = \arg\Phi$.

The equation for Φ can be written in the form of a conservation law by introducing the electromagnetic energy density w and the Poynting vector S in the manner used for homogeneous and stationary media [4.14]:

$$w = \frac{1}{16\pi}\left\{\frac{\partial\omega\epsilon_{\alpha\beta}}{\partial\omega}E_\alpha E_\beta^* + H_\alpha H_\alpha^*\right\}$$

$$= \frac{1}{16\pi}\left[\frac{\partial\omega^2\epsilon_{\alpha\beta}}{\omega\partial\omega}f_\alpha f_\beta^*\right]|\Phi|^2 \equiv \frac{\eta}{16\pi}|\Phi|^2 , \qquad (4.4.19)$$

$$S = \frac{c}{16\pi}(E\times H^* + E^*\times H) - \frac{\omega}{16\pi}\frac{\partial\epsilon_{\alpha\beta}}{\partial k}E_\alpha E_b^*$$

$$= \frac{|\Phi|^2}{16\pi}\left[c(f\times h^* + f^*\times h) - \omega\frac{\partial\epsilon_{\alpha\beta}}{\partial k}f_\alpha f_\beta^*\right] . \qquad (4.4.20)$$

The last term in (4.4.20) containing the derivative $\partial\epsilon_{\alpha\beta}/\partial k$ describes the energy flow associated with the coherent motion of charges [4.92] (also [4.14, 93]).

The frequency ω and the wave vector k in (4.4.19, 20) are functions of time and coordinates, $\omega = \omega(t,r)$ and $k = k(t,r)$, so one may verify directly that equating the real part of (4.4.18) to zero is equivalent to the following equation for $|\Phi|^2$ [4.85, 86]:

$$\frac{\partial w}{\partial t} + \mathrm{div}S = -\left[2\chi + \frac{1}{\eta}f_\alpha^* f_\beta \frac{\partial\epsilon_{\alpha\beta}}{\partial t}\right]w , \qquad \text{where} \qquad (4.4.21)$$

$$2\chi = -\frac{\omega}{\eta}\left[2i\epsilon_{\alpha\beta}^a + \left(\frac{\partial^2\epsilon_{\alpha\beta}}{\partial\omega\partial t} - \frac{\partial^2\epsilon_{\alpha\beta}}{\partial x_j\partial k_j}\right)f_\alpha^* f_\beta\right] . \qquad (4.4.22)$$

In turn, the equation for $\delta = \mathrm{arg}\Phi$ can be represented in the form

$$w\frac{\partial\delta}{\partial t} + S\nabla\delta = -wN , \qquad \text{where} \qquad (4.4.23)$$

$$N = \frac{1}{\eta}\mathrm{Im}\left\{cf\cdot\mathrm{curl}h^* + ch^*\cdot\mathrm{curl}f + f_\alpha^*\frac{\partial f_\beta}{\partial\omega}\frac{\partial\omega\epsilon_{\alpha\beta}}{\partial\omega} - \omega f_\alpha^*\frac{\partial f_\beta}{\partial x_j}\frac{\partial\epsilon_{\alpha\beta}}{\partial k_j}\right\} .$$

4.4.6 The Group Velocity Theorem

For a quasi-plane and quasi-monochromatic wave, the Poynting vector S and the density of electromagnetic energy w are related by [4.14]

$$\mathbf{S} = w\mathbf{g} \, , \tag{4.4.24}$$

where $\mathbf{g} = d\mathbf{r}/dt$ is the group velocity of the wave. This important relationship is allied with the names of Hamilton, Stokes, Rayleigh, Sommerfeld, and Brillouin. The group-velocity theorem can be extended into four dimensions [4.94, 95].

In the framework of the geometrical theory outlined in this text, the relationship $\mathbf{S} = w\mathbf{g}$ can be deduced invoking the eigenvalues of the matrix $\hat{Q}$ defined by (4.4.13):

$$\hat{Q}\mathbf{E} = \lambda\mathbf{E} \, . \tag{4.4.25}$$

Differentiating (4.4.25) first with respect to $\mathbf{k}$ and then to ω, premultiplying by $\mathbf{E}^*$ and letting $\lambda = 0$, we obtain

$$\mathbf{E}^* \frac{\partial\hat{Q}}{\partial\mathbf{k}}\mathbf{E} = \frac{\partial\lambda}{\partial\mathbf{k}}|\mathbf{E}|^2 \, ,$$

$$\tag{4.4.26}$$

$$\mathbf{E}^* \frac{\partial\hat{Q}}{\partial\omega}\mathbf{E} = \frac{\partial\lambda}{\partial\omega}|\mathbf{E}|^2 \, .$$

According to (4.4.13) we find for the derivatives of $Q_{\alpha\beta}$

$$\frac{\partial Q_{\alpha\beta}}{\partial k_l} = 2k_l\delta_{\alpha\beta} - k_\alpha\delta_{\beta l} - k_\beta\delta_{\alpha l} - \frac{\omega^2}{c^2}\frac{\partial\epsilon_{\alpha\beta}}{\partial k_l} \, ,$$

$$\frac{\partial Q_{\alpha\beta}}{\partial\omega} = \frac{1}{c^2}\frac{\partial\omega^2\epsilon_{\alpha\beta}}{\partial\omega} \, ,$$

and then express the left-hand parts of (4.4.26) through the $\mathbf{S}$ and w of (4.4.19, 20):

$$\mathbf{E}^* \frac{\partial\hat{Q}}{\partial\mathbf{k}}\mathbf{E} = \frac{16\pi c^2}{\omega}\mathbf{S} = \frac{\partial\lambda}{\partial\mathbf{k}}|\mathbf{E}|^2 \, ,$$

$$\tag{4.4.27}$$

$$\mathbf{E}^* \frac{\partial\hat{Q}}{\partial\omega}\mathbf{E} = -\frac{16\pi c^2}{\omega}w = \frac{\partial\lambda}{\partial\omega}|\mathbf{E}|^2 \, .$$

Finally we arrive at (4.4.24)

$$\frac{\mathbf{S}}{w} = -\left(\frac{\partial\lambda}{\partial\mathbf{k}}\right) \Big/ \left(\frac{\partial\lambda}{\partial\omega}\right) = -\left(\frac{\partial\mathscr{H}}{\partial\mathbf{k}}\right) \Big/ \left(\frac{\partial\mathscr{H}}{\partial\omega}\right) = \mathbf{g} \, .$$

A number of other relationships between the bilinear characteristics of a wave can be derived in a similar manner [4.94, 95].

It should be emphasized that provided the group velocity vector is defined by the kinematic relation $\mathbf{g} = d\mathbf{r}/dt$, as above, the expression $S = gw$ arises in the framework of the ray approximation only; that is, it will be valid only for almost-plane and almost-monochromatic waves. A more general statement for pulse and modulated-wave propagation without invoking the kinematic definition of the vector $\mathbf{g}$ has been considered in [4.96].

4.4.7 Integrating the Transfer Equation Along Space–Time Rays

Let us transform (4.4.21) observing (4.4.24), the equation[17]

$$\frac{1}{\eta} f_\alpha^* f_\beta \frac{\partial \epsilon_{\alpha\beta}}{\partial t} = - \frac{d\ln\omega}{dt} , \tag{4.4.28}$$

and the Liouville relation $\operatorname{div}\mathbf{g} = d\ln j(t)/dt$ to get

$$\frac{dw}{dt} + w \frac{d}{dt}\ln j(t) = \left(-2\chi + \frac{d\ln\omega}{dt}\right) w , \tag{4.4.29}$$

where the Jacobian $j(t)$ is defined by (2.7.42). This equation yields the energy density on the ray

$$w = w(t^0) \frac{j(t^0)\omega(t)}{j(t)\omega(t^0)} \exp\left(-2 \int_{t^0}^{t} \chi dt\right), \tag{4.4.30}$$

and the amplitude-modulus variation

$$|\Phi| = \sqrt{\frac{16\pi w}{\eta}} = |\Phi^0| \sqrt{\frac{\eta^0 j(t^0)\omega(t)}{\eta j(t)\omega(t^0)}} \exp\left(-\int_{t^0}^{t} \chi dt\right). \tag{4.4.31}$$

By virtue of (4.4.24), Eq.(4.4.23) for the argument δ becomes

$$\frac{\partial\delta}{\partial t} + \mathbf{g}\nabla\delta = \frac{d\delta}{dt} = - N , \quad \text{whence} \quad \delta = \delta^0 - \int_{t^0}^{t} N dt . \tag{4.4.32}$$

---

[17] In view of the Hamilton equations (4.4.15), $d\omega/dt = -(\partial\lambda/\partial t)/(\partial\lambda/\partial\omega)$. Differentiating (4.4.15) with respect to t we get

$$\frac{\partial\lambda}{\partial t}|\mathbf{E}|^2 = \mathbf{E}^* \frac{\partial Q}{\partial t}\mathbf{E} = - \frac{\omega^2}{c^2} f_\alpha^* f_\beta \frac{\partial\epsilon_{\alpha\beta}}{\partial t}|\mathbf{E}|^2 .$$

Inserting $\partial\lambda/\partial\omega$ from (4.4.27) we arrive at (4.4.28).

266

For a stationary formulation with no spatial dispersion, the solutions (4.4.31,32) convert, respectively, to (4.2.34,36); the proof is essentially the same as that in Sect.2.7).

4.4.8 Transverse Modes in an Isotropic Medium

The above discussion referred to anisotropic media where all eigenvalues λ_j are different, i.e., where polarization degeneracy is absent. Consider now the behavior of waves in an isotropic medium[18] whose permittivity tensor has the form [4.88, 89]

$$\epsilon_{\alpha\beta} = \epsilon_\perp(\omega,\mathbf{k};t,\mathbf{r})\left[\delta_{\alpha\beta} - \frac{k_\alpha k_\beta}{k^2}\right] + \epsilon_\parallel(\omega,\mathbf{k},t,\mathbf{r})\frac{k_\alpha k_\beta}{k^2} \,, \tag{4.4.33}$$

where $\epsilon_\perp$ is the transverse permittivity and $\epsilon_\parallel$ the longitudinal one.

In an isotropic medium the Hamiltonian (4.4.14) factors out:

$$\mathcal{H} = \lambda_1\lambda_2\lambda_3 = (k^2 - \epsilon_\perp \omega^2/c^2)^2(-\epsilon_\parallel\omega^2/c^2) = 0 \,, \tag{4.4.34}$$

where $\lambda_1 = \lambda_2 = -(k^2 - \epsilon_\perp\omega^2/c^2)$ are the multiple eigenvalues associated with polarization-degenerated transverse modes, and $\lambda_3 = -\epsilon_\parallel\omega^2/c^2$ corresponds to longitudinal oscillations.

Similar to the monochromatic problem, the leading terms of the electric and magnetic fields of a transverse wave may be represented in the form (4.1.13, 14), and the consistency conditions $\boldsymbol{\nu}\cdot\mathbf{Z} = 0$ and $\mathbf{b}\cdot\mathbf{Z} = 0$ of the equations of the first approximation lead to the following equations for the amplitudes Φ_ν and Φ_b:

$$\frac{d\Phi_b}{dt} - \kappa g_\perp \Phi_\nu + \alpha\Phi_b = 0 \,,$$

$$\tag{4.4.35}$$

$$\frac{d\Phi_\nu}{dt} + \alpha\Phi_\nu + \kappa g_\perp \Phi_b = 0 \,,$$

where κ is the ray torsion and $g_\perp$ is the group velocity of the transverse mode

$$g_\perp = c\left[2\sqrt{\epsilon_\perp'} - \frac{\omega}{c}\frac{\partial\epsilon_\perp}{\partial k}\right]1 / \eta_\perp \,,$$

$$\eta_\perp = \frac{\partial\omega^2\epsilon_\perp'}{\omega\partial\omega} \,,$$

---

[18] Describing waves in an isotropic medium one should keep in mind that the spatial dispersion due to inhomogeneities or nonuniform motion of the medium brings about weak anisotropy (both effects are discussed in Sect.4.5). Therefore, specifying a permittivity tensor in the form (4.4.33) implies that these effects are *negligibly small*.

$$\alpha = \frac{1}{2}\frac{d}{dt}\ln\left[\frac{\eta_\perp \, j(t)}{\omega}\right] + \chi_\perp \, , \tag{4.4.36}$$

$$\chi_\perp = \frac{1}{\eta_\perp}\left[2\omega\epsilon_\perp{}'' + \omega\left(\frac{\partial^2 \epsilon_\perp{}'}{\partial x_j \partial k_j} - \frac{\partial^2 \epsilon_\perp{}'}{\partial\omega\partial t}\right)\right] \, .$$

Here $\epsilon_\perp{}''$ is the imaginary part of the complex permittivity $\epsilon_\perp = \epsilon_\perp{}' + i\epsilon_\perp{}''$. In deriving (4.4.35) use was made of the equation $d\ln\omega/dt = -\eta_\perp^{-1}\partial\epsilon_\perp{}'/\partial t$, which can be proved similar to (4.4.28). In the absence of spatial dispersion when $\partial\epsilon_\perp{}'/\partial k = 0$, $g_\perp$ converts to (2.7.23).

For the field amplitude $A = |\mathbf{E}| = (|\Phi_\nu|^2 + |\Phi_b|^2)^{1/2}$, from (4.4.35) follows

$$dA/dt + \alpha A = 0, \tag{4.4.37}$$

the solution to which is

$$A = A^0\sqrt{\frac{\eta_\perp^0 \, j(t^0)\omega(t)}{\eta_\perp \, j(t)\omega(t^0)}}\exp\left(-\int_{t^0}^{t}\chi_\perp \, dt\right). \tag{4.4.38}$$

Equation (4.4.37) may also be represented as a conservation law

$$\frac{\partial w}{\partial t} + \mathrm{div}\mathbf{S} = \left(-2\chi_\perp - \frac{d\ln\omega}{dt}\right)w \, , \tag{4.4.39}$$

where $w = \eta_\perp |\mathbf{E}|^2/16\pi$ is the density of energy and

$$\mathbf{S} = \frac{c}{16\pi}\left[\mathbf{E}\times\mathbf{H}^* + \mathbf{E}^*\times\mathbf{H} - \frac{\omega}{c}\frac{\partial\epsilon_\perp}{\partial k}|\mathbf{E}|^2\right] = g_\perp w$$

is the density of energy flux (Poynting vector).

Setting in (4.4.35) $\Phi_\nu = \Phi\cos\theta$, $\Phi_b = \Phi\sin\theta$, we obtain for Φ the same equation as for A: $d\Phi/dt + \alpha\Phi = 0$. Consequently, the ratio Φ/A, defining the wave's polarization ellipse, remains unchanged over the entire ray and equals the initial value Φ^0/A^0. On the other hand, the angle $\theta = \arctan(\Phi_b/\Phi_\nu)$ describing the orientation of the polarization ellipse satisfies the equation

$$d\theta/dt = g_\perp \kappa \, . \tag{4.4.40}$$

In the stationary problem it transforms into Rytov's equation (4.1.26) because $g_\perp dt = d\sigma$.

4.4.9 Longitudinal Waves in an Isotropic Medium

When the longitudinal permittivity vanishes

$$\epsilon_{\parallel}(\omega,\mathbf{k};t,\mathbf{r}) = 0 \; , \tag{4.4.41}$$

from the zeroth approximation (4.4.11) it follows that

$$\left[k^2 - \frac{\omega^2}{c^2}\epsilon_{\perp}\right]\left[\delta_{\alpha\beta} - \frac{k_{\alpha}k_{\beta}}{k^2}\right]E_{\beta} = 0$$

or

$$\mathbf{E}_{\perp} = \mathbf{E} - \mathbf{k}(\mathbf{kE})/k^2 = 0 \; .$$

Hence, for $\epsilon_{\parallel} = 0$ the electric field is longitudinal and may be written as

$$\mathbf{E} = \Phi_{\parallel}\frac{\mathbf{k}}{k} = \Phi_{\parallel}\mathbf{1} \; . \tag{4.4.42}$$

To a zeroth approximation the magnetic field is zero:

$$\mathbf{H} = \frac{c}{\omega}\mathbf{k}\times\mathbf{E} = \frac{c}{\omega}\mathbf{k}\times\mathbf{1}\Phi_{\parallel} = 0 \; . \tag{4.4.43}$$

The consistency condition for the first-order approximation $\mathbf{1}\cdot\mathbf{Z} = 0$ leads us to the conservation law

$$\frac{\partial w}{\partial t} + \mathrm{div}\,\mathbf{S} = \left[-2\chi_{\parallel} - \frac{\mathrm{d}\ln\omega}{\mathrm{d}t}\right]w \; , \tag{4.4.44}$$

where

$$w = \frac{|\Phi_{\parallel}|^2\eta_{\parallel}}{16\pi} \; , \quad \eta_{\parallel} = \frac{\partial\omega^2\epsilon_{\parallel}}{\omega\partial\omega} = \omega\,\frac{\partial\epsilon_{\parallel}}{\partial\omega} \; ,$$

$$\mathbf{S} = -\frac{\omega}{16\pi}\frac{\partial\epsilon_{\parallel}}{\partial\mathbf{k}}|\Phi_{\parallel}|^2 = \mathbf{g}_{\parallel}\omega \; , \quad \mathbf{g}_{\parallel} = -\left[\frac{\partial\epsilon_{\parallel}}{\partial\mathbf{k}}\right]\bigg/\left[\frac{\partial\epsilon_{\parallel}}{\partial\omega}\right] \; ,$$

$$2\chi_{\parallel} = \frac{1}{\eta_{\parallel}}\left[2\omega\epsilon_{\parallel}'' + \omega\left(\frac{\partial^2\epsilon_{\parallel}}{\partial x_j\partial k_j} - \frac{\partial^2\epsilon_{\parallel}}{\partial\omega\partial t}\right)\right] \; .$$

From (4.4.44,45) it follows that the transport of energy by longitudinal waves is effected only by spatial dispersion; indeed, at $\partial\epsilon_{\parallel}/\partial\mathbf{k} = 0$ the Poynting vector vanishes along with the group velocity $\mathbf{g}_{\parallel}$.

The solution for $|\Phi_\parallel|$ is written in a form similar to (4.4.38), namely

$$\Phi_\parallel = \Phi_\parallel{}^0 \sqrt{\frac{\eta_\parallel{}^0 j(t^0)\omega(t)}{\eta_\parallel\, j(t)\omega(t^0)}}\, \exp\!\left(-\int_{t^0}^{t}\chi_\parallel dt\right). \tag{4.4.45}$$

4.4.10 Waves in Weakly Anisotropic Media

The quasi-isotropic approximation, described in Sect.4.3, may be invoked to describe the behavior of nonharmonic waves in weakly anisotropic media, in particular, those moving nonuniformly and possessing weak anisotropy. Equations (4.4.35) then acquire terms, associated with the anisotropic part of the permittivity tensor, which describe the conversion of modes in a weakly anisotropic, inhomogeneous, nonstationary medium [4.81]. A number of actual results bordering on the quasi-isotropic approximation have been obtained in [4.76-78].

4.5 Constitutive Equations for Nonstationary and Inhomogeneous Dispersive Media. Existence of Adiabatic Invariance

4.5.1 Corrections to the Quasi-Stationary Permittivity Tensor

We conducted the our consideration on the assumption that a permittivity tensor $\epsilon_{\alpha\beta}(\omega,\mathbf{k};t,\mathbf{r})$ is given. An actual calculation of this tensor for the general case of inhomogeneous, nonstationary, dispersive media presents formidable difficulties because deriving $\epsilon_{\alpha\beta}$ is equivalent to solving equations describing the response of the medium to the electromagnetic field. The derivation of $\epsilon_{\alpha\beta}$ markedly simplifies in the case of smoothly inhomogeneous and slowly nonstationary media, i.e., those for which the inequalities (4.4.4) are valid.

In essence, we are to handle a self-consistent problem on the behavior of a field-medium system by invoking equations for the microprocesses in the medium. The solution simplifies substantially for the case of smoothly inhomogeneous and slowly nonstationary media where one may apply the geometrical-optics method to a complete system of equations describing the field-medium system. Having no space to discuss this problem in more detail, we confine ourselves to qualitatively studying the new electromagnetic phenomena arising due to the smooth inhomogeneity and slow nonstationarity of dispersive media.

It is natural to expect for such media that $\epsilon_{\alpha\beta}$ will differ only insignificantly from its quasi-stationary value $\epsilon_{\alpha\beta}^0(\omega,\mathbf{k};t,\mathbf{r})$, which we define as the permittivity tensor for a homogeneous and stationary medium such that its macroscopic parameters γ_j (temperature, concentration, hydrodynamic velocity, etc., given in discrete or continuous form, as, say, a value set of the unperturbed electronic density distribution for a plasma) coincide with the $\gamma_j(t,\mathbf{r})$ of the considered inhomogeneous nonstationary medium at a given time t and at a point $\mathbf{r}$. In other words, if the electrodynamic properties of a homogeneous stationary medium are described by the tensor $\epsilon^h_{\alpha\beta}(\omega,\mathbf{k};\gamma_j)$, then for a medium of the variable parameters $\gamma_j = \gamma_j(t,\mathbf{r})$ the quasi-stationary tensor $\epsilon^0_{\alpha\beta}$ is introduced as

$$\epsilon^0_{\alpha\beta}(\omega,\mathbf{k},t,\mathbf{r}) = \epsilon^h_{\alpha\beta}(\omega,\mathbf{k},\gamma_j(t,\mathbf{r})) \ . \tag{4.5.1}$$

The actual tensor of a smoothly inhomogeneous and slowly nonstationary medium, $\epsilon_{\alpha\beta}$, which arises in equations of Sect.3.4, differs from its quasi-stationary value $\epsilon^h_{\alpha\beta}$ by a small quantity on the order of the small geometrical-optical parameters μ:

$$\epsilon_{\alpha\beta} = \epsilon^0_{\alpha\beta} + \delta\epsilon_{\alpha\beta} + O(\mu^2) \ , \quad \left|\delta\epsilon_{\alpha\beta}\right| \sim \mu \ll 1 \ . \tag{4.5.2}$$

In the limit of the homogeneous stationary medium the correction $\delta\epsilon_{\alpha\beta}$ reduces to zero; therefore we may anticipate that it should contain spatial and temporal derivatives of the parameters γ_j

$$\delta\epsilon_{\alpha\beta} \simeq B_0 \frac{\partial\epsilon_{\alpha\beta}^0}{\partial\gamma_j} \frac{\partial\gamma_j}{\partial t} + B_m \frac{\partial\epsilon_{\alpha\beta}^0}{\partial\gamma_j} \frac{\partial\gamma_j}{\partial x_m} \ . \tag{4.5.3}$$

The procedure of the actual derivation of $\delta\epsilon_{\alpha\beta}$ is facilitated by the fact that $\epsilon_{\alpha\beta}(\omega,\mathbf{k};t,\mathbf{r})$ describes, by definition, the medium response to a monochromatic plane wave $\mathbf{E}e^{i(\omega t-\mathbf{k}\mathbf{r})}$. This enables us to seek a solution of the microscopic equations defining the medium's response in the form

$$f(t,\mathbf{r}) = [F^0(t,\mathbf{r}) + F^1(t,\mathbf{r}) + ...] \ e^{i(\omega t-\mathbf{k}\mathbf{r})} \ ,$$

where $f(t,\mathbf{r})$ is one of parameters describing the wave processes in the medium, and $\left|F^1\right| \sim \mu\left|F^0\right| \ll \left|F^0\right|$, etc. Calculations of this sort can be found in [4.76-78, 97-99]. Some of these examples will be touched on in what follows.

4.5.2 Physical Phenomena Due to the Deviation of $\epsilon_{\alpha\beta}$ from Its Quasi-Stationary Value

The evaluation of $\delta\epsilon_{\alpha\beta}$ is justified only when it brings forth some novel phenomena and effects. Primarily, the inhomogeneous medium is able to support new modes absent in the homogeneous medium. The most spectacular example is the appearance of drift-dissipative waves in an inhomogeneous plasma embedded in a mangetic field which is also inhomogeneous. These waves often appear to be unstable because the electrons drifting in the inhomogenous magnetic field tend to give off their energy to the electromagnetic wave if their drifting velocity matches the wave speed in a definite way.

The geometrical-optics theory of drift-dissipative oscillations aims primarily at evaluating the imaginary (anti-Hermitian, to be more general) correction to the quasi-stationary value of permittivity. This correction should be proportional to the gradients of the particular plasma parameters and responsible (with a definite sign of the correction) for an increase in the oscillation amplitude. We shall not go further into this vast field, but note instead that the respective items have been elucidated in numerous reports, notably [4.83, 89, 100, 101].

Secondly, if a dispersive medium is nonstationary, the anti-Hermitian part of the correction $\delta\epsilon_{\alpha\beta}$ appears different from zero even when *no absorption* takes place. This effect is closely allied with the existence of the adiabatic invariant in transparent media and will be dealt with in Sects. 4.5.3-4.

Finally, electromagnetic processes in a nonstationary, inhomogeneous medium depend not only on the macroscopic parameter variation but also on the specific mechanism responsible for these variations [4.86, 99]. Some reports dealing with inhomogeneous, nonstationary dispersive media seem to lack an explicit understanding of these features.

4.5.3 Existence of the Adiabatic Invariant

If the condition

$$2\chi \equiv - \frac{\omega}{\eta}\left[2i\epsilon_{\alpha\beta}^{a} + \left(\frac{\partial^2 \epsilon_{\alpha\beta}^{h}}{\partial\omega\partial t} - \frac{\partial^2 \epsilon_{\alpha\beta}^{h}}{\partial x_j \partial k_j} \right) \right] f_\alpha^* f_\beta = 0 \tag{4.5.4}$$

is met (we reinstated the superscript ^{h} to the Hermitian part of the permittivity tensor), then (4.4.29) reduces to

$$\frac{d}{dt}\ln(wj/\omega) = 0 \ . \tag{4.5.5}$$

From this transfer equation it follows that

$$wj/\omega = \text{const} = w^0 j^0/\omega^0 \ , \tag{4.5.6}$$

which upon multiplying by $d\xi_1 d\xi_2 d\xi_3$ yields

$$d\mathcal{J} = wdV(t)/\omega = \text{const} = w^0 dV^0/\omega^0 \ , \tag{4.5.7}$$

where $dV(t) = j(t)d\xi_1 d\xi_2 d\xi_3 = dx_1 dx_2 dx_3$ is the elementary volume, and dV^0 is the elementary volume at $t = t^0$.

Therefore, with (4.5.4) satisfied, the combination $wdV/\omega = d\mathcal{J}$ appears to be a conserved quantity. Integrating $d\mathcal{J}$ over the whole volume of the wave packet leads us to the *adiabatic invariant* of the problem

$$\mathcal{J} = \int_{V(t)} \frac{wdV}{\omega} = \text{const} \ . \tag{4.5.8}$$

272

The ratio $w/\omega = d\mathcal{J}/dV$ is conveniently viewed as the *density of the adiabatic invariant*.

If the wave-packet volume $V(t)$ is so small that for all points within it the attendant values of frequency $\omega(t)$ are the same and vary as a whole as the volume travels, then we obtain for $\mathcal{J}$ an expression similar to that for the adiabatic invariant of a single oscillator:

$$\mathcal{J} \simeq \frac{\int w \, dV}{\omega} = \frac{W}{\omega} = \text{const.} \, , \tag{4.5.9}$$

where W is the total energy of the packet. Thus (4.5.8) may be thought of as the adiabatic invariant generalized over extended wave packets. Actual examples for conserving $\mathcal{J}$ can be found in [4.102-106] (also [4.86, 87]). Condition (4.5.4) has been formulated in [4.107] for a specific situation and generalized in [4.108].

4.5.4 Phenomenological Evaluation of the Anti-Hermitian Part of the Correction for the Quasi-Stationary Permittivity Tensor in Transparent Media

The argument of Sect. 4.5.3 can obviously be conducted in the reverse order; then the requirement for the adiabatic invariant to exist - natural for transparent (nonabsorbing) media - will entail (4.5.4) for which one can derive the anti-Hermitian correction $\delta\epsilon^a_{\alpha\beta}$ in a purely phenomenological manner [4.107, 108].

For nonabsorbing media, the quasi-stationary tensor $\epsilon^0_{\alpha\beta}$ is Hermitian, $\epsilon^0_{\alpha\beta} = \epsilon^h_{\alpha\beta}$, therefore in (4.5.2) $\epsilon^a_{\alpha\beta} = \delta\epsilon^a_{\alpha\beta} + 0(\mu^2)$. Observing this, substitute (4.5.2) into (4.5.4) to get the relation

$$\epsilon^a_{\alpha\beta} = \delta\epsilon^a_{\alpha\beta} = \frac{i}{2}\left[\frac{\partial^2 \epsilon^0_{\alpha\beta}}{\partial\omega\partial t} - \frac{\partial^2 \epsilon^0_{\alpha\beta}}{\partial x_j \partial k_j}\right] + 0(\mu^2) \, , \tag{4.5.10}$$

which provides a uniformly valid connection between the anti-Hermitian part of the permittivity tensor $\epsilon^a_{\alpha\beta}$ and the quais-stationary tensor $\epsilon^0_{\alpha\beta}$ subject to the existence of an adiabatic invariant. Equations similar to (4.5.10) are also valid for longitudinal and transverse waves in an isotropic media. Specifically, for an isotropic homogeneous nonstationary medium exhibiting only time dispersion, (4.5.10) yields the result of [4.107]

$$\delta\epsilon'' = \text{Im}\{\delta\epsilon\} = \frac{1}{2}\frac{\partial^2 \epsilon^0(\omega, t)}{\partial\omega\partial t} + 0(\mu^2) \, . \tag{4.5.11}$$

In stationary problems, the anti-Hermitian part of the permittivity tensor describes Joule's losses in the medium, and in active media the transport of energy from the medium to the wave. But in nonstationary transparent dispersive media the appearance of an anti-Hermitian part is not associated with Joule's losses; rather it is caused by an additioanl phase shift between **D** and **E**. For instance, when the concentration of electrons N_e in a plasma changes, the finite time it takes the polarization to set up in the dispersive medium causes oscillations of the electric displacement vector **D** to lag or lead in phase (depending on the sign of $\partial N_e / \partial t$) the oscillations of **E** at the same point. In a medium with spatial dispersion the phase shift to the type - owing its existence to transport of substance - can also be brought about by an inhomogeneity of the medium.

In stationary media where $dw/dt = 0$, so w is constant along the ray, from (4.5.6) it follows that the energy $dW = wdV$ is conserved in the elementary wave packet, but in a nonstationary medium the energy of a packet is not conserved. The change in W is caused by the transport of energy to and from the medium (the wave, e.g., acquires its energy from the kinetic energy of the directed motion of medium particles). This energy transfer does not, of course, change the net energy of the wave-medium system. Actually, the correction (4.5.10) has been introduced to provide for the net energy to be conserved in the case of transparent media, as it has been demonstrated, e.g., in [4.109].

When the properties of a medium change sharply as, say, in a fast motion of the interface between two media, the adiabatic invariant (and, consequently, the number of quanta) is not conserved [4.110-115].

4.6 Wave Processes in Nonstationary Media

4.6.1 One-Dimensional Problem. General Relationships

In considering wave processes in nonstationary media we shall attenpt to simplify the problem formulation as far as possible, confining ourselves to *one-dimensionally inhomogeneous* media. The most illustrative examples will be borrowed from the first reports on applying the geometrical-optics method to the study of waves in nonstationary media [4.102-104] and from surveys [4.86, 87].

Let the isotropic medium under investigation be inhomogeneous in the z direction with its refractive index $n = n(\omega; t, z)$. We consider waves traveling in the z direction only and adopt the simplest version by imposing the initial conditions - as in Sect.3.9.3 - that the field at $z = 0$ obeys the function of time:

$$E(t,0) = E^0(t) = A^0(t)e^{i\phi^0(t)} . \qquad (4.6.1)$$

In the one-dimensionally inhomogeneous isotropic medium, the space-time ray equations (4.4.17) corresponding to the eikonal equation $k = \omega n/c$ assume the form

$$\frac{dz}{dt} = c\left(\frac{\partial \omega n}{\partial \omega}\right)^{-1} \equiv g , \quad \frac{d\omega}{dt} = -\omega\left(\frac{\partial \omega n}{\partial \omega}\right)^{-1}\frac{\partial n}{\partial t} , \qquad (4.6.2a)$$

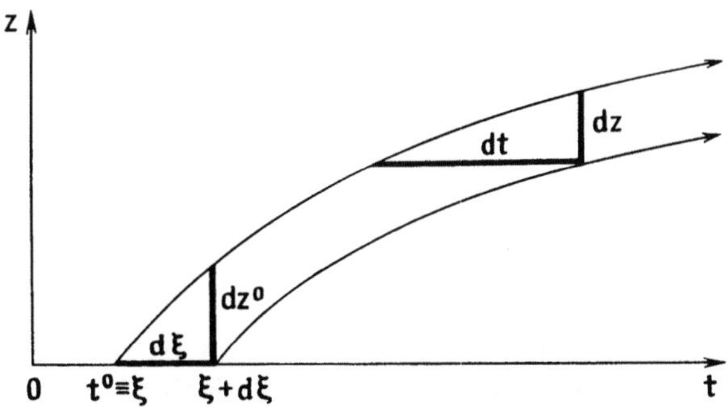

Fig.4.9. Evolution pulse duration and amplitude in a nonstationary medium

where $g = g_g$ is the group velocity of the wave. The equation

$$\frac{dk}{dt} = \omega \left(\frac{\partial \omega n}{\partial \omega} \right)^{-1} \frac{\partial n}{\partial z} \tag{4.6.2b}$$

for the wave number $k = k_z$ follows form (4.5.2a) and the eikonal equation $k = n\omega/c$.

It will be convenient to take the time instant $\xi \equiv t^0$ of the space-time ray $[t, z(t)]$ leaving the origin at $z = 0$ (Fig.4.9) as the parameter characterizing the family of solutions $z = z(t)$ and $\omega = \omega(t)$, as we did in Sect.4.9.3. With this choice of the parameter the desired solutions $z = Z(t, \xi)$ and $\omega = \Omega(t, \xi)$ should obey the following initial conditions:

$$Z(\xi, \xi) = 0 , \quad \Omega(\xi, \xi) = - \partial \phi^0(\xi)/\partial \xi . \tag{4.6.3}$$

The Jacobian of the one-dimensional problem is expressed as

$$j(t) = \partial Z(t, \xi)/\partial \xi . \tag{4.6.4}$$

The phase and amplitude are to found from the general formulas of Sects.2.7 and 4.4.

From (4.6.2a) it follows that the frequency ω increases on those segments of the path where the refractive index decreases in time ($\partial n/\partial t > 0$), and vice versa. As a result, even a strictly monochromatic initial oscillation develops frequency modulation which can be accompanied - especially in a dispersive medium - by the appearance of space-time caustics and this, in turn, can bring about substantial changes of the amplitude. These features of slowly nonstationary media provide a means of effectively controlling the frequency and amplitude of a wave. We shall consider several of these possibilities.

4.6.2 Nonstationary Nondispersive Media

In the absence of dispersion when the refractive index n depends on t and z only, the left equation of (4.6.2a) becomes independent of the right one, i.e.,

$$\frac{dz}{dt} = g = \frac{c}{n(t,z)} \; , \qquad \frac{d\ln\omega}{dt} = -\frac{\partial\ln[n(t,z)]}{\partial t} \; . \tag{4.6.5}$$

For arbitrary n(t,z) profiles, the equation for z = z(t) cannot be resolved analytically, yet in those specific situations which are analytically soluable for z(t) = Z(t,ξ) – some of the simplest examples are given below – the other parameters of the wave can be found as implicit functions.

In a nondispersive isotropic medium, the combination kg–ω vanishes and, by virtue of (2.7.33), the phase ϕ(t) on the ray [t,z(t)] does not change:

$$\phi(t) = \phi^0(\xi) \; , \qquad \xi = t^0 \; . \tag{4.6.6}$$

The phase ϕ(t,z) at a given point (t,z) can be derived by substituting ξ = ξ(t,z), implicitly given by z = Z(t,ξ), into (4.6.6).

The second equation of (4.6.5) yields the following law for the frequency variation ω(t) [4.103]:

$$\omega(t) = \omega^0(\xi)\exp\left\{-\int_\xi^t \frac{\partial\ln n}{\partial t}\, dt'\right\}, \qquad \omega^0(\xi) = -\frac{\partial\phi^0(\xi)}{\partial\xi} \; , \tag{4.6.7}$$

where the values n = n[t′,z(t′)] are taken along the ray [t′,z(t′)]. The quantities η and χ, appearing in (4.4.38) for the field amplitude A, become in a nondispersive isotropic medium (the subscript "$\perp$" is dropped)

$$\eta = 2n^2 \; , \qquad \chi = \omega\epsilon''/2\epsilon = \omega n''/n \; ,$$

where n″ = $\epsilon''/2\sqrt{\epsilon}$ is the small imaginary part of the refractive index, ϵ = ϵ' and n = $\sqrt{\epsilon'}$. Substituting these expressions into (4.4.38) yields

$$A(t) = A^0\,\frac{n^0}{n}\sqrt{\frac{j^0\omega}{j\omega^0}}\,\mathcal{P} \; , \tag{4.6.8}$$

where the superscript ^{0} refers to the initial values of A, n, j, and ω at the instant ξ of ray emission, and the factor

$$\mathcal{P} = \exp\left\{-\int_\xi^t \frac{\omega n''}{n}\, dt'\right\} \tag{4.6.9}$$

describes the reduciton of field intensity due to absorption. In stationary conditions, this factor coincides with (2.3.37) for dt = dz/g = ndz/c.

276

Equation (4.6.8) can be further simplified if we observe that by the Liouville formula

$$j(t) = j(\xi)\exp\left(\int_\xi^t \mathrm{div}\,g\,dt\right).$$

In a one-dimensional problem $\mathrm{div}\,g = \partial g/\partial z$, and in the absence of dispersion $g = c/n$; therefore

$$j(t) = j(\xi)\exp\left(-\int_\xi^t \frac{c}{n^2}\frac{\partial n}{\partial z}dt'\right).$$

The derivative $\partial n/\partial z$ can be expressed by $\partial n/\partial t$ and dn/dt since

$$\frac{dn}{dt} = \frac{\partial n}{\partial t} + g\frac{\partial n}{\partial z} = \frac{\partial n}{\partial t} + \frac{c}{n}\frac{\partial n}{\partial z}.$$

Accounting for (4.6.7), this leads to

$$j(t) = j(\xi)\frac{n^0}{n}\exp\left(\int_\xi^t \frac{\partial \ln n}{\partial t}dt\right) = j^0\frac{n^0\omega^0}{n\omega}. \tag{4.6.10}$$

Substituting (4.6.10) into (4.6.8) we finally arrive at

$$A = A^0\sqrt{\frac{n^0}{n}\frac{\omega}{\omega^0}\mathscr{P}}. \tag{4.6.11}$$

With no absorption ($\mathscr{P} = 1$) (4.6.11) yields follows the existence of the invariant $A^2 n/\omega^2 = \text{constant}$ or, what is the same,

$$\frac{S}{\omega^2} = \frac{c}{8\pi}\frac{A^2 n}{\omega^2} = \text{const}, \tag{4.6.12}$$

where $S = nc|E|^2/8\pi = cnA^2/8\pi$ is the density of energy flux in an isotropic medium. That the quantity S/ω^2 appears constant is equivalent to the existence of the adiabatic invariant (4.5.6) since in a nondispersive isotropic medium $w = n^2A^2/8\pi$, and $jn\omega = j^0n^0\omega^0$ by virtue of (4.6.10).

In a spatially homogeneous nonstationary medium of $n = n(t)$ we have from (4.6.7, 10)

$$\omega/\omega^0 = n^0/n, \quad A = A^0(n^0/n)^{3/2}\mathscr{P}, \tag{4.6.13}$$

and the dependence $z = Z(t,\xi)$ is given by

$$z = c \int_{\xi}^{t} \frac{dt'}{n(t')} \, .$$

From (4.6.13) it follows that, in order to alter markedly the wave frequency and amplitude, the refractive index should vary over a wide range. Systems with *traveling-wave refractive index* in the form

$$n = n(z - Vt) \equiv n(\zeta) \, , \tag{4.6.14}$$

where V is the velocity of the refractive-index traveling wave, are more sensitive to variations of n. In view of (4.6.5), $\zeta = z - Vt$ satisfies the equation

$$\frac{d\zeta}{dt} = g(\zeta) - V \, , \quad g(\zeta) = \frac{c}{n(\zeta)} \, . \tag{4.6.15}$$

The nature of the processes in the case under consideration depends on the relation between the wave velocity V and the group velocity g. In the non-synchronous mode ($g \neq V$), the solution of (4.6.15) is given by

$$t = \xi + \int_{-V\xi}^{z-Vt} \frac{d\zeta}{g(\zeta) - V} \, , \tag{4.6.16}$$

containing implicitly the solution $z = Z(t,\xi)$.

Using (4.6.14, 15) we transform the exponent in (4.6.7) as follows:

$$\int_{\xi}^{t} \frac{\partial \ln n}{\partial t} dt' = \int_{\zeta^0}^{\zeta} \frac{V}{n} \frac{dn}{d\zeta} \frac{d\zeta}{(g - V)} = \ln[1 - \beta n(\zeta)] \Big|_{\zeta^0 \, = \, - \, v\xi}^{\zeta \, = \, z-vt} \, ,$$

whereupon we obtain for the frequency

$$\omega = \omega^0 \frac{1 - \beta n(\zeta^0)}{1 - \beta n(\zeta)} \, , \tag{4.6.17}$$

with $\beta = V/c$. It is seen that approaching synchronism, as $g \rightarrow V$ and $1 - \beta n = 0$, the frequency deviate from ω^0 everywhere. However, in exact synchronism ($1 - \beta n = 0$) (4.6.17) fails.

Based on the following argument, we can establish how frequency varies in the *synchronous mode*. Let ζ_s be one of the synchronous points where $g(\zeta_s) = V$ and, consequently, $\beta n(\zeta_s) = 1$. In accordance with (4.6.15), $\zeta_s = z - Vt = \text{const} = \zeta_s^{\,0}$ so that the synchronous ray equation is

$$z = \zeta_s^{\,0} + Vt = V(t - \xi_s) \, ,$$

where $\xi_s = -\zeta_s/V$ is the time when the synchronous ray leaves the plane z = 0. On this ray the quantity

$$-\left.\frac{\partial \ln n}{\partial t}\right|_{\zeta=\zeta_s} = V\left.\frac{d\ln n(\zeta)}{d\zeta}\right|_{\zeta=\zeta_s} \equiv V\alpha$$

is constant, and by virtue of (4.6.7) the frequency ω varies exponentially:

$$\omega = \omega^0 e^{\alpha V(t-\xi_s)} = \omega^0 e^{\alpha z} \ .$$

For positive α the frequency rises exponentially, whereas for negative α it falls as z increases. Equation (4.6.18) makes it evident that even for small changes in n, the traveling wave in the synchronous mode can suffer very large changes in frequency provided that $|\alpha|z \gg 1$. For nonsynchronous points ($\zeta \neq \zeta_s$) the frequency assumes values intermediate between the maximum and minimum values attained at the synchronous rays.

Detailed analyses for both synchronous and nonsynchronous modes for a sinusoidal refractive index profile, $n(\zeta) = n_0(1+\delta\cos K\zeta)$, have been outlined in [4.103, 104]. For various applications of nonstationary wave systems (frequency and phase modulators in distributed diodes and nonstationary lines, traveling-wave modulators in laser engineering, amplifiers and generators of short pulses, ultrasound and elastic wave modulators, etc.), see [4.86, 87].

4.6.3 Nonstationary Dispersive Media

In the presence of dispersion, the left equation of (4.6.2) cannot be solved independently of the right one, so we have to resort to discussing specific situations allowing for analytic solutions. One of them is a nonstationary but homogeneous dispersive medium of $n = n(\omega, t)$ [4.116]. Since the medium is homogeneous ($\partial n/\partial z = 0$) the wave number $k = k_z$ is constant (4.6.2b):

$$k = n(\omega, t)\omega/c = \text{const} = k^0 = n^0 \omega^0/c \ . \tag{4.6.18}$$

Solving (4.6.18) for ω we may express it as a function of time and time origin $t^0 \equiv \xi$, $\omega = \Omega(t,\xi)$. Substituting this expression for the frequency in (4.6.2a), we obtained the path $z = Z(t,\xi)$ as the quadrature

$$z = \int_\xi^t g[t',\Omega(t',\xi)]dt' \ . \tag{4.6.19}$$

The Jacobian j(t) results from differentiating the quadrature with respect to ξ at t = constant:

$$j(t) = \frac{\partial z}{\partial \xi} = g[t, \Omega(t,\xi)] + \int_\xi^t \frac{\partial g}{\partial \Omega} \frac{\partial \Omega}{\partial \xi} dt' . \qquad (4.6.20)$$

As a simple example, consider the variation of frequency in a transverse electromagnetic wave traveling in a nonstationary cold plasma of $\epsilon(\omega, t) = 1 - \omega_p^2(t)/\omega^2$, where $\omega_p = [4\pi e^2 N(t)/m]^{1/2}$ is the plasma frequency.[19] In this case the dispersion equation (4.6.18) takes the form[20]

$$k = \frac{\omega}{c} \sqrt{1 - \frac{\omega_p^2(t)}{\omega^2}} = k^0 = \frac{\omega^0(\xi)}{c} \sqrt{1 - \frac{\omega_p^2(\xi)}{[\omega^0(\xi)]^2}} , \qquad (4.6.21)$$

from which it follows the frequency variation

$$\omega(t) \equiv \Omega(t,\xi) = \sqrt{\omega_p^2(t) - \omega_p^2(\xi) + [\omega^0(\xi)]^2} . \qquad (4.6.22)$$

Now that $\omega(t)$ is known, the ray path may be found by quadrature (for plasma, $g = c\sqrt{\epsilon}$):

$$z(t) = \int_\xi^t g(t')dt' = c \int_\xi^t \sqrt{1 - \frac{\omega_p^2(t')}{\Omega^2(t',\xi)}} dt' .$$

Analysing integrals for the type of a linear concentration profile $\omega_p^2(t) = A + Bt$ has revealed that, as the electronic density decreases, it can give rise to space-time caustics corresponding to pulse compression [4.86, 87].

In considering the evolution of waves in dispersive media of refractive-index traveling waves, the ray equations may be integrated in quadratures by introducing the variable $\zeta = z - Vt$, where V is the wave speed of the respective parameter. The variation of frequency due to nonstationarity of the dispersive media cause changes in the group velocity, thus facilitating - in comparison to a nondispersive media - the formation of space-time caustics with the corresponding geometrical-optical singularities of the field.

---

[19] Recall that contrary to the amplitude behavior the variation of frequency is independent of the electron-density variation mechanism. Therefore we do not attach a physical cause to the nonstationarity.

[20] Note that the plasma dispersion law (4.6.21), which can be represented as $k^2 c^2 = \omega^2 - \omega_p^2$, also involves longitudinal waves in isotropic plasma (to this end, c should be understood as the speed of thermal electronic motion) and waves in a cylindrical waveguide loaded by a nondispersive dielectric [in this case $v_0 = c\sqrt{\epsilon\mu}$ should be substituted for c, and $v_0 \kappa$ for ω_p, where κ is the transverse wave number for a given mode]. There are reports on systems of more sophisticated laws of dispersion, such as magnetoactive plasma and plasma-filled waveguide [4.86, 87].

A salient feature of a dispersive medium, worth emphasizing is that synchronism $g \simeq V$ is reached for a *finite interval* of the values of V. By the same token, the synchronism can be realized not only for a part of a wave packet but for the entire wave: to class there correspond the so-called stationary waves of an envelope [4.86,87,116]. One may obtain a general insight into the motion of rays in a system of traveling wave by the graphic method, as was done in [4.117] for a uniformly moving plasma of an exponential electron density profile.

The quality of a geometrical-optics approximation can for a *number of cases* be assessed by comparing the results with the respective exact solutions. For media with a traveling parameter, the latter can be derived by converting to variables in which the parameter wave becomes "frozen-in" [4.112,114].

4.6.4 Evolution of Short Pulses

A short pulse whose width $\ell = g\tau$ and duration τ are small as compared with the inhomogeneity scale L and the nonstationarity time T may be thought of as a *homogeneous* wave packet, since its frequency ω and wave number k, being constant originally, do not change practically within the pulse. The variations of ℓ and τ obey simple laws stated in [4.90].

Let dz and dz^0 be spatial cross sections of the ray tube corresponding to the time interval $d\xi = dt^0$ (Fig.4.9). The short pulse width ℓ varies proportional to dz:

$$\frac{\ell}{\ell^0} \simeq \frac{dz}{dz^0} = \frac{\partial z/\partial t^0}{\partial z/\partial t^0} = \frac{j(t)}{j(\xi)} \equiv \frac{j}{j^0} , \tag{4.6.23}$$

where ℓ is the initial pulse width and $\xi = t^0$ the initial moment of time. For the pulse duration $\tau = \ell/g$, we have

$$\frac{\tau}{\tau^0} \simeq \frac{dt}{d\xi} = \left(\frac{dz}{g}\right) : \left(\frac{dz^0}{g^0}\right) = \frac{jg^0}{j^0 g} . \tag{4.6.24}$$

We may use the pulse duration τ to express the signal spectrum width $\Delta\omega = 1/\tau$ and the relative spectrum width $\delta = \Delta\omega/\omega = 1/\omega\tau$ so that

$$\Delta\omega/\Delta\omega^0 = \tau^0/\tau = j^0 g/jg^0 ,$$

$$\delta/\delta^0 = \omega^0\tau^0/\omega\tau = j\omega^0 g/j^0 \omega g^0 . \tag{4.6.25}$$

It is an easy matter to verify that in a stationary inhomogeneous medium the duration of a homogeneous wave packet does not change (the "collimated beam zone" of a pulse, Sect.3.9):

$$\tau = \tau^0 . \tag{4.6.26}$$

Letting in (4.6.24) $\tau = \tau^0$, we obtain $j/j^0 = g/g^0$, which implies[21]

$$\ell/\ell^0 = g/g^0 \ . \tag{4.6.27}$$

Observing that in a stationary medium $\omega = \omega^0$, we obtain

$$\Delta\omega = \Delta\omega^0 \ , \quad \delta = \delta^0 \ .$$

4.6.5 Reflection from Moving Interfaces

Following [4.110, 113] we consider a one-dimensional problem of the reflection and refraction of waves at a moving interface Q, given by the equation $z = Z_b(t)$. Let an incident wave $E_i = A_i \exp(i\phi_i)$ strike the interface between a homogeneous medium 1 of permittivity ϵ_1 and a homogeneous medium 2 of permittivity ϵ_2, moving from 1 into 2 toward increasing values of z so that $k_1 = \partial\phi_i/\partial z > 0$.

The phase of the incident wave at the moving boundary, $\phi_{i,Q}$, imposes the initial conditions for the phases of the reflected and refracted waves:

$$\phi_{r,Q} = \phi_{t,Q} = \phi_{i,Q} \equiv \phi_i(t, Z_b(t)) \ . \tag{4.6.28}$$

Differentiating (4.6.29) with respect to t we obtain the equations

$$\frac{\partial\phi_i}{\partial t} + \frac{\partial\phi_i}{\partial z} \frac{dZ_b}{dt} = -\omega_i + k_i V_b = -\omega_r + k_r V_b = -\omega_t + k_t V_b$$

$$V_b = dZ_b/dt \ ,$$

which, along with the dispersion equation $k^2 c^2 = \omega^2 \epsilon_{1,2}(\omega)$, yield the initial conditions on ω and k for both reflected and refracted waves. Hence we may now construct a system of reflected and refracted space-time rays to derive not only the phases but also the amplitudes of the waves. The initial amplitudes $A_{r,Q}$ and $A_{t,Q}$ are then to be expressed in terms of $A_{i,Q}$ from the known boundary conditions at the moving interface [4.113]:

$$\{E + \beta \times B\}_{tan} = \{H - \beta \times D\}_{tan} = 0 \ ,$$

where $\beta = V_b/c$, and V_b is the speed of the boudnary; the braces denote the difference taken on either side of the boundary.[22]

The effects arising at a moving interface allow for an intuitively appealing ray interpretation on the $\{t,z\}$ plane. Figure 4.10 illustrates typical situations arising in uniform motion of the boundary $Z_b(t) = Z_b^0 + V_b t$, ($V_b < 0$) counter to the incident wave. It is assumed that both media 1 and 2 are nondispersive so that the group velocities g_1 and g_2 are independent

---

[21] The obtained result agrees with that of the direct computation, which yields $j = -g$ under the given circumstances.

[22] The salient features of the field behavior at moving interfaces have been discussed in [4.113].

282

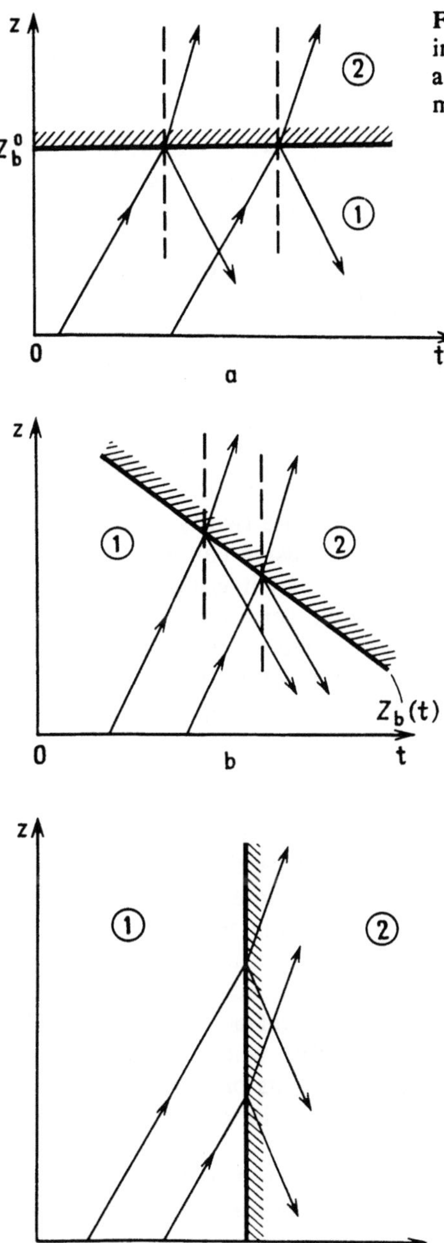

Fig.4.10. Ray interpretation of plane wave incidence on (a) a stationary interface, (b) a uniformly moving interface, and (c) a medium with instantly altering permittivity

of frequency (to be definite, assume that $g_2 > g_1$). Figure 4.10a shows the simplest situation of a plane wave incident on a stationary interface ($V_b = 0$, $Z_b = Z_b^0$), and Fig.4.10b illustrates the configuration of space-time rays for a relatively slow motion of the boundary when V_b does not exceed g_1. Comparing Figs.4.10a,b makes it obvious that in the case of the moving boundary the reflected pulse is shorter than the incident one, corresponding to an increasing field amplitude. The amplitude of the reflected wave becomes infinite at $|V_b| = g_1$ when the interface "pushes" the reflected wave ahead of itself. A "bulldozing" effect of a similar kind takes place also for $g_1 < |V_b| < g_2$.

For $|V_b| > g_2$ the second medium propagates two waves at the group velocity g_2 traveling in opposite directions. Figure 4.10c plots the rays in the limiting case of $|V_b| = \infty$ when the entire space simultaneously experiences a sudden change in permittivity form ϵ_1 to ϵ_2 at the instant t_0. The appearance of two oppositely traveling waves in medium 2 is necessitated by the initial condtions at $t = t_0$ [4.115].

In dispersive media, the pattern of reflections and refractions become complicated in two respects. First, the group velocity of the reflected and refracted waves depends now on the speed of the boundary (in terms of frequency which varies with V_b due to the Doppler effect). Therefore, when the boundary moves nonuniformly ($V_b \neq$ const.) some groups of waves leaving it can catch up and overtake other groups, thus giving rise to space-time foci and caustics. Figures 4.11a,b illustrate two types of possible focusing - for reflected and refracted waves. Focusing can also take place when the boundary moves uniformly, provided that it is illuminated by a *non-monochromatic* wave packet corresponding, for instance, to the far field of the space-time diffraction of a pulse. Figure 4.11c depicts such a packet as a diverging bundle of rays (*Felsen* [4.23, 118] has treated the focusing of rays at $V_b = \infty$, i.e., when the medium properties undergo a sudden change). In all the situations mentioned, the moving boundary plays the role of a *space-time lens* [4.86, 119].

Secondly, moving media can generate new types of natural modes which are absent in a medium at rest. The possibility has been discussed in [4.113] for a medium consisting of oscillators to infer that the boundary is followed by two new waves whose frequencies are close to the natural frequency of the oscillators. These waves appear owing to transients in the oscillators following the passage of the wave front of a parameter.

At a sharp interface of two *anisotropic* media, an incident wave induces, in general, all the types of reflected and refracted normal modes consistent with the dispersion law. Specifically, when the anisotropic medium all at once undergoes a sudden change in its properties, this gives rise to two pairs of normal modes propagating in opposite directions [4.120].

Obviously, the adiabatic invariant cannot be expected to be conserved at a moving interface. An elaborate analysis has indicated that the quanta increase in number [4.86, 87, 110]; hence wave *amplification* is possible when the frequency of one of the waves, ω_j, turns out to be negative - the area of the anomalous Doppler effect corresponding to the interface moving faster than the phase velocity $v_{ph}(\omega_j)$.

It is worth noting that in dealing with reflection at moving boundaries in 3-D space, the reflection and refraction laws for a ray assume a familiar form only in a reference frame tied with the interface, whereas in the laboratory frame of reference the angle of reflection differs from the angle of

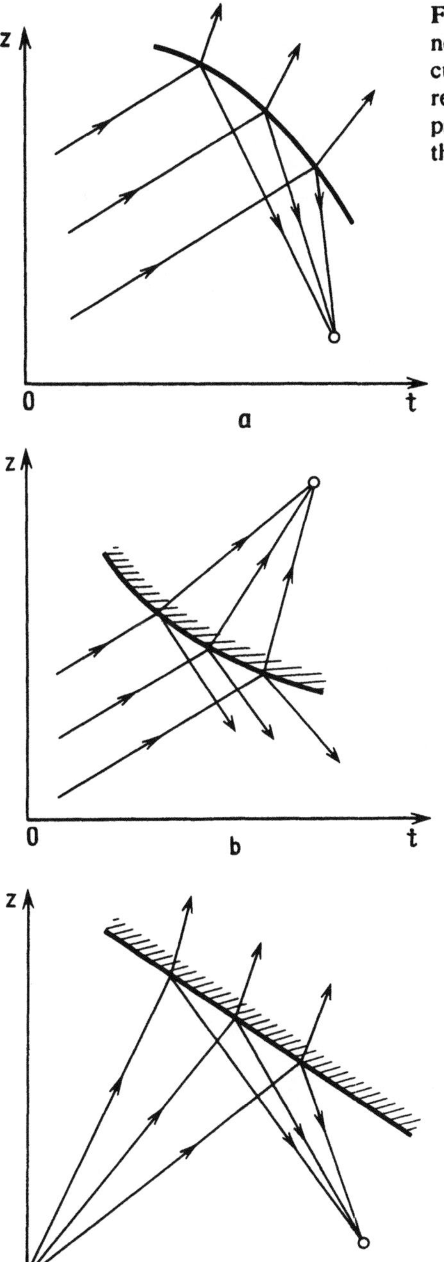

Fig.4.11. Space-time lenses produced by a nonuniformly advancing interface: (**a**) focussing of reflected waves; (**b**) focussing of refracted waves. (**c**) Focussing of a wave packet (a bundle of rays fanning out from the origin) on a uniformly moving interface

incidence. In many situations, though, an apparent interface may be selected such that, on the one hand, it keeps the familiar properties of rays untouched and, on the other hand, renders the scattered fields amenable to efficient handling by the geometrical-optics approximation [4.121].

4.6.6 Perturbation Theory in Nonstationary Problems

When the medium parameters deviate from their means only slightly, then the geometrical-optics equations may be handled by perturbation techniques. Many situations allow for a *quasi-stationary account* of the temporal variations of the refractive index, whereby the dependence of n on time is reestablished at the last computational stage only. The quasi-stationary approach is feasible provided that the wave passes an inhomogeneity of size ℓ for a time $\Delta t \simeq \ell/g$ smaller than the characteristic time of medium properties variation τ, so that $\ell/g \ll \tau$. The calculations of this sort are often employed to determine the Doppler shift of a radio wave passed through a nonstationary ionospheric plasma [4.39, 122].

More general calculations under the conditions when the quasi-stationary approximation is invalid, i.e., when $\ell/g \geq \tau$, have been undertaken in [4.86, 123-125] as applied to the fluctuations of electromagnetic waves in a turbulent plasma.

5. Conclusion

Having presented the fundamentals of ray theory and examined its applications to various problems of real value, it would seem only natural for the book to conclude with a general description of the method of geometrical optics.

The most attractive feature of the ray method is that it is intuitively appealing. Notably, the ray patterns agree well with our intuitive representations of propagating waves in general and light waves in particular. This is precisely why geometrical optics has remained the most useful approximation of wave theory.

A second important feature explaining the method's popularity is connected with its universality, and in many cases it is the only possible approximation for calculating wave fields. This is particularly true of smoothly inhomogeneous media, to which a large part of the book is devoted. Here geometrical optics is virtually without peer among both analytic and numerical methods. A special function of geometrical optics as applied to inhomogeneous media is due to the fact that the parameters of the media, as a rule, are known approximately, and attempts to solve initial equations exactly lead to a superfluous precision.

The versatility of the approximation is supplemented by the equally universal, heuristic conditions of the method's applicability (Sect.2.10). These criteria take care of the method's applicability since they employ the ray constructions of Fresnel volumes. In a number of cases these criteria help to evaluate the field intensity even where geometrical optics is inapplicable. The last possibility paves the way for a geometrical-optics analysis of a wave pattern in its entity. Such an approach is valuable both in the correct formulation of theoretical studies and a preliminary assessment of fields in engineering.

The popularity of geometrical optics is also favored by the existence of an extremely wide range of actual physical and technical problems which the theory is able to describe adequately. This multiplicity of problems arises due to the variability of inhomogeneous media propagating waves of various physical natures and the implementation in physics and technology of an ever broader spectrum of wavelengths. Under these circumstances, for a given inhomogeneous medium, one can almost always pinpoint a class of short-wavelength problems amenable to ray treatment and yielding tolerably accurate wave fields.

Numerous reported calculations have demonstrated enough of the strength of geometrical optics in various disciplines of physics and egineering that the effectiveness of the method does not require special proof.

However, with the student in mind, we have included in this book several typical examples that illustrate the actual strenght of the method.

Geometrical optics should also be appraised as the heuristic base for many approximate methods in the theory of wave propagation and diffraction. In the short-wavelength limit, asymptotic methods covered by geometrical optics, the rays lose their original physical sense of the lines along which energy of the wave propagates, although they continue to fulfill the role of "a skeleton that carries the wave flesh". This important implication of rays contributed to the larger share of applications of the ray method, already extensively employed in radio engineering, acoustics, and optics.

The *method of complex geometrical optics* is utilized for describing fields in strongly absorbing media and in the region of caustic shadows [5.1,2]). The generalization of this method over the space-time complex rays has ben given in [5.3,4] and over anisotropic media in [5.5].

A number of generalizations about geometrical optics have been designed for describing fields in the presence of caustics. One of them, the *method of canonic functions*, is a 3-D analog of the related-equation technique which is applied in solving one-dimensional problems [5.6,7]. The method yields the asymptotic expansion of the wave field in terms of special functions such as Airy, Bessel, or Weber. *Uniformly valid* field asymptotics so derived have been reported in [5.8-12], whereas local asymptotics were given in [5.13,14]. A uniformly valid asymptotic expansion can also be derived by *Maslov's canonic operator technique* [5.15,16] (see the reviews [5.10,17,18].

To determine the wave fields where interfaces are involved, *Keller* [5.19] devised the *geometrical theory of diffraction*, which has been extended in numerous subsequent works [5.20-22]. The contemporary state of this efficient method is outlined in [5.23]. Similar ideas form the basis of the *vertex-wave method* developed by *Ufimtsev* [5.24]. The generalization of the geometrical theory of diffraction over the pulse propagation in dispersive media is dealt with in [5.25,26].

Penumbra asymptotic expansion are used to define wave fields near the shadow boundary [5.27-29] (also [5.23]), while an alternative approach of generalized penumbra asymptotics must be employed where broken caustics are involved [5.12].

The *method of parabolic equation* by *Leontovich* and *Fock* [5.27,30] is a quasi-optical generalization of geometrical optics designed to account for energy transfer between ray tubes. An alternative generalization of the quasi-optical type has been supplied by the smooth-perturbation technique developed by *Rytov* [5.31]. The Kirchhoff-Huygens diffraction integral for smoothly inhomogeneous media has been used in [5.32]. The interference-integral (virtual ray) method has been suggested in [5.33,34] to describe the superposition of locally plane normal modes.

Finally, several generalizations of the ray method for nonlinear problems have been developed.

The generalizations of geometrical optics substantially widened the possibilities of the method so that now wave fields may be evaluated even in regions where the method is no longer valid, such as shadow and penumbra regions, the vicinity of caustics, foci, etc., as it was demonstrated in excellent surveys of *Berry* and *Mount* [5.11], *Berry* and *Upstill* [5.35], *Theocaris* and *Michopoulos* [5.36], *Cornbleet* [5.37], and *Stamnes* [5.38]. These generalizations as well as geometrical optics itself become markedly more efficient if computers can be employed in their applications. The short-wavelength asymptotic methods will undoubtedly retain their key position in wave theory. We are going to illustrate this fact in depth in our forthcoming book *Caustics, Catastrophes, and Wave Fields*, to appear in the present Springer Series on Wave Phenomena.

References

Chapter 1

1.1   W.R. Hamilton: *Mathematical Papers* (Cambridge Univ. Press, London 1932, 1940)

1.2   J.L. Singe: *Geometrical Optics. An Introduction to Hamilton's Method* (Cambridge Univ. Press, London 1954)

1.3   J.L. Singe: *Mechanics and De Broglie Waves* (Cambridge Univ. Press, London 1954)

1.4   L.D. Landau, E.M. Lifshitz: *Mechanics*, 3rd ed. (Pergamon, Oxford 1976)

1.5   G. Goldstein: *Classical Mechanics* (Addison-Wesley, Reading, MA 1950)

1.6   G.G. Slyusarev: *Geometricheskaya Optika* (geometrical optics) (Izd. AN SSSR, Moscow 1946)

1.7   G.G. Slyusarev: *Aberration and Optical Design Theory* (Hilger, Bristol 1984)

1.8   B.N. Begunov: *Geometricheskaya Optika* (geometrical optics) (MGU, Moscow 1966)

1.9   A.I. Tudorovsky: *Teoriya Opticheskikh Priborov* (The Theory of Optical Instruments), Vols.1,2 (Izd. AN SSSR, Moscow 1952)

1.10   M. Herzberger: *Modern Geometrical Optics* (Interscience, New York 1958)

1.11   W. Brouwer: *Matrix Methods in Optical Instrument Design* (Benjamin, New York 1964)

1.12   H.A. Buchdahl: *Optical Aberration Coefficients* (Dover, New York 1968)

1.13   O.N. Stavroudis: *The Optics of Rays, Wavefronts and Caustics* (Academic, New York 1972)
O.N. Stavroudis: *Modular Optical Design*, Springer Ser. Opt. Sci., Vol. 28 (Springer, Berlin, Heidelberg 1982)

1.14   P. Debye: (A note to A. Sommerfeld, J. Runge: Anwendung der Vektorenrechnung auf die Grundlagen der geometrischen Optik): Ann. Phys. **35**, 277 (1911). Also P. Debye: *Polar Molecules* (Chemical Catalog Co., New York 1929, reprinted by Dover, New York 1947)

1.15   V.I. Smirnov: *A Course of Higher Mathematics* (Pergamon, Oxford 1964) Vol. 4, Chap. 3

1.16   S.M. Rytov: Modulated oscillations and waves. Tr. Fiz. Inst. Akad. Nauk SSSR **2**, 3 (1940)

1.17   S.M. Rytov: On the transition from wave to geometrical optics. Dokl. Akad. Nauk SSSR **18**, 263 (1938)

1.18   R.K. Luneburg: *Mathematical Theory of Optics* (University of California Press, Berkeley, CA 1964)

1.19   R. Courant: *Partial Differential Equations* (Interscience, New York 1962)

1.20   J.B. Keller, R.M. Lewis, B.D. Seckler: Asymptotic solution of some diffraction problems. Commun. Pure Appl. Math. **9**, 207 (1956)

1.21   V.M. Babich, V.S. Buldyrev: *Asymptotic Methods in Short-Wave Diffraction Theory*, Springer Ser. Wave Phenom., Vol. 4 (Springer, Berlin, Heidelberg 1989)

1.22   C.G. Mikhlin (ed.): *Lineinye Uravneniya Matematicheskoi Fiziki* (Linear Equations of Mathematical Physics) (Nauka, Moscow 1964)

1.23 Yu.A. Kravtsov, L.A. Ostrovsky, N.S. Stepanov: Proc. IEEE **62**, 1492 (1974)

1.24 M. Born, E. Wolf: *Principles of Optics*, 5th edn. (Pergamon, Oxford 1975)

1.25 L.M. Brekhovskikh: *Waves in Layered Media* 2nd edn. (Academic, New York 1980)
L.M. Brekhovskikh, O.A. Godin: *Acoustics of Layered Media I. Plain Wave Theories*, Springer Ser. Wave Phen., Vol.5 (Springer, Berlin, Heidelberg 1990)

1.26 L.D. Landau, E.M. Lifshitz: *The Classical Theory of Fields*, 4th edn. (Pergamon, Oxford 1975)

1.27 V.L. Ginzburg: *The Propagation of Electromagnetic Waves in Plasmas*, 2nd edn. (Pergamon, Oxford 1970)

1.28 E.L. Feinberg: *Rasprostraneniye Radiovoln vdol Zemnoi Poverkhnosti* (Radiowave Propagation Along the Earth's Surface) (Izd. AN SSSR, Moscow 1961)

1.29 M. Kline, I.W. Kay: *Electromagnetic Theory and Geometrical Optics* (Wiley, New York 1965)

1.30 L.B. Felsen, N. Marcuvitz: *Radiation and Scattering of Waves*, Vols. 1,2 (Prentice-Hall, Englewood Cliffs, NJ 1973)

1.31 G.L. James: *Geometrical Theory of Diffraction for Electromagnetic Waves* (Peregrinus, London 1976)

1.32 P. Frank, R. von Mises: *Die Differential- und Integralgleichungen der Mechanik und der Physik*, 2 Vols. (Vieweg, Braunschweig 1930)

Chapter 2

2.1 S.M. Rytov: Trudy FIAN 2(1), 3 (1940)

2.2 S.M. Rytov: Dokl. Akad. Nauk SSSR **18**(2), 263 (1938)

2.3 R.P. Feynman, A.R. Hibbs: *Quantum Mechanics and Path Integrals* (McGraw-Hill, New York 1965)

2.4 V.S. Buslaev: In *Problemy Mathematicheskoi Fiziki* (problems of mathematical physics), Vol.2 (Izd. LGU, Leningrad 1967) p.85

2.5 M.V. Berry, K.E. Mount: Rep. Prog. Phys. **35**(4), 315 (1972)

2.6 G.B. Whitham: J. Fluid Mech. **22**(2), 273 (1965)

2.7 G.B. Whitham: *Linear and Nonlinear Waves* (Wiley, New York 1974)

2.8 L.A. Ostrovsky, N.S. Stepanov: Izv. VUZ Radiofiz. **14**(4), 489 (1971)

2.9 Yu.A. Kravtsov, L.A. Ostrovsky, N.S. Stepanov: Proc. IEEE **62**(11), 1492 (1974)

2.10 P. Debye (A note to A. Sommerfeld, J. Runge): Ann. Phys. **35**, 277 (1911)
P. Debye: *Polar Molecules* (Chemical Catalog Co., N.Y. 1929, reprinted by Dover, New York 1947)

2.11 W.R. Hamilton: *Mathematical Papers* (Cambridge Univ. Press, London 1932, 1940)

2.12 J.L. Synge: *Geometrical Optics. An Introduction to Hamilton's Method* (Cambridge Univ. Press, London 1954)

2.13 M. Born, E. Wolf: *Principles of Optics*, 5th edn. (Pergamon, Oxford 1975)

2.14 L.D. Landau, E.M. Lifshits: *Mechanics* 3rd. edn. (Pergamon, Oxford 1976)

2.15 G. Goldstein: *Classical Mechanics* (Addison-Wesley, Reading, Mass. 1950)

2.16 R. Courant: *Partial Differential Equations* (Interscience, New York 1962)

2.17 V.V. Stepanov: *Kurs Differentsiel'nykh Uravnenii* (a course of differential equations) (Fizmatgiz, Moscow 1958)

2.18 G.A. Korn, T.M. Korn: *Mathematical Handbook for Scientists and Engineers* (McGraw Hill, New York 1961)

2.19 V.M. Babich, V.S. Buldyrev: *Asymptotic Methods in Short-Wavelength Diffraction Theory*, Springer Ser. Wave Phen., Vol.4 (Springer, Berlin, Heidelberg 1989)

2.20 V.P. Maslov: *Théorie des Perturbations et Méthods Asymptotique* (Dunod, Paris 1972)

V.P. Maslov, M.V. Fedoryuk: *Semi-Classical Approximation in Quantum Mechanics* (Reidel, Hingham, MA 1981)

2.21 G.L. James: *Geometrical Theory of Diffraction for Electromagnetic Waves* (Peregrinus, London 1976)

2.22 V.A. Borovikov, B.E. Kinber: *Geometricheskaye Teoria Difraktsii* (the geometrical theory of diffraction) (Radio i Svjaz, Moscow 1978)

2.23 L.B. Felsen: In *Quasi-Optics*, ed. by J. Fox (Polytechnic Press, Brooklyn, NY 1964)

2.24 R.M. Lewis, J.B. Keller: NY Univ. Res. Rep. EM-194 (1964)

2.25 L.M. Brekhovskikh: *Waves In Layered Media*, 2nd edn. (Academic, New York 1980)
L.M. Brekhovskikh, O.A. Godin: *Acoustics of Layered Media I. Plane-Wave Theories*, Springer Ser. Wave Phen., Vol.5 (Springer, Berlin, Heidelberg 1990)

2.26 Yu.A. Kravtsov: Izv. VUZ Radiofiz. 10(9,10), 1283 (1967) [Engl. transl.: Radiophys. Quantum Electron. 10(9,10), 719 (1967)]

2.27 D. Wey-Yi Wang, G.A. Deschamp: Proc. IEEE 62, 1541 (1974)

2.28 Yu.I. Orlov: In *Pryamye i Obratnye Zadachi Teorii Difraktsii* (direct and inverse problems of diffraction theory) (Izd. IRE AN SSSR, Moscow 1979) p.5

2.29 M.V. Berry: Adv. Phys. 25, 1 (1976)

2.30 A.I. Tudorovsky: *Teoriya Opticheskikh Priborov* (the theory of optical instruments), Vol.1, Vol.2 (Izd. AN SSSR, Moscow 1948, 1952)

2.31 M. Herzberger: *Modern Geometrical Optics* (Interscience, New York 1958)

2.32 O.N. Stavroudis: *The Optics of Rays, Wavefronts and Caustics* (Academic, New York 1972)

2.33 Yu.I. Orlov: Izv. VUZ Radiofiz 13(5), 760 (1970) [Engl. transl.: Radiophys. Quant. Electron. 13(1970)]

2.34 Yu.I. Orlov: Izv. VUZ Radiofiz 9(3), 497; 9(4), 657 (1966) [Engl. transl.: Radiophys. Quant. Electron. 9(1966)]

2.35 Yu.I. Orlov: Izv. VUZ Radiofiz. 20(11), 1669 (1977) [Engl. transl.: Radiophys. Quant. Electron. 20, 1148 (1977)]

2.36 Yu.I. Orlov: Izv. VUZ Radiofiz. 11(2), 317 (1968) [Engl. transl.: Radiophys. Quant. Electron. 11(1968)]

2.37 Yu.I. Orlov: Radiotekh. Elektron. 11(6), 1125 (1966)

2.38 Yu.I. Orlov: Izv. VUZ Radiofiz. 10(1), 30 (1967) [Engl. transl.: Radiophys. Quantum Electron. 10(1), 13 (1967)]

2.39 L.M. Brekhovskikh (ed.): *Akustika Okeana* (ocean acoustic) (Nauka, Moscow 1974)
L. Brekhovskokh, Yu. Lysanov: *Fundamentals of Ocean Acoustics*, Springer Ser. Electrophys., Vol.8 (Springer, Berlin, Heidelberg 1982)

2.40 Yu.A. Kravtsov: Zh. Eksp. Teor. Fiz. 55(3), 798 (1968) [Engl. transl.: Sov. Phys.-JETP 28(3), 413 (1969)]

2.41 M.A. Miller, G.V. Permitin, A.A. Fraiman: Izv. VUZ Radiofiz. 21(11), 1603 (1978) [Engl. transl.: Radiophys. Quant. Electron. 21, 1114 (1978)]

2.42 V.I. Arnold: *The Mathematical Methods of Classical Mechanics* (Springer, Berlin, Heidelberg 1978)

2.43 V.I. Arnold: Usp. Mat. Nauk 30(5), 3 (1975)

2.44 V.I. Arnold: *Additional Chapters to the Theory of Ordinary Differential Equations* (Springer, Berlin, Heidelberg 1983)

2.45 V.I. Arnold: *Catastrophe Theory* (Springer, Berlin, Heidelberg 1983)

2.46 R. Gilmore: *Catastrophe Theory for Scientists and Engineers* (Wiley, New York 1981)

2.47 T.H. Bröcker, L. Lander: *Differential Germs and Catastrophes* (Cambridge Univ. Press, Cambridge 1974)

2.48 V.I. Arnold: Commun. Pure Appl. Math. 29(6), 557 (1976)

2.49 J.N.L. Connor: Molec. Phys. 31(1), 33 (1976)

2.50 V.I. Arnold, A.N. Varchenko, S.M. Gusein-Zade: *Singularities of Differentiable Mappings*, Vols.1 and 2 (Birkhäuser, Boston 1985 and 1988)
2.51 J.J. Duistermaat: Commun. Pure Appl. Math. 27, 207 (1974)
2.52 V.I. Arnold: Usp. Mat. Nauk 28(5), 17 (1973)
2.53 M.V. Berry, C. Upstill: Catastrophe optics: morphologies of caustics and their diffraction patterns. *Progress in Optics* 18, 257-346 (North-Holland, Amsterdam 1980)
2.54 V.I. Arnold: Sov. Phys.- Uspekhi 26(12), 1024 (1983)
2.55 Yu.A. Kravtsov, Yu.I. Orlov: Sov. Phys.-Uspekhi 26(12), 1038 (1983)
2.56 L. Levey, L.B. Felsen: Radio Sci. 4(10), 959 (1969)
2.57 Yu.I. Orlov: Radiotekh. Elektron. 20(2), 242 (1975)
2.58 Yu.I. Orlov: Radiotekh. Elektron. 21(4), 730 (1976) [Engl. transl.: Radio Eng. Electron (USSR) 21(4), 50 (1976)]
2.59 L. Levey, L.B. Felsen: J. Inst. Math. Appl. 3(1), 76 (1967)
2.60 Yu.I. Orlov: Cand. Science Thesis, Moscow Power Institute (1969)
2.61 N.S. Orlova, Yu.I. Orlov: Radiotekh. Elektron. 22(9), 1811 (1977) [Engl. transl.: Radio Eng. Electron (USSR) 22(9), 35 (1977)]
2.62 Yu.I. Orlov: Radiotekh. Elektron. 21(1), 62 (1976) [Engl. transl.: Radio Eng. Electron (USSR) 21(1), 50 (1976)]
2.63 Yu.A. Kravtsov: Izv. VUZ Radiofiz. 8(4), 659 (1965) [Eng. transl.: Radiophys. Quant. Electron. 22(4), 40 (1965)]
2.64 V.A. Kaloshin, Yu.I. Orlov: Radiotekh. Elektron. 18(10), 2028 (1973)
2.65 R.M. Lewis: Arch. Ration Mechan. Anal. 20(3), 191 (1965)
2.66 L.D. Landau, E.M. Lifshits: *The Classical Theory of Fields*, 4th ed. (Pergamon, Oxford 1975)
2.67 Yu.A. Kravtsov: Akust. Zh. 14(1), 1 (1968) [Engl. transl.: Sov. Phys. - Acoustics 14(1), 1 (1968)]
2.68 D. Ludwig: SIAM Rev. 12(3), 325 (1970)
2.69 V.M. Babich: In *Voprosy Dinamicheskoi Teorii Rasprostranenia Seismicheskikh Voln* (topics of the dynamic theory of seismic wave propagation), Vol.5 (Leningrad State Univ. - Izd. LGU 1961) p.145
2.70 J.B. Keller, S. Rubinov: Ann. Phys. 9(1), 24 (1960)
2.71 V.P. Bykov: In *Power Electroncis*, Vol.4 (Nauka, Moscow 1965) p.66
2.72 L.A. Vainstein: *Open Resonators and Waveguides* (Golem, Boulder 1969)
2.73 A.S. Alekseyev, V.M. Babich, B.Ya. Gel'chinsky: In *Voprosy Dinamicheskoi Teorii Resprostranenia Seismicheskikh Voln* (topics of the dynamic theory of seismic wave propagation), Vol.5 (Izd. LGU, Leningrad 1961) p.3
2.74 Yu.A. Kravtsov, Yu.I. Orlov: Usp. Fiz. Nauk 132(3), 476 (1980) [Engl. transl.: Sov. Phys.-Usp. 23(11), 750 (1980)]
      Yu.A. Kravtsov, Yu.I. Orlov: Radio Sci., 16(6), 975 (1981)
2.75 J.B. Keller, F.C. Karal: J. Appl. Phys. 31(6), 1039 (1960)
2.76 B.R. Levy, J.B. Keller: Comm. Pure Appl. Math. 12(1), 159(1959)
2.77 R.M. Lewis, N. Bleistein, D. Ludwig: Commun. Pure Appl. Math. 20(2), 295 (1967)
2.78 V.M. Babich: In *Voprosy Dinamicheskoi Teorii Resprostranenia Seismicheskikh Voln* (topics of the dynamic theory of seismic wave propagation) Vol.6 (Izd. LGU 1962) p.60
2.79 I.G. Kondratjev, G.V. Permitin: Izv. VUZ Radiofiz. 13(12), 1795 (1970) [Engl. transl.: Radiophys. Quant. Electron. 13(12), 130 (1970)]
2.80 Yu.A. Kravtsov: Izv. VUZ Radiofiz. 11(10), 1582 (1968) [Engl. transl.: Radiophys. Quant. Electron. 11(10), 120 (1968)]
2.81 L.B. Felsen, N. Marcuvitz: *Radiation and Scattering of Waves*, Vols. 1,2 (Prentice-Hall, Englewood Cliffs, NJ 1973)
2.82 R.M. Lewis: In *Quasi-Optics*, ed. by J. Fox (Polytechnic Press, Brooklyn, NY 1964)

2.83 L.B. Felsen: IEEE Trans. **AP-17**, 191 (1969)

2.84 L.B. Felsen: SIAM Review 12(3), 424 (1970)

2.85 F. Friedlander: *Sound Pulses* (Cambridge Univ. Press, Cambridge 1958)

2.86 L.A. Vainstein: Usp. Fiz. Nauk 118(2), 339 (1976) [Engl. transl.: Sov. Phys. - Usp. 19(1976)]

2.87 L.B. Felsen (ed.): *Transient Electromagnetic Fields*, Topics Appl. Phys., Vol.10 (Springer, Berlin, Heidelberg 1976)

2.88 S.L. Sobolev: *The Wave Equation in an Inhomogeneous Medium*, Trudy Seismol. Inst. Acad. Sci. USSR, Vol.6 (1930);
V.I. Smirnov: *A Course of Higher Mathematics*, Vol.4 (Pergamon, Oxford 1964)

2.89 M. Kline, I.W. Kay: *Electromagnetic Theory and Geometrical Optics* (Wiley, New York 1965)

2.90 V.A. Borovikov: *Difraktsiya na Mnogougol'nikakh i Mnogogrannikakh* (diffraction by polygons and polyhedrons) (Nauka, Moscow 1966)

2.91 V.M. Babich: In *Voprosy Dinamicheskoi Teorii Resprostranenia Seismicheskikh Voln*, Vol.5 (topics of the dynamic theory of seismic wave propagation) (Izd. LGU 1961) p.25

2.92 P. Wolfe, R.M. Lewis: J. Diff. Equat. 2(3), 328 (1966)

2.93 R.M. Lewis, W. Pressman: J. Diff. Equat. 3(3), 360 (1967)

2.94 L.A. Ostrovsky: Izv. VUZ Radiofiz. 18(4), 618 (1975)

2.95 S.G. Mikhlin (ed.): *Lineinye Uravneniya Matematicheskoi Fiziki* (Linear Equations of Mathematical Physics) (Nauka, Moscow 1964)

2.96 P. Frank, R. von Mises: *Die Differential- und Integralgleichungen der Mechanik und Physik*, 2 Vols. (Vieweg, Braunschweig 1930)

2.97 E. Kamke: *Gewöhnliche Differentialgleichungen* (Akademische Verlagsgesellschaft Geest und Portig, Leipzig 1959)

2.98 P.M. Morse, H. Feshback: *Methods of Theoretical Physics*, Vol.2 (McGraw-Hill, New York 1953)

2.99 E. Madelung: *Die mathematischen Hilfsmittel des Physikers* (Springer, Berlin, Heidelberg 1957)

2.100 Ya.N. Feld, L.S. Benenson: *Antenna Feeder Assemblies* (Izd. VVIA, Moscow 1959) Pt.2

2.101 M.A. Kolosov, A.V. Shabel'nikov: *Refraktsiya Elektromagnitnykh Voln v Atmosferakh Zemli, Venery i Marsa* (refraction of electromagnetic waves in the atmospheres of Earth, Venus and Mars) (Sov. Radio, Moscow 1976)

2.102 V.G. Burkov, Yu.Ya. Yashin: Izv. VUZ Radiofiz. 10(12), 1631 (1967) [Engl. transl.: Radiophys. Quantum Electron. 10(12), 911 (1967)]

2.103 D.L. Nielson: Rad. Sci. 3(1), 101 (1968)

2.104 Yu.I. Orlov, A.P. Anyutin: In *Trudy MEI*, Vol.119 (Power Institute, Moscow 1972) p.92

2.105 E.L. Boyarintsev, Yu.Ya. Yashin: Geomagn. Aeronomija 14(6), 1019 (1974)

2.106 S.J. Maurer, L.B. Felsen: Proc. IEEE 55(10), 1718 (1967)

2.107 N.P. Mar'in: Radiotekh. Elektron. 10(2), 235 (1965)

2.108 N.P. Mar'in: Geomagn. Aeronomia 5(2), 260 (1965)

2.109 Yu.M. Zhidko: Izv. VUZ Radiofiz. 11(6), 876 (1968)

2.110 A.P. Yarygin: Radiotekh. Elektron. 14(5), 912 (1969)

2.111 N.P. Mar'in: Radiotekh. Elektron. 10(10), 1765 (1965)

2.112 N.P. Mar'in: Geomagn. Aeronomia 5(3), 568 (1965)

2.113 R. Woo, A. Ishimaru: Radio Sci. 6(5), 583 (1971)

2.114 B.I. Semenov: Radiotekh. Elektron. 17(2), 1725 (1972)

2.115 B.I. Semenov: Radiotekh. Elektron. 19(1), 51 (1974)

2.116 Yu.Y. Yashin: Izv. VUZ Radiofiz. 11(4), 491 (1968)

2.117 Yu.Ya. Yashin: Some Topics of EM Wave Propagation Theory for Inhomogeneous Anisotropic Media. Cand. Science Thesis, Gorky State University (GGU, Gorky 1969)

2.118 G.G. Slyusarev: *Geometricheskoye Optika* (geometrical optics) (Izd. AN SSSR, Moscow, Leningrad 1946)

2.119 G.G. Slyusarev: *Aberration and Optical Design Theory* (Hilger, Bristol 1984)

2.120 B.N. Begunov: *Geometricheskaya Optika* (geometrical optics) (Izd. MGU, Moscow 1966)

2.121 W. Brouwer: *Matrix Methods in Optical Instrument Design* (Benjamin, New York 1964)

2.122 H.A. Buchdahl: *Optical Aberration Coefficients* (Dover, New York 1968)

2.123 R.K. Luneburg: *Mathematical Theory of Optics* (Univ. California Press, Berkley 1964)

2.124 L.A. Chernov: *Wave Propagation in Random Media* (McGraw-Hill, New York 1960)

2.125 V.I. Tatarski: *The Effect of the Turbulent Atmosphere on Wave Propagation* (US Dep. Commerce, Springfield 1971)

2.126 Yu.N. Barabanenkov, Yu.A. Kravtsov, S.M. Rytov, V.I. Tatarski: Usp Fiz. Nauk 102(1), 3 (1970) [Engl. transl.: Sov. Phys.-Usp. 13(5), 551–78 (1971)]

2.127 S.M. Rytov, Yu.A. Kravtsov, V.I. Tatarski: *Principles of Statistical Radiophysics*, Vol.4, *Wave Propagation through Random Media* (Springer, Berlin, Heidelberg 1989)

2.128 A. Ishimaru: *Wave Propagation and Scattering in Random Media*, 2 Vols. (Academic, New York 1978)

2.129 R.S. Lawrence, G.G. Little, J.A. Chivers: Proc. IEEE 52(1), 4 (1964)

2.130 Yu.A. Kravtsov, Z.I. Feizulin, A.G. Vinogradov: *Prokhozhdenie Radiovoln Cherez Atmosferzu Zemli* (radiowave propagation through the Earth atmosphere) (Radio i Svjaz, Moscow 1983)

2.131 Yu.A. Kravtsov, Z.I. Feizulin: Radiotekh. Elektron. 16(2), 1777 (1971)

2.132 V.A. Baranov, Yu.A. Kravtsov: Izv. VUZ Radiofiz. 18(1), 52 (1975)

2.133 M.V. Tinin: In *Issledovaniya po Geomagnetizmu, Aeronomii i Fizike Solntsa* (studies in geomagnetism, aeronomy and solar physics), Vol.39 (Nauka, Moscow 1976) p.166

2.134 V.A. Baranov, A.V. Popov: In *Rasprostraneniye Dekametrovykh Radiovoln* (propagation of decametric radio waves) (IZMIRAN, Moscow 1975) p.14

2.135 T.N. Duboshin: *Solar Mechanics. Basic Problems and Methods* (Nauka, Moscow 1968)

2.136 M.F. Subbotin: *Introduction to Theoretical Astronomy* (Nauka, Moscow 1968)

2.137 H.L. Bertoni, L.B. Felsen, A. Hessel: IEEE Trans. AP-19(3), 226 (1971)

2.138 F.B. Cherny: *Rasprostraneniye Radiovoln* (radiowave propagation) (Sov. Radio, Moscow 1972)

2.139 E.L. Feinberg: *Rasprostranenie Radiovoln vdol Zemnoy Poverkhnosti* (the propagation of radio waves above the Earth's surface) (Izd. AN SSSR, Moscow 1961)

2.140 Yu.A. Kravtsov, Z.I. Feizulin: Izv. VUZ Radiofiz. 12(6), 886 (1969)

2.141 G.S. Gorelik: *Kolebaniya i Volny* (oscillations and waves), 2nd. ed. (Fizmatgiz, Moscow 1959)

2.142 P.V. Bliokh: Izv. VUZ Radiofiz. 7(3), 460 (1964)

2.143 A.G. Litvak, V.I. Talanov: Izv. VUZ Radiofiz. 10(4), 539 (1967) [Engl. transl.: Radiophys. Quant. Electron. 10(4), 296 (1967)]

2.144 Yu.L. Gazaryan: In *Voprosy Dinamicheskoy Teorii Rasprostraneniya Seismicheskikh Voln* (topics of the dynamic theory of seismic wave propagation), Vol.5 (Izd. LGU, Leningrad 1961) p.73

2.145 V.Ya. Groshev, Yu.A. Kravtsov: Izv. VUZ Radiofiz. 11(12), 1812 (1968)

2.146 Yu.A. Kravtsov: Izv. VUZ Radiofiz. 7(4), 664 (1964)

2.147 D. Ludwig: Commun. Pure Appl. Math. 19(2), 215 (1966)

2.148 Yu.I. Orlov: In *Proc. MEI*, Vol.194 (Power Institute, Moscow 1974) p.103

2.149 Yu.A. Kravtsov: Rays and caustics as physical objects. *Progress in Optics* **26**, 227–348 (North-Holland, Amsterdam 1988)
2.150 A.A. Asatryan, Yu.A. Kravtsov: Wave Motion **10**(1), 45 (1988)

Chapter 3

3.1 M. Herzberger: *Modern Geometrical Optics* (Interscience, New York 1958)
3.2 A.V. Pogorelov: *Lektsii po Differentsial'noi Geometrii* (lectures in differential geometry) (Izd. KhGU, Kharkov 1967)
3.3 A.I. Tudorovsky: *Teoriya Opticheskikh Priborov* (the theory of optical instruments) Vol.1, Vol.2 (Izd. AN SSSR, Moscow, 1948, 1952)
3.4 P.K. Rashevsky: *Kurs Differentsyalnoi Geometrii* (a course of differential geometry) (GITTL, Moscow 1956)
3.5 M.V. Berry: Adv. Phys. **25**, 1 (1976)
3.6 G. Stroke: *An Introduction to Coherent Optics and Holography* (Academic, New York 1966)
3.7 V.A. Borovikov, B.E. Kinber: *Geometricheskaya Teoria Difraktsii* (the geometrical theory of diffraction) (Svjaz, Moscow 1978)
3.8 G.T. Markov, A.F. Chaplin: *Vozbuzhdeniye Electromagnytnykh Voln* (excitation of electromagnetic waves) (Energia, Moscow 1967)
3.9 Yu.I. Orlov, A.P. Anyutin, S.K. Tropkin: In *Diffraction and Wave Propagation Theory*, Proc. VI All-Union Symposium, Vol.1 (Izd. AN SSSR, Moscow 1977) p.183
3.10 E.L. Feinberg: *Rasprostraneniye Radiovoln vdol Zemnoi Poverkhnosti* (radio-wave propagation along the Earth's surface) (Izd. AN SSSR, Moscow 1961)
3.11 Yu.I. Orlov, S.K. Tropkin: In *Trudy MEI*, No.341 (Power Institute, Moscow 1977) p.37
3.12 Yu.I. Orlov, S.K. Tropkin: In *Mashinnoye Proyektirovanie Ustroistv i Sistem SVCh* (computer aided design of microwave systems) (Izd. TGU, Tbilisi 1979) p.215
3.13 Ya.N. Feld (ed.): *Antenny SVCh* (microwave antennas), Vols.1,2 (Sov. Radio, Moscow 1950)
3.14 Ya.N. Feld, L.S. Benenson: *Antenna Feeder Assemblies*, Pt.2 (Izd. VVIA, Moscow 1959)
3.15 Yu.A. Kravtsov, Yu.I. Orlov: Usp. Fiz. Nauk **132**(3), 476 (1980) [Engl. transl.: Sov. Phys.-Usp. **23**(11), 750 (1980)]
3.16 L.M. Brekhovskikh: *Waves in Layered Media*, 2nd edn. (Academic, New York 1980)
L.M. Brekhovskikh, O.A. Godin: *Acoustics of Layered Media I. Plane-Wave Theories*, Springer Ser. Wave Phen., Vol.5 (Springer, Berlin, Heidelberg 1990)
3.17 T. Pearcey: Phil. Mag. **37**, 311 (1946)
3.18 V.A. Fock: Zh. Eksp. Teor. Fiz. **20**(11), 961 (1950)
3.19 V.A. Fock: *Electromagnetic Diffraction and Propagation Problems* (Pergamon, New York 1965)
3.20 B.Ya. Gel'chinsky: Dokl. Akad. Nauk SSSR **118**(3), 458 (1958)
3.21 I.M. Fuks: Izv. VUZ Radiofiz. **8**(6), 1078 (1965)
3.22 F.G. Bass, I.M. Fuks: *Wave Scattering from a Statistically Rough Surface* (Pergamon, Oxford 1979)
3.23 G.A. Deschamps: Proc. IEEE **60**(9), 1022 (1972)
3.24 G.L. James: *Geometrical Theory of Diffraction for Electromagentic Waves* (Peregrinus, London 1976)
3.25 F. Friedlander: *Sound Pulses* (Cambridge Univ. 1958)
3.26 J.B. Keller, H.B. Keller: J. Opt. Soc. Amer. **40**(1), 48 (1950)
3.27 M.I. Kantorovich, Yu.K. Muravjev: Zh. Tekh. Fiz. **22**(3), 394 (1952)

3.28 S.W. Lee: IEEE Trans. **AP-23**(2), 184 (1975)

3.29 G.G. Slyusarev: *Geometricheskaya Optika* (geometrical optics) (Izd. AN SSSR, Moscow, Leningrad 1946)

3.30 G.G. Slyusarev: *Aberation and Optical Design Theory* (Hilger, Bristol 1984)

3.31 B.N. Begunov: *Geometricheskaya Optika* (geometrical optics) (Izd. MGU, Moscow 1966)

3.32 W. Brouwer: *Matrix Methods in Optical Instrument Design* (Benjamin, New York 1964)

3.33 H.A. Buchdahl: *Optical Aberration Coefficients* (Dover, New York 1968)

3.34 O.N. Stavroudis: *The Optics of Rays, Wavefronts and Caustics* (Academic, New York 1972)

3.35 J.B. Keller, R.M. Lewis, B.D. Seckler: Comm. Pure. Appl. Math. **9**(2), 207 (1956)

3.36 R.M. Lewis: In *Quasi-Optics* (Polytechnic, Brooklyn, New York 1964)

3.37 V.I. Ivanov: Zh. Vych. Mat. Mat. Fiz. **8**(5), 1141 (1968)

3.38 V.I. Ivanov: Zh. Vych. Mat. Mat. Fiz. **10**(2), 490 (1970)

3.39 R.G. Kouyoumjian, P.G. Pathak: Proc. IEEE **62**(11), 1448 (1974)

3.40 E.A. Stager, E.V. Chaevsky: *Resseyaniye na Telakh Slozhnoi Formy* (wave scattering by complex-shaped bodies) (Sov. Radio, Moscow 1974)

3.41 V.O. Kobak: *Radiolokatsionnye Otrazhateli* (Radar Reflectors) (Sov. Radio, Moscow 1975)

3.42 J.R. Mentzer: *Scattering and Diffraction of Radio Waves* (Pergamon, London, New York 1955)

3.43 Yu.I. Orlov: Izv. VUZ Radiofiz. **13**(5), 760 (1970) [Engl. transl.: Radiophys. Quant. Electron. **13**(5), 534 (1970)]

3.44 R.C. Spencer, G. Hyde: IEEE Trans. **AP-16**, 317 (1968)

3.45 M.S. Zhuk, Yu.B. Molochkov: *Proyektirovaniye Antenno-Fidernykh Ustroistv* (design of antenna-feeder assemblies) (Energia, Moscow, Leningrad 1966)

3.46 V.M. Babich, V.S. Buldyrev: *Asymptotic Methods in Short-Wavelength Diffraction Theory*, Springer Ser. Wave Phen., Vol.4 (Springer, Berlin, Heidelberg 1989)

3.47 L.A. Vainstein: *Open Resonators and Waveguides* (Golem, Boulder 1969)

3.48 B.E. Kinber: Radiotekh. Elektron. **6**(8), 1273 (1961)

3.49 R. Burridge: Proc. Roy. Soc. **A-276**(1366), 367 (1963)

3.50 Z. Bartkowski: Optik **18**(1), 22 (1961)

3.51 V.P. Bykov: Izv. VUZ Radiofiz. **19**(1), 85 (1966)

3.52 E.I. Kheifets, E.L. Shenderov: Akust. Zh. **18**(3), 456 (1972)

3.53 G.S. Landsberg: *Optics* (Nauka, Moscow 1976)

3.54 F.B. Cherny: *Rasprostreneniye Radiovoln* (Radiowave Propagation) (Sov. Radio, Moscow 1972)

3.55 L.M. Brekhovskikh: Usp. Fiz. Nauk **32**(4), 464 (1947)

3.56 L.M. Brekhovskikh (ed.): *Akustika Okeana* (ocean acoustics) (Nauka, Moscow 1974)

3.57 Yu.A. Kravtsov: In *Progress in Optics* **26**, 227 (North-Holland, Amsterdam 1988)

3.58 G.D. Malyuzhinets: Usp. Fiz. Nauk **69**(2), 321 (1959)

3.59 V.A. Fock, L.A. Vainstein: Radiotekh. Elektron. **8**(3), 363, 377 (1963)

3.60 L.B. Felsen: In *Quasi-Optics* (Polytechnic, Brooklyn, New York 1964)

3.61 L.B. Felsen, N. Marcuvitz: *Radiation and Scattering of Waves*, Vols.1,2 (Prentice-Hall, Englewood Cliffs 1973)

3.62 R.M. Lewis, J.B. Keller: New York University Res. Rep. EM-194 (1964)

3.63 Yu.I. Orlov: Izv VUZ Radiofiz.**9**(3), 497; **9**(4), 657 (1966) [Engl. transl.: Radiophys. Quantum Electron. **9**(3), 348; **9**(4), 460 (1966)]

3.64 Yu.I. Orlov: Izv. VUZ Radiofiz. **20**(11), 1669 (1977) [Engl. transl.: Radiophys. Quant. Electron. **20**(11), 1148 (1977)]

3.65   Yu.I. Orlov: Izv. VUZ Radiofiz. 11(2), 317 (1968) [Engl. transl.: Radiophys. Quant. Electron. 11(2), 221 (1968)]

3.66   V.V. Zheleznyakov: *Elektromagnitnye Volny v Kosmicheskoi Plazme* (electromagnetic waves in interstellar plasma) (Nauka, Moscow 1977)

3.67   A.N. Barkhatov, I.I. Shmelev: Akust. Zh. 4(1), 100 (1958)

3.68   A.N. Barkhatov: *Modelirovaniye Rasprostraneniya Zvuka v More* (modeling sound propagation in sea) (Hidrometeoizdat, Leningrad 1968)

3.69   I. Tolstoy, C.S. Clay: *Ocean Acoustics* (McGraw-Hill, New York 1966)

3.70   Yu.I. Orlov: Izv. VUZ Radiofiz. 13(3), 412 (1970)

3.71   M.A. Miller, G.V. Permitin, A.A. Fraiman: Izv VUZ Radiofiz. 21(11), 1603 (1978) [Engl. transl.: Radiophys. Quant. Electron. 21(11), 1114 (1978)]

3.72   Yu.I. Orlov, S.K. Tropkin: Izv. VUZ Radiofiz. 23(12), 1473 (1980) [Engl. transl.: Radiophys. Quant. Electron. 23(12), 1030 (1980)]

3.73   M.A. Pedersen, D.F. Gordin: J. Acoust. Soc. Amer. 47(2), 419 (1967)

3.74   Yu.A. Kravtsov: Izv. VUZ Radiofiz. 8(4), 659 (1965) [Engl. transl.: Radiophys. Quant. Electron. 8, 453 (1965)]

3.75   Yu.A. Kravtsov: Izv. VUZ Radiofiz. 10(9,10), 1283 (1967) [Engl. transl.: Radiophys. Quant. Electron. 10(2,10), 719 (1967)]

3.76   V.A. Kaloshin, Yu.I. Orlov: Radiotekh. Elektron. 18(10), 2028 (1973)

3.77   M.V. Tinin: In *Studies in Geomagnetism, Aeronomy and Solar Physics*, Vol.29 (Nauka, Moscow 1973) p.157

3.78   Yu.A. Kravtsov, M.V. Tinin, Yu.N. Cherkashin: Geomagn. Aeronomia 19(5), 769 (1979)

3.79   A.V. Gurevich, E.E. Tsedilina: *Sverkhdal'neye Resprostraneniye Korotkikh Radiovoln* (long-range propagation of short radio waves) (Nauka, Moscow 1979)

3.80   K.G. Budden: *Radio Waves in the Ionosphere* (Cambridge Univ. Press, Cambridge 1961)

3.81   J.M. Kelso: *Radio Ray Propagation in the Ionosphere* (McGraw-Hill, London 1964)

3.82   Ya.L. Al'pert: *Rasprostraneniye Radiovoln i Ionosfera* (radio wave propagation and the ionosphere) (Izd. AN SSSR, Moscow 1960)

3.83   K. Davies: *Ionospheric Radio Waves* (Blaisdell, Div. of Ginn, Wallham, Mass. 1969)

3.84   T.S. Kerblay, E.M. Kovalevskaya: *O Trayektoriyakh Korotkikh Radiovoln v Ionosfere* (on the short wave propagation paths in the ionosphere) (Nauka, Moscow 1974)

3.85   Yu.I. Orlov: Izv. VUZ Radiofiz. 10(1), 30 (1967) [Engl. transl.: Radiophys. Quantum Electron. 10(1), 23 (1967)]

3.86   V.L. Ginzburg: *The Propagation of Electromagnetic Waves in Plasmas*, 2nd edn. (Pergamon, Oxford 1970)

3.87   A. Erdélyi: *Asymptotic Expansions* (Dover, New York 1956)

3.88   A.H. Nayfeh: *Perturbation Methods* (Wiley, New York 1973)

3.89   J. Heading: *An Introduction to Phase-Integral Methods* (Methuen - Wiley, London 1962)

3.90   N. Fröman, P.O. Fröman: *JWKB Approximation* (North Holland, Amsterdam 1965)

3.91   V.B. Gil'denburg, Yu.M. Zhidko, I.G. Kondratjev, M.A. Miller: Izv. VUZ Radiofiz. 10(9,10), 1358 (1967) [Engl. transl.: Radiophys. Quant. Electron. 10(9,10), 761 (1967)]

3.92   S.D. Zhernosek, I.G. Kondratjev: Izv. VUZ Radiofiz. 13(2), 656 (1970) [Engl. transl.: Radiophys. Quant. Electron. 13(2), 450 (1970)]

3.93   V. Cerveny, I.A. Molotkov, I. Psencik: *Ray Method in Seismology* (Univ. Karlova, Praha 1977)

3.94   Yu.A. Kravtsov: Akust. Zh. 14(1), 1 (1968) [Engl. transl.: Sov. Phys.-Acoust. 14(1), 1 (1968)]

3.95 Yu.A. Kravtsov: Izv. VUZ Radiofiz. 7(4), 664 (1964) [Engl. transl.: Radiophys. Quant. Electron. 7(4), 104 (1964)]

3.96 D. Ludwig: Comm. Pure Appl. Math. 19(2), 215 (1966)

3.97 R.H. Lang, J. Shmoys: J. Acoust. Soc. Am. 48(1), 242 (1970)

3.98 I.A. Molotkov: In *Topics of the Dynamic Theory of Seismic Wave Propagation*, Vol.6 (Izd. LGU – State Univ., Leningrad 1962) p.92

3.99 N.V. Tsepelev: In *Mathematical Problems of Wave Propagation Theory*, Vol.3 (Nauka, Leningrad 1970) p.209

3.100 B.D. Seckler, J.B. Keller: J. Acoust. Soc. Am. 31(2), 192 (1959)

3.101 B.D. Seckler, J.B. Keller: J. Acoust. Soc. Am. 31(2), 206 (1959)

3.102 J.R. Wait: *Electromagnetic Waves in Stratified Media* (Pergamon, Oxford 1962)

3.103 P.E. Krasnushkin: *Metod Normalnykh Voln v Primenenii k Probleme Dal'nikh Radiosvyazei* (the method of normal modes as applied to long-range radio communications) (Izd. MGU, Moscow 1947)

3.104 Yu.A. Kravtsov, Z.I. Feizulin: Izv. VUZ Radiofiz. 12(6), 886 (1969) [Engl. transl.: Radiophys. Quant. Electron. 12(6), 597 (1969)]

3.105 I.G. Kondratjev, G.V. Permitin: Izv. VUZ Radiofiz. 13(12), 1795 (1970) [Engl. transl.: Radiophys. Quant. Electron. 13(12), 1240 (1970)]

3.106 G.V. Permitin: Izv. VUZ Radiofiz. 16(2), 254 (1973) [Engl. transl.: Radiophys. Quant. Electron. 16(2), 188 (1973)]

3.107 M.A. Kolosov, A.V. Shabel'nikov: *Refraktsiya Electromagnitnykh Voln v Atmosferakh Zemli, Venery i Marsa* (refraction of electromagnetic waves in the atmosphere of Earth, Venus and Mars) (Sov. Radio, Moscow 1976)

3.108 R. Woo, A. Ishimaru: Radio Sci. 6(5), 583 (1971)

3.109 L.D. Landau, E.M. Lifshits: *Mechancis* (Pergamon, Oxford 1969)

3.110 G. Goldstein: *Classical Mechanics* (Addison-Wesley, Reading, Mass. 1950)

3.111 J.L. Synge: *Geometrical Optics. An Introduction to Hamilton's Method* (Cambridge Univ. Press, London 1954)

3.112 J.L. Synge, B.A. Griffith: *Principles of Mechanics* (McGraw-Hill, New York 1942)

3.113 T.N. Duboshin: *Solar Mechanics, Basic Problems and Methods* (Nauka, Moscow 1968)

3.114 E.L. Stiefel, G. Scheifele: *Linear and Regular Celestial Mechanics. Perturbed Two-Body Motion. Numerical Methods. Canonical Theory* (Springer, Berlin, Heidelberg 1971)

3.115 N.G. Alexopoulos: IEEE Trans. AP-22(2), 242 (1974)

3.116 N.F. Mott, H.S.W. Massey: *The Theory of Atomic Collisions* (Clarendon, Oxford 1965)

3.117 R.G. Newton: *Scattering Theory of Waves and Particles* (McGraw-Hill, New York 1965)

3.118 N.P. Mar'in: Radiotekh. Elektron. 10(2), 235 (1965)

3.119 J. Hasselgrove: In Proc. Cambridge Conf. on Phys. Ionosphere (Phys. Soc., London 1955) p.355

3.120 E.G. Zelkin, R.A. Petrova: *Linzovye Antenny* (Lens Antennas) (Sov. Radio, Moscow 1974)

3.121 E.E. Voik (Khvoikova): Geomagnetism i Aeronomia 14(1), 57 (1974)

3.122 P.E. Krasnushkin: Dokl. Akad. SSSR 200(6), 1313 (1971)

3.123 L.C. Humphrey: Radio Sci. 3(11), 1074 (1968)

3.124 D.S. Lukin, E.A. Palkin: In *Difraktsionnye Effecty Dekametrovykh Radiovoln v Ionosfere* (diffraction effects of decametric radio waves in the ionosphere) (Nauka, Moscow 1977) p.12

3.125 S.V. Khudyakov: Zh. Eksp. Teor. Fiz. 56(3), 938 (1969) [Engl. transl.: Sov. Phys.-JETP 29(3), 507-512 (1969)]

3.126 S.V. Khudyakov: Zh. Eksp. Teor. Fiz. 57(3), 927 (1969)

3.127 V.P. Maslov: *Théorie des Perturbation et Méthodes Asymptotique* (Dunod, Paris 1972)
V.P. Maslov, M.V. Fedoryuk: *Semi-classical Approximation in Quantum Mechnics* (Reidel, Hingham, Mass. 1981)
3.128 B.I. Semenov: Radiotekh. Elektron. 17(2), 1725 (1972)
3.129 R.S. Lawrence, G.G. Little, J.A. Chivers: Proc. IEEE 52(1), 4 (1964)
3.130 B.I. Semjonov: Radiotekh. Electron 19(1), 51 (1974)
3.131 V.P. Maslov: Dokl. Akad. Nauk SSSR 151(2), 306 (1963)
3.132 I.T. Selezov, V.V. Yakovlev: *Difraktsiya Voln na Simmetrichnykh Neodnorodnostyakh* (diffraction of waves at symmetrical inhomogeneities) (Naukova Dumka, Kiev 1979)
3.133 V.A. Permyakov: Izv. VUZ Radiofiz. 11(4), 531 (1968)
3.134 E.S. Birger, L.A. Vainstein, N.B. Konyukhova: Zh. Vych. Mat. Mat. Fiz. 16(6), 1526 (1976)
3.135 L.A. Vainstein, E.S. Birger, N.B. Konyukhova, E.L. Kosarev, G.P. Prudkovsky: Fizika Plasmy 2(4), 658 (1976)
3.136 V.A. Permyakov: Izv. VUZ Radiofiz. 19(10), 1556 (1976) [Engl. transl.: Radiophys. Quant. Electron. 19(10), 1090 (1976)]
3.137 A.L. Gutman, V.A. Chesnokov: Radiotekh. Elektron. 14(2), 335 (1969)
3.138 Yu.I. Orlov, A.V. Demin: Trudy MEI, Vol.497 (Power Inst., Moscow 1980) p.10
3.139 V.A. Permyakov: Izv. VUZ Radiofiz. 23(9), 780 (1980) [Engl. transl.: Radiophys. Quant. Electron. 23(9), 716 (1980)]
3.140 Yu.I. Orlov: Radiotekh. Elektron. 11(6), 1125 (1966)
3.141 N.P. Mar'in: Geomagn. Aeronomia 5(2), 260 (1965)
3.142 Yu.M. Zhidko: Izv. VUZ Radiofiz, 11(6), 876 (1968)
3.143 A.P. Yarygin: Radiotekh. Elektron. 14(5), 912 (1969)
3.144 D.L. Nielson: Rad. Sci. 3(1), 101 (1968)
3.145 Yu.I. Orlov, A.P. Anyutin: Trudy MEI, Vol.119 (Power Inst., Moscow 1972) p.92
3.146 E.L. Boyarintsev, Yu.Ya. Yashin: Geomagn. Aeronomia 14(6), 1019 (1974)
3.147 M. Skolnik (ed.): *Radar Handbook* (McGraw-Hill, New York 1970)
3.148 G.L. Gdalevich, K.I. Gringauz, V.A. Rudakov, S.M. Rytov: Radiotekh. Elektron, 8(6), 942 (1963)
3.149 Yu.A. Kravtsov, Z.I. Feizulin, A.G. Vinogradov: *Prokhozhdeniye Radiovoln Cherez Atmosferu Zemli* (radiowave propagation through the Earth's atmosphere) (Radio i Syjaz, Moscow 1983)
3.150 V.A. Baranov, Yu.A. Kravtsov: Izv. VUZ Radiofiz. 18(1), 52 (1975)
3.151 V.A. Krasilnikov: Dokl. Akad. Nauk SSSR 47(7), 486 (1945)
3.152 P.G. Bergman: Phys. Rev. 70, 486 (1946)
3.153 L.A. Chernov: *Wave Propagation in Random Media* (McGraw-Hill, New York 1960)
3.154 V.I. Tatarski: *The Effect of the Turbulent Atmosphere on Wave Propagation* (US Dep. Commerce, Springfield 1971)
3.155 Yu.N. Barabanenkov, Yu.A. Kravtsov, S.M. Rytov, V.I. Tatarski: Usp. Fiz. Nauk 102(1), 3 (1970) [Engl. transl.: Sov. Phys.-Usp. 13(5), 551–578 (1971)]
3.156 S.M. Rytov, Yu.A. Kravtsov, V.I. Tatarski: *Principles of Statistical Radiophysics. Vol.4 Wave Propagation through Random Media* (Springer, Berlin, Heidelberg 1989)
3.157 Yu.A. Kravtsov: Zh. Eksp. Teor. Fiz. 55(3), 798 (1968) [Engl. transl.: Sov. Phys.-JETP 28(3), 413 (1969)]
3.158 V.A. Baranov, Yu.A. Kravtsov: Izv. VUZ Radiofiz. 12(10), 1500 (1969)
3.159 S.M. Flatte, R. Dashen, W.H. Munk, K.M. Watson, F. Zachariasen: *Sound Transmission through a Fluctuating Ocean* (Cambridge Univ. Press, London 1979)
3.160 S.J. Maurer, L.B. Felsen: Proc. IEEE 55(10), 1718 (1967)

3.161 V.S. Buldyrev: *The Field of a Point Source in a Waveguide*, Proc. Leningrad Dept. of Math. Inst., LOMI, Vol. CXV (Nauka, Leningrad 1971) p.78

3.162 A.V. Vagin, N.E. Maltsev: In *Problems of Ship Building. Acoustics*, Vol. 9 (Rumb, Leningrad 1977) p.61

3.163 P.E. Krasnushkin: Zh. Eksp. Teor. Fiz. 18(4), 431 (1948)

3.164 H. Bremmer: *Terrestrial Radio Waves. Theory of Propagation* (Elsevier, New York 1949)

3.165 K.G. Budden: *The Wave-Guide Mode Theory of Wave Propagation* (Logos, London 1961)

3.166 J.B. Keller, J.S. Papadakis (eds.): *Wave Propagation and Underwater Acoustics* (Springer, Berlin, Heidelberg 1979)

3.167 B.Z. Katsenelenbaum: *Teoriya Neregulyarnykh Volnovodov s Medlenno Menyayushchimisya Parametrami* (the theory of nonregular waveguides with slowly varying parameters) (Izd. AN SSSR, Moscow 1961)

3.168 M. Kruskal: *Asymptotic Theory of Hamiltonian and Other Systems with All Solutions Nearly Periodic* (Princeton Univ. Press, Princeton 1961)

3.169 V.A. Borovikov, B.E. Kinber, Yu.A. Kravtsov, A.V. Popov, P.Ya. Ufimtsev: Waves and Rays in Nonregular Waveguides, Preprint No.13 (212) (IZMIRAN, Moscow 1978)

3.170 B.E. Kinber, Yu.A. Kravtsov: Radiotekh. Elektron. 22(12), 2470 (1977)
B.E. Kinber, Yu.A. Kravtsov: In Proc. URSI Symp. on electromagnetic wave theory, Stanford University (1977) p.55

3.171 B.E. Kinber: Radiotech. Elektron. 21(6), 1314 (1976) [Engl. transl.: Radio Eng. Electron Phys. (USSR) 21(6), 127 (1976)]

3.172 B.E. Kinber, Yu.A. Kravtsov, M.P. Salganik: Radiotekh. Elektron. 24(9), 1721 (1979) [Engl. transl.: Radio Eng. Electron. Phys. (USSR) 24(9), 10 (1979)]

3.173 B.E. Kinber, N.E. Maltsev, A.I. Tokatly: Radiotekh. Elektron. 15(12), 2512 (1970)

3.174 V.A. Borovikov, A.V. Popov: In *Pryamye i Obratnye Zadachi Teorii Diffraktsii* (direct and inverse problems of diffraction theory) (Izd. IRE AN SSSR, Moscow 1979) p.157

3.175 A.V. Gurevich: Geomagn. Aeronomy 11(6), 961 (1971)

3.176 L.M. Brekhovskikh: Akust. Zh. 11(2), 148 (1965)

3.177 D. Marcuse: *Theory of Dielectric Waveguides* (Academic, New York 1978)

3.178 V.A. Borovikov: Radiotekh. Elektron. 23(7), 1365 (1978)

3.179 B.E. Kinber, H.H. Komissarova, Yu.A. Kravtsov: Izv. VUZ Radiofiz. 22(4), 414 (1979) [Engl. transl.: Radiophys. Quant. Electron. 22(4), 285 (1979)]

3.180 V.A. Kaloshin: Radiotekh. Elektron 19(12), 2623 (1974)

3.181 J.B. Keller, S. Rubinov: Ann. Phys. 9(1), 24 (1960)

3.182 V.P. Bykov: In *Power Electroncis*, Vol.4 (Nauka, Moscow 1965) p.66

3.183 V.I. Arnold: Funk. Anal. Prilozh. 6(2), 12 (1972)

3.184 P.M. Morse, H. Feshbach: *Methods of Theoretical Physics*, Vol.2 (McGraw-Hill, New York 1953)

3.185 N.P. Mar'in: Radiotechn. Electron. 10(10), 1765 (1965)

3.186 N.P. Mar'in: Geomagn. Aeronomia 5(3), 568 (1965)

3.187 I.G. Kondratjev: Izv. VUZ Radiofiz. 20 (1), 118 (1977) [Engl. transl.: Radiophys. Quant. Electron. 20(1), 77 (1977)]

3.188 Yu.M. Zhidko: Izv. VUZ Radiofiz. 12(8), 1205 (1969)

3.189 G.V. Permitin: Izv. VUZ Radiofiz. 16(1), 62 (1973) [Engl. transl.: Radiophys. Quant. Electron. 16(1), 45 (1973)]

3.190 Yu.A. Kravtsov, S.A. Namazov: Radiotekh. Elektron 25(3), 459 (1980) [Engl. transl.: Radio Eng. Electron Phys. (USSR) 25(3), 8 (1980)]
Yu.A. Kravtsov, S.A. Namasov: In Proc. URSI Spring Meeting and USA Nat'l Radio Sci. Meeting, Seattle, Wash. (1979) (Comm.G, p.262)

3.191 Yu.A. Kravtsov, A.I. Saichev: Sov. Phys.-Uspekhi 25(7), 494 (1982); J. Opt. Soc. Am. A 2(12), 2100 (1985)

3.192 M.V. Berry, K.E. Mount: Rep. Prog. Phys. 35(4), 315 (1972)

3.193 Yu.A. Kravtsov, Z.I. Feizulin: Radiotekh. Elektron. 16(2), 1777 (1971)

3.194 Yu.A. Kravtsov, L.A. Ostrovsky, N.S. Stepanov: Proc. IEEE 62(11), 1492 (1974)

3.195 Yu.I. Orlov: In *Pryamye i Obratnye Zadachi Teorii Difraktsii* (direct and inverse problems of diffraction theory) (Izd. IRE AN SSSR, Moscow 1979) p.5

3.196 L.A. Vainstein, G.M. Vakman: *Metod Razdeleniya Chastot* (frequency separation method) (Nauka, Moscow 1983)

3.197 L.A. Ostrovsky, M.I. Rabinovich: *Nelineinye Nestatsionarnye Volny* (nonlinear and nonstationary waves) (Radiotechnical Inst., Ryasan 1975)

3.198 L.B. Felsen: IEEE Trans. AP-19(3), 424 (1971)

3.199 P.V. Bliokh: Izv. VUZ Radiofiz. 7(3), 460 (1964)

3.200 A.P. Anyutin, Yu.I. Orlov: Izv. VUZ Radiofiz. 19(4), 495 (1976) [Engl. transl.: Radiophys. Quant. Electron. 19(4), 347 (1976)]

3.201 L.A. Ostrovsky, N.S. Stepanov: Izv. VUZ Radiofiz. 14(4), 489 (1971)

3.202 L.A. Ostrovsky: Izv. VUZ Radiofiz 12(9), 133 (1969)

3.203 L.A. Vainstein: Usp. Fiz. Nauk 118(2), 339 (1976) [Engl. transl.: Sov. Phys.-Usp. 19(2), 259 (1976)]

3.204 V.A. Zverev: *Radiooptika* (radio optics) (Sov. Radio, Moscow 1975)

3.205 J.M. Kelso: Rad. Sci. 3(1), 1 (1968)

3.206 M.P. Kiyanovsky: In *Studies in Geomagnetism, Aeronomy and Solar Physics*, Vol.25 (Nauka, Moscow 1972) p.87

3.207 V.Yu. Kim, L.N. Solodovnikova, L.D. Shoya: In *Rasprostraneniye Dekametroviykh Radiovoln* (propagation of decametric radio waves) (Nauka, Moscow 1978) p.56

3.208 N.E. Maltsev: In *Teoriya Difraktsii i Rasprostroneniya Voln* (theory of diffraction and wave propagation), Vol.2 (Izd. AN SSSR, Moscow 1973) p.122

3.209 V.A. Eliseyevnin: Akust. Zh. 10(3), 284 (1964)

3.210 A.V. Belonosova, S.S. Tadzhimukhametova, A.S. Alekseyev: In *Metody i Algoritmy Interpretatsii Geofizicheskikh Dannykh* (methods and algorithms for handling geophysical data) (Nauka, Moscow 1967) p.124

3.211 Yu.A. Burmakov, T.I. Oblogina: Izv. Akad. Nauk SSSR, Fiz. Zemli 12, 81 (1968)

3.212 M.V. Tinin: In *Issledovaniya po Geomagnetizmu, Aeronomii i Fizike Solntsa* (studies in geomagnetism, aeronomy, and solar physics), Vol. 39 (Nauka, Moscow 1976) p.166

3.213 V.A. Baranov, A.V. Popov: In *Rasprostraneniye Dekametrovykh Radiovoln* (propagation of decametric radio waves) (IZMIRAN, Moscow 1975) p.14

3.214 K.V. Svistunov, M.V. Tinin: In *Studies in Geomagnetism, Aernomy, and Solar Physics*, Vol. 45 (Nauka, Moscow 1979) p.178

3.215 A.L. Karpenko: In *Difraktsionnye Effecty Decametrovykh Radiovoln v Ionosfere* (diffraction effects of decametric radio waves in the ionosphere) (Nauka, Moscow 1977) p.54

3.216 D.S. Lukin, Yu.G. Spiridonov: In *Luchevoye Priblizheniye i Voprosy Rasprostraneniya Radivoln* (ray approximation and problems of radio wave propagation) (Nauka, Moscow 1971) p.265

3.217 A.N. Kazantsev, D.S. Lukin, Yu.G. Spiridonov: Kosmich. Issled. 5(4), 593 (1967)

3.218 Yu.G. Spiridonov, D.S. Lukin: Izv. VUZ Radiofiz. 12, 1769 (1969)

3.219 T.A. Croft: J. Geophys. Res. 72(9), 2343 (1967)

3.220 P. Ugincius: J. Acoust. Soc. Amer. 45(1), 206 (1969); ibid 47(1), 339 (1969)

3.221 R.F. MacKinnon, J.S. Partridge, S.H. Toole: J. Acoust. Soc. Amer. 52(5), 1471 (1972)

3.222 K.G. Chen, D. Ludwig: J. Acoust. Soc. Amer. 54(2), 431 (1973)

3.223 V.V. Stepanov: *Kurs Differentsial'nykh Uravnenii* (a course of differential equations) (Fizmatgiz, Moscow 1958)

3.224 S.B. Stechkin, Yu.N. Subbotin: *Splainy v Vychislitel'noy Matematike* (splines in computational mathematics) (Nauka, Moscow 1976)

3.225 C.B. Mohler, L.P. Solomon: J. Acoust. Soc. Amer. **48**(3), 739 (1970)

3.226 M.V. Tinin: In *Teoriya Diffraktsii i Rasprastraneniye Voln* (theory of diffraction and wave propagation) (7th All-Union Symposium), Vol.1 (Izd. AN SSSR, Moscow 1977) p.58

3.227 B.R. Julian, D. Gubbins: Trans. AGU **56**(12), 1027 (1975)

3.228 W. Hauf, V. Grigull: *Optical Methods in Heat Transfer*, Advances in Heat Transfer, Vol.6 (Academic, New York 1970)

3.229 B.E. Kinber: Akust. Zh. **1**(3), 221 (1955)

3.230 J.B. Keller: IRE Trans. AP-7(2), 146 (1959)

3.231 M.R. Weiss: J. Opt. Soc. Amer. **58**(11), 1524 (1968)

3.232 R.M. Lewis: IEEE Trans. AP-17(3), 308 (1969)

3.233 R. Kühn: *Mikrowellenantennen* (Technik, Berlin 1964)

3.234 A.S. Alekseyev: Fiz. Zemli **11**, 1514 (1962)

3.235 A.S. Alekseyev: In *Nekotorye Metody i Algoritmy Interpretatsii Geofizicheskikh Dannykh* (methods and algorithms for geophysical data analysis) (Nauka, Moscow 1967) p.9

3.236 I.S. Berzon, I.P. Pasechnik: *Stroyeniye Zemly po Dynamicheskim Kharakteristikam Seismicheskikh Voln* (Earth build analysis by the dynamic characteristics of seismic waves) (Nauka, Moscow 1976)

3.237 M.M. Lavrentyev, V.G. Vasilyev, V.G. Romanov: *Mnogomernye Obratnye Zadachi dlya Differentsialnykh Uravnenii* (multidimensional inverse problems for differential equations) (Nauka, Novosibirsk 1969)

3.238 Yu.E. Anikonov: *Nekotorye Metody Issledovaniya Mnogomernykh Obratnykh Zadach dlye Differentsialnykh Uravnenii* (some methods to study multidimensional inverse problems for differential equations) (Nauka, Novosibirsk 1978)

3.239 A.S. Barashkov, V.I. Dmitriev: In *Studies in Geomagnetism, Aeronomy and Solar Physics*, Vol.25 (Nauka, Moscow 1972) p.3

3.240 A.S. Barashkov, V.I. Dmitriev, D.P. Kostomarov: In *Studies in Geomagnetism, Aeronomy, and Solar Physics*, Vol.25 (Nauka, Moscow 1972) p.37

3.241 E.L. Boyarintsev, L.S. Gotsakova, Yu.Ya. Yashin: In *Studies in Geomagnetism, Aeronomy and Solar Physics*, Vol.25 (Nauka, Moscow 1972) p.76

3.242 R.G. Newton: SIAM Review **12**(3), 346 (1970)

3.243 U. Buck: Rev. Mod. Phys. **46**(2), 369 (1974)

3.244 U. Buck: J. Chem. Phys. **54**(5), 1923 (1971)

3.245 V.E. Golant: *Sverkhvysokochastotnye methody issledovaniya plasmy* (microwave methods of plasma diagnostics) (Nauka, Moscow 1971)

3.246 J. Sheffield: *Plasma Scattering of Electromagnetic Radiation* (Academic, New York 1975)

3.247 Z.S. Agranovich, V.A. Marchenko: *Obratnye zudachi teorii rasseyaniya* (inverse problem of scattering theory) (Izd. Kharkov Univ., Kharkov 1960)

3.248 A.I. Tikhonov. V.Y. Arsenin: *Metody Resheniya Nekorrektnykh Zadach* (solution techniques for ill-posed problems) (Nauka, Moscow 1974)

3.249 A.N. Tikhonov, V.B. Glasko: Zh. Vych. Mat. Mat. Fiz. **5**(3), 93 (1965)

3.250 A.S. Alekseyev, E.N. Bessonova, N.N. Matveyeva: *Inverse Kinematic Problems of Blast Seismology* (Nauka, Moscow 1979)

3.251 A.O. Blagoveshchensky: In *Pryamye i Obratnye Zadachi Teorii Diffraktsii* (direct and inverse problems of diffraction theory) (Izd. IRE AN SSSR, Moscow 1979) p.115

3.252 L.P. Nizhnik: *Obratnaya Nestatsionarnaya Zadacha Rasseyeniya* (inverse nonstationary scattering problem) (Naukova Dumka, Kiev 1973)

3.253 U.K. Nigul (ed.): *Ekho-signaly ot Uprugikh Obektov* (echo signals from elastic objects), Vols.1,2 (Valgus, Tallinn 1974, 1976)

3.254 E.M. Kennaugh, D.L. Moffatt: Proc. IEEE 53(8), 893-901 (1965)

3.255 S.A. Namazov, Yu.I. Orlov, S.K. Tropkin: In *Proc. XII All-Union Conf. on Radio Wave Propagation*, Pt.1 (Nauka, Moscow 1978) p.42

3.256 A.A. Lukin, Yu.N. Cherkashin: Radiotekh. Elektron. 20(5), 898 (1975)

3.257 Yu.V. Timoshin: *Impul'snaya Seismicheskaya Golografiya* (seismic pulse holography) (Nedra, Moscow 1978)

3.258 V.V. Klyuev (ed.): *Pribory dlya Nerazrushayushchego Kontrolya Materialov i Izdelii* (instruments for nondestructive control of materials and products. A handbook) (Mashinostrojenije, Moscow 1976)

3.259 A.K. Beketova, A.F. Belozerov, A.N. Berezkin: *Golograficheskaya Interferometriya Fazovykh Obyektov* (holographic interferometry of phase objects) (Nauka, Leningrad 1979)

3.260 R.K. Mueller, M. Kaveh, J.F. Greenleaf (eds.): *Acoustical Imaging*, Vol.13 (Plenum, New York 1984)

3.261 W.H. Munk, C. Wunsch: Deep Sea Res. A 26, 123 (1979)

Chapter 4

4.1 V.S. Ignatovsky: Proc. State Opt. Inst. 1(3), 1 (1919);
V.A. Fock: ibid 3(27), 1 (1924)

4.2 S.M. Rytov: Trudy FIAN 2(1), 3 (1940)

4.3 S.M. Rytov: Dokl. Akad. Nauk SSSR 18(2), 263 (1938)

4.4 M. Kline, I.W. Kay: *Electromagnetic Theory and Geometrical Optics* (Wiley, New York 1965)

4.5 R.M. Lewis: IEEE Trans. AP-14(1), 100 (1966)

4.6 R. Courant, D. Hilbert: *Methods of Mathematical Physics*, 2 vols. (Wiley, New York 1953, 1962)

4.7 V.V. Vladimirskii: Dokl. Akad. Nauk SSSR 31(3), 222 (1941)

4.8 M.V. Berry: Proc. Roy. Soc. A 392, 45 (1984); Sci. Am. 259(6), 26 (1988)

4.9 Yu.A. Kravtsov: Izv. VUZ Radiofiz. 13(2), 281 (1970)

4.10 A.L. Gutman, A.P. Yarygin: Radiotekh. Elektron. 16(1), 3 (1971)

4.11 Yu.I. Orlov, S.A. Vlasov: Radiotekh. Elektron. 23(1), 17 (1978)
Yu.I. Orlov, S.A. Vlasov: Izv. VUZ Radiofiz. 21(3), 290 (1978) [Engl. transl.: Radiophys. Quant. Electr. 21(3), 290 (1978)]

4.12 L.A. Apresyan: Izv. VUZ Radiofiz. 16(3), 461 (1973) [Engl. transl.: Radiophys. Quant. Electr. 16(3), 348 (1973)]

4.13 V.L. Ginzburg: *Propagation of Electromagnetic Waves in Plasma* (Pergamon, Oxford 1970)

4.14 L.D. Landau, E.M Livshits: *Electrodynamics of Continuous Media* (Pergamon, Oxford 1960)

4.15 W. Pauli: Helv. Phys. Acta 5(1), 179 (1932)

4.16 A.D. Galanin: J. Phys. (USSR) 6(1), 35 (1942)

4.17 M.L. Levin, S.M. Rytov: Akust. Zh. 2(2), 173 (1956)

4.18 V.M. Babich: Dokl. Akad. Nauk SSSR 110(3), 355 (1956)

4.19 F.C. Karal, J.B. Keller: J. Acoust. Soc. Amer. 31(6), 694 (1959)

4.20 L. Brekhovskikh, V. Goncharov: *Mechanics of Continua and Wave Dynamics*, Springer Ser. Wave Phen., Vol.1 (Springer, Berlin, Heidelberg 1985)

4.21 V. Cerveny, I.A. Molotkov, I. Psencik: *Ray Method in Seismology* (Univerzita Karlova, Praha 1977)

4.22 M. Born, E. Wolf: *Principles of Optics*, 5th edn. (Pergamon, Oxford 1975)

4.23 L.B. Felsen, N. Marcuvitz: *Radiation and Scattering of Waves*, Vols. 1,2 (Prentice-Hall, Englewood Cliffs 1973)

4.24 F.I. Fedorov: *Optika Anizotropnikh Sred* (optics of anisotropic media) (Izd. AN BSSR, Minsk 1958)

4.25 V.V. Zheleznyakov: *Elektromagnitnye Volny v Kosmicheskoi Plazme* (Electromagnetic Waves in Interstellar Plasma) (Nauka, Moscow 1977)

4.26 K.G. Budden: *Radio Waves in the Ionosphere* (Cambridge Univ. 1961)

4.27 J.M. Kelso: *Radio Ray Propagation in the Ionosphere* (McGraw-Hill, London 1964)

4.28 J.A. Ratcliff: *An Introduction to the Ionosphere and Magnetosphere* (Cambridge Univ. Press, Cambridge, MA 1972)

4.29 V.V. Zheleznyakov: *Radio Emission of the Sun and Planets* (Pergamon, Oxford 1970)

4.30 R. Courant: *Partial Differential Equations* (Interscience, New York 1962)

4.31 R.M. Lewis: Arch. Ration Mechan. Anal. **20**(3), 191 (1965)

4.32 S. Veinberg: Phys. Rev. **126**(5), 1829 (1962)

4.33 H. Poeverlein: Phys. Rev. **128**(3), 956 (1962)

4.34 J. Bazer, J. Harley: J. Geophys. Res. **68**(1), 147 (1963)

4.35 K. Suchy: Ann. Phys. 11(2.3), 113 (1952); Pt.2 ibid 13(1), 178 (1953)

4.36 Yu.A. Zaitsev, Yu.A. Kravtsov, Yu.Ya. Yashin: Izv. VUZ Radiofiz. 11(12), 1802 (1968)

4.37 Yu.A. Kravtsov: In *Analytical Methods in the Theory of Diffraction and Propagation* (Scient. Council on Acoust. with the Ministry of Radio Industry, Moscow 1970) p.257

4.38 V. Ellis, S. Buksbaum, L. Bers: *Waves in Anisotropic Plasma* (MIT Press, Cambridge, Mass. 1963)

4.39 K. Davies: *Ionospheric Radio Waves* (Blaisdell, Wallham, Mass., Toronto, London 1969)

4.40 F.I. Fedorov. V.V. Fillipov: *Otrazheniye i Prelomleniye Sveta Prozrachnymi Kristallami* (light reflection and refraction by transparent crystals) (Nauka, Minsk 1976)

4.41 P.L. Christiansen: *Comparative Studies of Diffraction Processes in the Geometrical Theory of Diffraction* (Polyteknick, Lyngby 1975)

4.42 L.B. Felsen: IEEE Trans. **AP-12**(5), 624 (1964)

4.43 E. Arbel, L.B. Felsen: In *Electromagnetic Theory and Antennas* (Pergamon, Oxford 1963)

4.44 Yu.Y. Yashin: Izv. VUZ Radiofiz. 11(4), 491 (1968)

4.45 A.E. Krupina: Izv. VUZ Radiofiz. 19(12), 1810 (1976) [Engl. transl.: Radiophys. Quant. Electr. 19(12), 1254 (1976)]

4.46 Yu.A. Kravtsov, Yu.Ya. Yashin: Izv. VUZ Radiofiz 12(8), 1175 (1969)

4.47 V.A. Baranov, Yu.A. Kravtsov: Radiotekh. Elektron. 23(8), 1588 (1978)

4.48 V.M. Babich: *The Ray Method of Wavefront's Intensity Evaluation in the Case of an Elastic Inhomogeneous Anisotropic Medium*, Topics of the Dynamic Theory of Seismic Wave Propagation, Vol.5 (Izd. LGU, Leningrad 1961) p.36

4.49 R.K. Luneburg: *Mathematical Theory of Optics* (Univ. California Press, Berkley 1964)

4.50 V.B. Berestetsky, E.M. Lifshits, L.P. Pitaevsky: *Relativistic Quantum Theory*, Pt.1 (Pergamon, Oxford 1971)

4.51 V.P. Maslov, M.V. Fedoryuk: *Semiclassical Approximation in Quantum Mechanics* (Reidel, Dordrecht 1981)

4.52 V.A. Dubrovsky: Dokl. Akad. Nauk SSSR 150(3), 527 (1963)

4.53 Yu.A. Kravtsov: Dokl. Akad. Nauk SSSR 183(1), 74 (1968)

4.54 O.N. Naida: Radiotekh. Elektron. 23(12), 2489 (1978)

4.55 O.N. Naida: Izv. VUZ Radiofiz. 14(12), 1843 (1971)

4.56 R.V. Khokhlov: IRE Trans. CT-7, 398 (1960)

4.57 O.N. Naida: Radiotekh. Elektron. 23(12), 2489 (1978)

4.58 V.V. Zheleznyakov, V. V. Kocharovsky, Vl. V. Kocharovsky: Sov. Phys. - Uspekhi 141(2), 257 (1983)

4.59 V.E. Golant, A.D. Pilia: Usp. Fiz. Nauk 104(3), 413 (1971)

4.60 N.S. Erokhin, S.S. Moiseyev: *Wave Processes in an Inhomogeneous Plasma*, Topics in Plasma Theory, Vol.7 (Atomizdat, Moscow 1973) p.146

4.61 L.D. Landau, E.M. Livshits: *Quantum Mechanics* (Pergamon, Oxford 1974)

4.62 E.E. Nikitin, S.Ya. Umansky: *Nonadiabatic Transitions in Slow Atomic Collisions* (Atomizdat, Moscow 1979)

4.63 I.I. Sobelman, L.A. Vainstein, E.A. Yukov: *Excitation of Atoms and Broadening of Spectral Lines*, Springer Ser. Chem. Phys., Vol.7 (Springer, Berlin, Heidelberg 1981)

4.64 V.V. Kucherenko: Izv. Akad. Nauk SSSR, Ser. Mat. 38(3), 625 (1974)

4.65 C. Zener: Proc. Roy. Soc. London Ser. A 137, 696 (1932)

4.66 W. Wasow: *Asymptotic Expansions for Ordinary Differential Equations* (Wiley, New York 1965)

4.67 P.E. Krasnushkin: Dokl. Akad. Nauk SSSR 239(4), 815 (1978)

4.68 Yu.A. Kravtsov, O.N. Naida: Zh. Eksp. Teor. Fiz. 71(1), 237 (1976) [Engl. transl.: Sov. Phys.-JETP 44(1), 122 (1976)]

4.69 N.S. Bellyustin: Izv. VUZ Radiofiz. 21(4), 487 (1978) [Engl. transl.: Radiophys. Quant. Electr. 21(4), 332 (1978)]

4.70 N.G. Denisov: Izv. VUZ Radiofiz. 21(7), 1921 (1978) [Engl. transl.: Radiophys. Quant. Electr. 21(7), 647 (1978)]

4.71 O.N. Naida: Izv. VUZ Radiofiz. 15(5), 751 (1972)

4.72 O.N. Naida: Izv. VUZ Radiofiz. 17(6), 896 (1974)

4.73 L.L. Goryshnik, Yu.A. Kravtsov, L.Ya. Tomashuk, B.V. Fomin: Geomagnetism i Aeronomia 9(5), 873 (1969)

4.74 V.V. Zheleznyakov, E.Ya. Zlotnik: Izv. VUZ Radiofiz. 20(9), 1444 (1977) [Engl. transl.: Radiophys. Quant. Electr. 20(9), 997 (1977)]

4.75 L.A. Apresyan: Astronom. Zh. 53(1), 53 (1976)

4.76 N.S. Stepanov, V.G. Gavrilenko: Dokl. Akad. Nauk SSSR 201(3), 577 (1971)

4.77 V.G. Gavrilenko, G.A. Lupanov, N.S. Stepanov: Izv. VUZ Radiofiz. 15(2), 183 (1972)

4.78 V.G. Gavrilenko, N.S. Stepanov: Astron. Zh. 53(2), 291 (1976)

4.79 O.N. Naida: Dokl. Akad. Nauk SSSR 236(4), 842 (1977); Akust. Zh. 24(5), 731 (1978)

4.80 L.A. Apresyan, Yu.A. Kravtsov, Yu.Ya Yashin, V.A., Yashnov: Izv. VUZ Radiofiz. 19(9), 1296 (1976)

4.81 O.N. Naida: Izv. VUZ Radiofiz. 20(3), 383 (1977) [Engl. transl.: Radiophys. Quant. Electr. 20(3), 261 (1977)]

4.82 F.R. Gantmacher: *The Theory of Matrices*, Vols.1,2 (Chelsea, New York 1959)

4.83 A.A. Rukhadze, V.P. Silin: Usp. Fiz. Nauk 96(1), 87 (1968)

4.84 C. Chen: J. Math. Phys. 12(5), 743 (1971)

4.85 Yu.A. Kravtsov: Zh. Eksp. Teor. Fiz. 55(4), 1470 (1968) [Engl. transl.: Sov. Phys.-JETP 28(4), 769 (1969)]

4.86 Yu.A. Kravtsov, L.A. Ostrovsky, N.S. Stepanov: Proc. IEEE 62(11), 1492 (1974)

4.87 L.A. Ostrovsky, N.S. Stepanov: Izv. VUZ Radiofiz. 14(4), 489 (1971)

4.88 V.L. Ginzburg, A.A. Rukhadze: In *Handbuch der Physik*, Vol.49/2 (Springer, Berlin, Heidelberg 1972)

4.89 A.F. Aleksandrov, L.S. Bogdankevich, A.A. Ruhadze: *Principles of Plasma Electrodynamics*, Springer Ser. Electrophys., Vol.9 (Springer, Berlin, Heidelberg 1984)

4.90 N.S. Stepanov: Izv. VUZ Radiofiz. 12(2), 283 (1969)

4.91 L.A. Ostrovsky: Radiotekh. Elektron. 10(7) 1176 (1965)

4.92 M.E. Gertsenshtein: Zh. Eksp. Teor. Fiz. 26(6), 680 (1954)

4.93 T.H. Stix: *The Theory of Plasma Waves* (McGraw-Hill, New York 1962)

4.94 V.G. Polevoy, S.M. Rytov: Usp. Flz. Nauk 125(3), 549 (1978) [Engl. transl.: Sov. Phys.-Usp. 21(7), 630 (1978)]

4.95 S.M. Rytov: Zh. Eksp. Teor. Fiz. 17(10), 930 (1947)

4.96 L.A. Vainstein: Usp. Fiz. Nauk 118(2), 339 (1976) [Engl. transl.: Sov. Phys.-Usp. 19(1976)]

4.97 Yu.A. Kravtsov, A.I. Kugushev, A.V. Chernykh: Zh. Eksp. Teor. Fiz. 59(5), 2160 (1970) [Engl. transl.: Sov. Phys. JETP 29(3), 1168 (1970)]

4.98 V.D. Pikulin, N.S. Stepanov: Izv. VUZ Radiofiz. 16(8), 1138 (1973) [Engl. transl.: Radiophys. Quant. Electr. 16(8), 877 (1973)]

4.99 N.S. Stepanov: Izv. VUZ Radiofiz. 19(7), 960 (1976) [Engl. transl.: Radiophys. Quant. Electr. 19(7), 683 (1976)]

4.100 A.A. Rukhadze, V.P. Silin: Usp. Fiz. Nauk 82(3), 499 (1964)

4.101 A.B. Mikhailovsky: *Teoriya Plazmenykh Neustoichivostey* (theory of plasma instability) (Atomizdat, Moscow 1970, 1971)

4.102 S.I. Averkov, L.A. Ostrovsky: Izv. VUZ Radiofiz. 1(1), 46 (1958)

4.103 S.I. Averkov, N.S. Stepanov: Izv. VUZ Radiofiz. 2(2), 203 (1959)

4.104 N.S. Stepanov: Izv. VUZ Radiofiz. 3(4), 672 (1960)

4.105 N.S. Stepanov: Izv. VUZ Radiofiz. 6(1), 112 (1963)

4.106 N.S. Stepanov: Izv. VUZ Radiofiz. 11(5), 700 (1968)

4.107 L.P. Pitaevskii: Zh. Eksp. Teor. Fiz. 39(11), 1450 (1960) [Engl. transl.: Sov. Phys.-JETP 12(5), 1008 (1960)]

4.108 Yu.A. Kravtsov, N.S. Stepanov: Zh. Eksp. Teor. Fiz. 57(11), 1730 (1969) [Engl. transl.: Sov. Phys.-JETP 30(5), 935 (1970)]

4.109 V.I. Karpman: *Nelineinye Volny v Dispergiruyushchikh Sredakh* (nonlinear waves in dispersive media) (Nauka, Moscow 1973)

4.110 L.A. Ostrovsky: Zh. Eksp. Teor. Fiz. 61(2), 551 (1971) [Engl. transl.: Sov. Phys.-JETP 34(2), 293 (1971)]

4.111 Yu.M. Sorokin: Izv. VUZ Radiofiz. 15(1), 51 (1972)

4.112 N.S. Stepanov, Yu.M. Sorokin: Zh. Tekh. Fiz. 42(3), 578 (1972)

4.113 L.A. Ostrovsky: Usp. Fiz. Nauk 116(2), 315 (1975)

4.114 M.A. Miller, Yu.M. Sorokin, N.S. Stepanov: Usp. Fiz. Nauk 121(3), 525 (1977)

4.115 S.N. Stolyarov: In *Einsteinovskii Sbornik* (Einstein Collective Works of 1975-76) (Nauka, Moscow 1978) p.152

4.116 L.A. Ostrovsky: Izv. VUZ Radiofiz. 4(2), 293 (1961)

4.117 J.B. Whithman, L.B. Felsen: J. Math. Phys. 13(5), 760 (1972)

4.118 L.B. Felsen: SIAM Review 12(3), 424 (1970)

4.119 L.A. Ostrovsky: Izv. VUZ Radiofiz. 18(4), 618 (1975)

4.120 B.E. Rok: Izv. VUZ Radiofiz. 17(11), 1728 (1974)

4.121 V.N. Krasilnikov, L.N. Lutchenko: In *Radio Wave Diffraction and Propagation Problems*, Vol.12 (Izd. LGU, Leningrad 1973) p.150

4.122 Yu.A. Kravtsov, Z.I. Feizulin, A.G. Vinogradov: *Radiowave Propagation through the Earth Atmosphere* (Radio i Svyaz', Moscow 1983)

4.123 V.G. Gavrilenko, N.S. Stepanov: Izv. VUZ Radiofiz. 16(1), 69 (1973) [Engl. transl.: Radiophys. Quant. Electr. 16(1), 50 (1973)]

4.124 V.G. Gavrilenko, N.S. Stepanov: Radiotekh. Elektron. 18(6), 1105 (1973)

4.125 V.G. Gavrilenko, M.N. Krom, N.S. Stepanov: Izv. VUZ Radiofiz. 20(8), 1181 (1977) [Engl. transl.: Radiophys. Quant. Electr. 20(8), 820 (1977)]

Chapter 5

5.1 Yu.A. Kravtsov: Izv. VUZ Radiofiz. 10 (9,10), 1283 (1967) [Engl. transl.: Radiophys. 10 (9,10), 719 (1967)]

5.2 Wey-Yi D. Wang, G.A. Deschamp: Proc. IEEE 62, 1541 (1974)

5.3 K.A. Connor, L.B. Felsen: Proc. IEEE 62 (11), 1586 (1974)

5.4   Yu.I. Orlov: In *Pryamye i Obratnye Zadachi Teorii Difraktsii* (Direct and Inverse Problems of Diffraction Theory) (Izd. IRE AN SSSR, Moscow 1979) p.5

5.5   Yu.A. Kravtsov, Yu.Ya. Yashin: Izv. VUZ Radiofiz. 12 (5), 674 (1969)

5.6   A. Erdélyi: *Asymptotic Expansions* (Dover, New York 1956)

5.7   A.H. Nayfeh: *Perturbation Methods* (Wiley, New York 1973)

5.8   Yu.A. Kravtsov: Izv. VUZ Radiofiz. 7 (4), 664 (1964)

5.9   D. Ludwig: Comm. Pure Appl. Math. 19 (2), 215 (1966)

5.10 Yu.A. Kravtsov: Akust. Zh. 14 (1), 1(1968)

5.11 M.V. Berry, K.E. Mount: Rep. Prog. Phys. 35 (2), 313 (1972)

5.12 Yu.A. Kravtsov, Yu.I. Orlov: Sov. Phys. - Uspekhi 26 (12), 1038 (1983)

5.13 R.N. Buchal, J.B. Keller: Comm. Pure Appl. Math. 13 (1), 85 (1960)

5.14 Yu.L. Gazaryan: In *Voprosy Dinamicheskoy Teorii Rasprostraneniya Seismicheskikh Voln* (Topics of the Dynamic Theory of Seismic Wave Propagation), Vol.5 (Leningrad State Univ., Leningrad 1961) p.73

5.15 V.P. Maslov: *Théorie des Perturbations et Méthodes Asymptotiques* (Dunod, Paris 1972)

5.16 V.P. Maslov, M.V. Fedoryuk: *Semiclassical Approximation in Quantum Mechaniscs* (Reidel, Dordrecht 1981)

5.17 J.M. Arnold: Rad. Sci. 17 (5), 1181 (1982)

5.18 R. Ziolkowski, G.A. Deschamps: Rad. Sci. 19 (4), 1001 (1984)

5.19 J.B. Keller: In *Calculus of Variations and its Applications*, Proc. Symp. Appl. Math., Vol.8 (McGraw-Hill, New York 1958) p.27

5.20 B.R. Levy, J.B. Keller: Comm. Pure Appl. Math. 12 (1), 159 (1959)

5.21 J.B. Keller: J. Opt. Soc. Amer. 52 (2), 116 (1962)

5.22 P.H. Pathak, R.G. Kouyoumjian: Proc. IEEE 62 (11), 1438 (1974)

5.23 G.L. James: *Geometrical Theory of Diffraction for Electromagnetic Waves* (Peter Peregrinus 1976)

5.24 P.Ya. Ufimtsev: *Method of Edge Waves in the Physical Theory of Diffraction* (US AF, Forlin Technology Div., Wrigth Patterson AFB, Ohyo 1971)

5.25 R.M. Lewis: *Asymptotic Theory of Transients. - Electromagnetic Wave Theory*, Pt.2 (Pergamon, New York 1967) p.845

5.26 A.P. Anyutin, Yu.I. Orlov: Izv. VUZ Radiofiz. 19 (4), 495 (1976) [Engl. transl.: Radiophys. Quant. Electr. 19 (4), 347 (1976)]

5.27 V.A. Fock: *Electromagnetic Diffraction and Propagation Problems* (Pergamon Press, New York 1965)

5.28 R.M. Lewis, J. Boersma: J. Math. Phys. 10 (12), 2291 (1969)

5.29 D.S. Ahluwalia: SIAM J. Appl. Math. 18 (2), 287 (1970)

5.30 M.A. Leontovich, V.A. Fock: Zh. Eksp. Teor. Fiz. 16 (7), 557 (1946)

5.31 S.M. Rytov, Yu.A. Kravtsov, V.I. Tatarski: *Principles of Statistical Radiophysics*, Vol.4 (Springer, Berlin, Heidelberg 1989)

5.32 Yu.A. Kravtsov, Z.I. Feizulin: Izv. VUZ Radiofiz. 12 (6), 886 (1969)

5.33 Yu.I. Orlov: Izv. VUZ Radiofiz. 17 (7), 1035 (1974)

5.34 V.B. Avdeev, A.V. Demin, Yu.A. Kravtsov, M.V. Tinin, A.P. Yarygin: Izv. VUZ Radiofiz. 31 (11), 1279 (1988)

5.35 M.V. Berry, C. Upstill: Catastrophe optics: Morphologies of caustics and their diffraction patterns, in *Progress in Optics* 18, 257 (North-Holland, Amsterdam 1980)

5.36 P.S. Theocaris, J.G. Michopoulos: Generalization of the theory of far-field caustics by the catastrophe theory. Appl. Opt. 21 (6), 1080 (1982)

5.37 S. Cornbleet: Geometrical optics reviewed: A new light on old subject. Proc. IEEE 71(4), 471 (1983)

5.38 J.J. Stamnes: Diffraction, asymptotics and catastrophes. Opt. Acta 29 (6), 823 (1982)

Subject Index